T0281035

Lecture Notes in Business Information Processing 501

Series Editors

LNBIP reports state-of-the-art results in areas related to business information systems and industrial application software development – timely, at a high level, and in both printed and electronic form.

The type of material published includes

- Proceedings (published in time for the respective event)
- Postproceedings (consisting of thoroughly revised and/or extended final papers)
- Other edited monographs (such as, for example, project reports or invited volumes)
- Tutorials (coherently integrated collections of lectures given at advanced courses, seminars, schools, etc.)
- Award-winning or exceptional theses

LNBIP is abstracted/indexed in DBLP, EI and Scopus. LNBIP volumes are also submitted for the inclusion in ISI Proceedings.

Maria Papadaki · Marinos Themistocleous ·
Khalid Al Marri · Marwan Al Zarouni
Editors

Information Systems

20th European, Mediterranean, and Middle Eastern Conference, EMCIS 2023
Dubai, United Arab Emirates, December 11–12, 2023
Proceedings, Part I

Editors
Maria Papadaki
British University in Dubai
Dubai, United Arab Emirates

Marinos Themistocleous
University of Nicosia
Nicosia, Cyprus

Khalid Al Marri
British University in Dubai
Dubai, United Arab Emirates

Marwan Al Zarouni
Dubai Blockchain Center
Dubai, United Arab Emirates

ISSN 1865-1348 ISSN 1865-1356 (electronic)
Lecture Notes in Business Information Processing
ISBN 978-3-031-56477-2 ISBN 978-3-031-56478-9 (eBook)
https://doi.org/10.1007/978-3-031-56478-9

This Springer imprint is published by the registered company Springer Nature Switzerland AG
The registered company address is: Gewerbestrasse 11, 6330 Cham, Switzerland

Paper in this product is recyclable.

Preface

The European, Mediterranean, and Middle Eastern Conference on Information Systems (EMCIS) is an annual research event that explores the field of Information Systems (IS) from practical, theoretical, regional and global perspectives. Renowned for its success in bringing together researchers from across the globe, EMCIS fosters a friendly atmosphere that encourages the exchange of innovative ideas. Recognized as one of the premier conferences in Europe and the Middle Eastern region for IS academics and professionals, EMCIS comprehensively addresses technical, organizational, business, and social issues in the application of information technology. The conference is committed to defining and establishing IS as a discipline of high impact for professionals and practitioners in the field. EMCIS places a strong emphasis on facilitating the identification of innovative research that holds significant relevance to the IS discipline.

EMCIS 2023 marked the celebration of two significant milestones: (a) the conference's 20th edition and (b) the return to an in-person format. Held in December and hosted by the British University in Dubai, this year's conference was particularly special. Out of the submissions received, 43 papers were accepted, representing an acceptance rate of 35%. These papers spanned various tracks, including:

- Artificial Intelligence
- Big Data and Analytics and Machine Learning
- Blockchain Technology and Applications
- Climate Change and Emerging Technologies
- Cloud Computing
- Digital Services and Social Media
- Digital Governance
- Emerging Computing Technologies and Trends for Business Process Management
- Enterprise Systems
- Information Systems Security and Information Privacy Protection
- Healthcare Information Systems
- Managing Information Systems
- Innovative Research Projects
- Metaverse
- Smart Cities

The submitted papers underwent a double-blind review process with a minimum of two reviewers. Additionally, submissions from track chairs were assessed by a member of the EMCIS Executive Committee and a member of the international committee. Finally, contributions from conference chairs underwent review by two senior external reviewers.

EMCIS has once again demonstrated its truly international nature, with authors originating from 22 different countries and participants representing 29 countries. The diversity of contributions is highlighted in the summary below, organized by the country of origin of authors / participants:

- Austria
- Brazil
- Bulgaria
- Cameroon
- Canada
- Croatia
- Cyprus
- Estonia
- Finland
- France
- Germany
- Greece
- The Netherlands
- India
- Italy
- Norway
- Pakistan
- Poland
- Romania
- Russia
- Saudi Arabia
- Slovakia
- South Africa
- Sweden
- Switzerland
- Tunisia
- United Arab Emirates
- UK
- USA

The papers were accepted for their theoretical and practical excellence and promising results. We hope the readers will find them interesting and consider joining us at the next edition of the conference.

December 2023

Maria Papadaki
Marinos Themistocleous
Khalid Al Marri
Marwan Al Zarouni

Organization

Conference Chairs

Khalid Almarri	British University in Dubai, UAE
Maria Papadaki	British University in Dubai, UAE

Conference Executive Committee

Richard Kirkham (Publications Chair)	University of Manchester, UK
Marinos Themistocleous (Program Chair)	University of Nicosia, Cyprus
Nikolay Mehandjiev (Public Relations Chair)	University of Manchester, UK

International Committee

Marwan Al Zarouni	Dubai Blockchain Center, UAE
Charalampos Alexopoulos	University of the Aegean, Greece
Nikolaos Bakas	GRNET, Greece
Paulo Henrique de Souza Bermejo	Universidade de Brasília, Brazil
Lasse Berntzen	University of South-Eastern Norway, Norway
Yannis Charalabidis	University of the Aegean, Greece
Savvas Chatzichristofis	Neapolis University Pafos, Cyprus
Klitos Christodoulou	University of Nicosia, Cyprus
Paulo Rupino Cunha	University of Coimbra, Portugal
Vasiliki Diamantopoulou	University of the Aegean, Greece
Irenee Dondjio	Hague University, The Netherlands
Catarina Ferreira da Silva	University Institute of Lisbon, Portugal
Besart Hajrizi	University of Mitrovica, Kosovo
Elias Iosif	University of Nicosia, Cyprus
Muhammad Kamal	Coventry University, UK
Angeliki Kokkinaki	University of Nicosia, Cyprus
Przemysław Lech	University of Gdańsk, Poland
Euripidis N. Loukis	University of the Aegean, Greece
Nikolay Mehandjiev	University of Manchester, UK

Paulo Melo	University of Coimbra, Portugal
Vincenzo Morabito	Bocconi University, Italy
Andriana Prentza	University of Piraeus, Greece
Maribel Yasmina Santos	University of Minho, Portugal
Alan Serrano	Brunel University London, UK
António Trigo	Coimbra Business School, Portugal
Horst Treiblmaier	Modul University Vienna, Austria
Aggeliki Tsohou	Ionian University, Greece
Piotr Soja	Cracow University of Economics, Poland
Gianluigi Viscusi	Linköping University, Sweden

Contents – Part I

Digital Governance

Healthcare Information Systems

Contents – Part II

Big Data and Analytics

Digital Services and Social Media

Innovative Research Projects

Managing Information Systems

Metaverse

Pros and Cons of Integrating the Metaverse into Education: A Comprehensive Analysis

Soulla Louca(✉) and Saadya Chavan

University of Nicosia, 1700 Nicosia, Cyprus
louca.s@unic.ac.cy, chevan.s@live.unic.ac.cy

Abstract. The Metaverse holds the promise of profoundly transforming education by reshaping how courses are delivered and how students learn and engage, thanks to the novel possibilities it offers for immersive and interactive learning experiences. This study seeks to pinpoint both the challenges and advantages of integrating the metaverse into education, based on a systematic review of existing literature from both a technological and human viewpoint. It provides an in-depth analysis of not only the benefits of employing the Metaverse in education, but also the obstacles that educational institutions, designers, and implementers encounter when creating or utilizing these digital realms. The study's findings underscore the need to develop and deploy metaverse systems to improve educational methods and maximize learner satisfaction. Furthermore, they will provide researchers with insights to tackle the multifaceted challenges associated with the use of metaverse tools and environments in education.

Keyword: Metaverse in Education

1 Introduction

The most recent advancement in online learning is the rise of the metaverse, an interactive digital environment that combines elements of both the physical and virtual realms. This innovative space allows learners and educators to engage in immersive and collaborative experiences. The metaverse introduces exciting opportunities for online learning, including the creation of lifelike simulations, interactive social interactions, exploration of diverse virtual worlds, and the cultivation of digital competencies. Nonetheless, the metaverse also presents certain obstacles and potential hazards for online learning, including ethical considerations, privacy worries, disparities in digital access, and the potential for cyber-bullying [63].

The word "metaverse" was first coined by Neal Stephenson in 1992, in his book "Snow Crash". In the book, the metaverse is portrayed as a virtual reality-based successor to the internet, where users can engage in immersive, three-dimensional experiences and interact with others in a shared virtual space. Stephenson coined the term by combining "meta" (meaning beyond or transcending) with "universe", indicating a digital realm that goes beyond the physical world [7, 75]. Since then, the concept of the metaverse has gained traction and has been widely discussed in various fields, including technology,

M. Papadaki et al. (Eds.): EMCIS 2023, LNBIP 501, pp. 3–17, 2024.
https://doi.org/10.1007/978-3-031-56478-9_1

gaming, and futurism, as a potential evolution of the internet and virtual reality. It has become a popular term to describe a future interconnected digital realm where people can engage, create, and interact in a virtual environment.

The metaverse is an immersive, online environment that enables multi-user interactions, transcending physical boundaries. It incorporates three-dimensional elements, personified agents, and simulations, fostering social and economic engagement. Users can explore diverse contexts and lifestyles through their avatars [11, 29] emphasize the merging of physical and virtual realms in the metaverse, where everyday activities and economic endeavors extend from the real world. Avatars in the metaverse are tightly linked to users' real identities, actively engaging in social, economic, and cultural experiences within the virtual realm. Sarıtaş and Topraklıkoğlu [49] propose an alternative definition of metaverse technology, encompassing digital reality platforms and three-dimensional computer software-based platforms. In this definition, the metaverse comprises multiple virtual worlds where individuals interact through avatars, leveraging blockchain [75, 76] and digital reality technologies. Liu [34] points out to metaverse as being the integration of the virtual and the real world but does not consider it as a video game. Yilmaz, et al., [66] highlight terminological confusion and varying definitions in literature. Jiang and Xu [24] note the influence of the metaverse's evolving nature on its definition. Talan and Kalinkara [53] observe that definitions of the metaverse change over time, and Yue [67] emphasizes that many definitions envision the future of the metaverse rather than capturing its current state.

The concept of the metaverse in education has gained significant attention as a potential transformative tool for teaching and learning. In the context of education, metaverse offers numerous possibilities for immersive and collaborative learning experiences. Students can explore virtual environments, engage in simulations, and interact with peers and instructors from anywhere in the world. Studies show that metaverse in education may enhance student engagement, foster creativity, and improve critical thinking skills. Furthermore, the metaverse can provide opportunities for personalized learning, allowing students to tailor their educational experiences according to their individual needs and interests [25]. With the rapid advancements in virtual reality and augmented reality technologies, the metaverse holds immense potential to revolutionize the education landscape, creating inclusive and immersive educational environments [18].

The aim of this study is to explore the challenges and benefits of incorporating the metaverse into educational settings. This exploration is carried out through a comprehensive review of existing literature, considering both technological and human factors. The outcomes underscore the importance of designing and deploying metaverse systems to enhance educational methodologies and boost learner satisfaction. Additionally, these insights will assist researchers in tackling the diverse obstacles related to the use of metaverse tools and environments in education. The methodology is detailed in Sect. 2, followed by Sect. 3 that presents the results and study limitations. Section 4 offers a discussion based on the obtained results, and the paper concludes with future research directions.

2 Methodology

A systematic literature review (SLR) [55] was adopted in this study with the aim of identifying and analyzing relevant peer-reviewed articles on metaverse to identify the benefits and challenges of the use of metaverse in education. A SLR is a type of review that uses repeatable methods to find, select, and synthesize all available evidence on a specific research question. It involves systematically searching for, selecting, critically analyzing, and synthesizing peer-reviewed articles that meet pre-defined criteria [20]. This systematic process ensures a comprehensive exploration of existing literature and provides a foundation for understanding the current state of knowledge. To maintain thoroughness and openness, the SLR follows a distinct procedure that details the research queries, database selection, exclusion standards, and quality criteria for examining only relevant documents. The SLR protocol lessens the likelihood of random decisions during the review phase and limits redundancy. SLRs are extensively employed in diverse sectors, such as healthcare, social sciences, and engineering, to discover, assess, and amalgamate relevant evidence [27].

The SLR process generally encompasses five stages [1], that are described below. Upon completion of these stages, the outcomes are categorized and scrutinized, succeeded by a discourse on the discoveries and a recognition of the challenges and constraints. The categorization and analysis, discussion, challenges, and limitations are elaborated in Sect. 3 and Sect. 4 respectively. Similar methods have been used by [5, 9] and [10].

2.1 The Systematic Literature Review

Phase 1: Research Question. Research questions in a SLR utilize a clear and concise form of inquiry that guides the search for the relevant literature and the analysis of the data to help identify and evaluate the current state of knowledge on a particular topic [9, 10]. The main research questions this study deals with the identification of the advantages and the challenges of using metaverse in education based on the various experiments and other related studies performed so far.

Phase 2: Searching for Studies. A comprehensive search of multiple databases was conducted to find studies that met a set of selection criteria. The search strategy included the use of specific keywords and Boolean operators. This stage entailed identifying and choosing pertinent databases likely to house articles related to our research question. Our review was conducted using various computer science and education databases. The databases examined were ERIC, IEEE Xplore, Professional Development Collection, ScienceDirect, and Scopus. In order to identify the relevant articles, two main keywords were used: metaverse and education. These were processed as (TITLE-ABS-KEY(metaverse, education) AND PUBYEAR > 2013 AND PUBYEAR < 2024 AND (LIMIT-TO (LANGUAGE, "English"))). No other keywords were used as words such as Virtual Reality, Extended Reality, Online Learning, and Immersive Learning, are part of what is considered to be the metaverse. Restrictions were imposed on the date and on the language. We screened articles published in the year 2014 or later and we limited it only to English written articles.

Phase 3: Eligibility Criteria. Eligibility criteria pertain to a pre-established set of guidelines that dictate the selection of studies to be incorporated or excluded in the SLR [9, 56]. Exclusion criteria are specific elements that result in a study's exclusion from the review. Conversely, inclusion criteria are elements that a study must exhibit to be included in the review. These criteria are formulated based on the research questions and objectives, aiming to ensure that the studies included in the review are pertinent and of high quality, thereby reducing the risks of bias [10].

The inclusion criteria were (a) documents related to the implementation of metaverse educational environments or documents related to metaverse in education in any field; (b) articles that were already published; (c) year of publication greater that 2013; (d) documents written in the English language. All other articles were excluded.

Phase 4: Quality Criteria. The quality criteria are used to ensure that the included studies are of high quality and that their findings can be trusted [10]. The criteria established by the study for ensuring the quality and relevance of the selected documents were peer reviewed conference or journal articles related to the use of metaverse in education. By employing these selection criteria, the study aimed to identify quality documents that provided valuable insights in terms of the advantages and disadvantages of the use of metaverse in education based on either qualitative or quantitative research, practical or experimental, as well as the theoretical aspects of education in the metaverse.

Phase 5: Screening and Data Extraction. For choosing relevant studies and extracting data, the PRISMA (Preferred Reporting Items for Systematic Reviews and Meta-Analysis) [43] protocol was employed to ensure the clarity and precision of the screening and data extraction procedure. The PRISMA protocol encompasses several primary steps. The application of the PRISMA protocol ensured that the screening and data extraction procedure was carried out systematically and transparently, thereby boosting the rigor and credibility of the SLR. Specifically, adhering to a reference protocol helps reduce biases [38].

The study began with a total of 1313 records, which underwent a four-phase filtering process. In the first phase, duplicate records were removed, resulting in 1189 unique records. During the second phase, the titles and abstracts of the papers were screened, leading to the exclusion of 1101 papers. Subsequently, 88 records remained eligible for the analysis of metaverse and education. Out of these 88 records, 63 were deemed suitable for the analysis on the on the benefits and challenges associated with utilizing metaverse in education.

3 Results from Systematic Literature Review

This section provides an overview of the benefits and obstacles associated with integrating the metaverse into education, as identified in the SLR.

3.1 Benefits of Metaverse in Education

Improving the Educational Experience and Fostering Effective Learning. Ho (2022) [20] examines metaverse technology's impact on teaching, enhancing competitiveness.

Sánchez-López et al. (2022) [48] suggest avatars, multimodal literacy, game mechanics, and transhuman capabilities create a unique learning experience. Jiang and Xu (2022) [24] emphasize the metaverse's role in improving education quality and driving transformative practices. Ruwodo et al. (2022) [46] propose "metaversity", challenging traditional universities with extended reality for exploration and problem-solving. Yue (2022) [67], Yilmaz et al. (2023) [66], and Ng (2022) [40] share similar ideas. The metaverse revolutionizes learning with convenience and accessibility, particularly in astrophysics [14, 31, 64]. Khansulivong et al. (2022) [26] find integrated content benefits agriculture students while Watanabe (2023) [61] through his experiments has identified the metaverse allows for certain activities to be performed effectually and effectively. Integrating the metaverse with Learning Management Systems (LMS) connects physical and virtual realms [45]. The metaverse offers an immersive experience for visualizing challenging concepts like molecules [6, 42], recreating historical events, and overcoming geographical limitations [6, 39].

New Pedagogical Models. Virtual worlds offer educational opportunities aligned with digital pedagogical models, supporting diverse teaching methods like flipped classrooms and collaborative learning. They are scalable, incorporating ICT resources and interactive elements for versatile educational experiences [3, 11]. The metaverse provides an immersive, authentic, and innovative environment that fosters personalized learning, gamified learning, collaboration, and the development of new pedagogical models [24, 36]. This immersive space is created through an integration of various technologies such as virtual reality, multi-user virtual environments, mixed reality, and augmented reality allowing for psychological immersion and facilitating "situational learning" based on students' experiences within the metaverse [74].

Promote Participation of Women in STEM. Research indicates the need to increase women's representation in STEM (Science, Technology, Engineering, and Mathematics) to address the gender gap in technology [13, 57]. Elementary school girls show a positive attitude towards life logging in the metaverse, suggesting its potential to bridge the gender gap in technology [52], and thus enhance women's involvement in STEM fields.

Increased Learner Collaborations. Metaverse immersive environments require students to collaborate in teams, solving puzzles or riddles and bringing augmented and virtual realities into the classroom [59, 60].

Convenience. Yue (2022) [67] suggests that metaverse technology enables students to overcome institutional limitations and attend classes at prestigious educational institutions. It streamlines access to learning materials, allowing individuals to access extensive libraries from home. Additionally, while students in video livestream format may experience virtual fatigue, those in the virtual reality space report reduced effects [18].

Inclusion - Equal Opportunities to Education. Yue (2022) [67] suggests the metaverse can democratize education, providing equal opportunities regardless of knowledge, finances, social status. Sá et al. (2023) [47] emphasize its potential for accessibility across economic, cultural, and social backgrounds. Additionally, the metaverse supports individuals with specific needs, including Autism Spectrum Disorder, social anxiety disorders, and physical disabilities, further democratizing education [14, 47].

Personalization and Empowerment. Metaverse-based learning goes beyond being a technological tool, becoming an authentic reality. Learners shape their own reality, constructing inclusive learning environments through community building. This constructivist approach promotes social change and active knowledge construction [6, 12, 36, 71]. Benefits include realistic experiences, multisensory sensations, digitized representations, optimized learning paths, improved concrete thinking, and learner-defined virtual space [69]. Jovanović et al. (2022) [25] suggest using 3D avatars for personalized experiences and problem-solving in education.

Bridging the Gap Between Physical and Virtual Realms in Time and Space. Guo, H., and Gao, W. (2022) [15] state that the metaverse provides an immersive and interactive learning environment, bridging the gap between physical and virtual realms. Metaverse education seamlessly integrates time, connecting past, present, and future, a feat not possible in the real world. Logical time integration through virtual avatars enables this time fusion in learning [70, 72].

Student and Teacher Engagement. Using metaverse in education enhances active class participation, immersion, student interaction, customization, creativity, motivation, and engagement [30]. Immersive and interactive teaching fosters a nearly realistic social space, cultivating a sense of community [50]. Virtual worlds make learning easier and more captivating by providing a video game-like experience, offering a significant advantage over traditional classroom sessions [25]. The metaverse enables diverse learning activities enabling learners to actively engage through avatars in social and educational activities with others [30–32] while enabling at the same time intelligent teaching evaluation [36]. Immersive learning and interactive tools like role-playing, gaming consoles, real-time strategy, and multiplayer online gamification enhance class participation [3, 17].

Access to Limited/Restricted Equipment. Metaverse provides real-time access to lab resources, including virtual labs as digital twins, and restricted resources like conducting tests on Uranium within the metaverse [60].

Decreased Costs. In certain subjects such as Chemistry and Physics, experiments can be simulated, as highlighted by [6]. The Internet of Everything facilitates convenient information search and knowledge acquisition, thereby reducing the overall cost of education for individuals [6, 17].

Ecological Education. The use of metaverse in education allows for ecological education assisting in improving environmental quality. The immersive experience of the users offers unique benefits in stimulating learners' ecological emotions. This can help to enhance the content and format of traditional ecological education and optimize the actual outcomes of such education. This is highly significant for ecological governance and the construction of an ecological civilization [35].

3.2 Challenges of Metaverse in Education

Technology Related. Technology maturity and requirements: Yang et al. (2022) [65] examined the use of metaverse in basketball, highlighting its focus on representation and evolving infrastructure. The complexity of metaverse poses challenges in development and promotion due to its reliance on diverse digital technologies [6]. Weak Artificial intelligence limits personalized learning guidance, and handling data from student interactions overwhelms existing analytics tech. Building a realistic metaverse requires skilled personnel and expensive 3D modeling tech, presenting practical obstacles [68, 71]. Wang et al. (2022) [60] emphasize the importance of maintaining user interactivity bandwidth for user retention. However, current user inputs on mobile mixed reality and virtual reality headsets are unsatisfactory, leading to low throughput rates and hindering expression in virtual-physical environments.

Additionally, current interactivity technologies struggle to capture subtle body movements, necessitating more precise peripherals like sensors and haptics for accurate user input and feedback. Furthermore, enabling a seamless metaverse experience requires handling large amounts of data generated by simultaneous and widespread online user interactions. This demands robust storage, computing capabilities, and a stable, low-latency, high-bandwidth connection [2, 8, 60].

Realistic rendering of digital avatars for physical participants is a concern, as advanced avatars from classroom sensing may exceed virtual reality headset capabilities. Real-time video synchronization with avatar actions and audio transmission is crucial for many courses. Addressing latency while maximizing video quality has been explored in cloud gaming, using joint source coding and forward error correction. However, networking limitations currently cause delayed haptic feedback and impact user experiences [60].

Accessing the metaverse world at reasonable costs necessitates affordable smart wearables for learners and teachers. The cost is amplified by the inclusion of simulations [60, 64, 68, 73]. Furthermore, the inclusion of various simulations adds to the overall cost [64].

Governance. Governance poses a major challenge in metaverse development. Balancing conservative and radical rules is crucial to meet expectations without intensifying real-world conflicts [21]. The ideal metaverse promotes inclusivity and personal freedom by reducing central privileges and bypassing local restrictions. Decentralized autonomous organizations (DAOs) establish regulations, facilitate e-commerce, and enable user participation in decision-making. However, key questions remain regarding rule makers, metaverse management, ownership, control mechanisms, ethical standards, human rights, freedom of speech, and preventing user-generated content violations [2, 65]. Yang et al. (2022) [65] studied metaverse basketball and emphasized addressing classroom management challenges in online sports. Community governance, including spreading conduct standards, is crucial to prevent moral issues like insults and bullying [33]. Metaverse applications in education raise additional concerns about inclusivity, accessibility, lifelong learning data management, analysis, and controlling interactions among social, technical, and economic factors [37, 62].

Learning Difficulties and Disruption. Using the metaverse for learning brings challenges like increased difficulty, distractions, disconnection from real-life experiences, and potential disruptions to class discipline [4, 53]. Furthermore, Themistocleous et al. [54] examine how the mobility of avatars can cause disruptions and divert the attention of students during class.

Privacy and Security. Metaverse companies will gather extensive personal data from users to gain insights into their thoughts and behavior [4, 6, 22, 29, 35, 39, 68, 71], raising concerns about potential unethical uses [64]. Creators and developers must provide solutions to manage data usage and protect users' privacy and security in accordance with established policies [2].

Not Everything Can be Taught Online. Disciplines such as complex surgery and civil engineering face challenges in comprehensive learning due to practical limitations and real-world complexities. Metaverse applications offer valuable examples and scenarios, but real-world practice remains essential due to unforeseen variables, moral considerations, and safety concerns [8, 14].

Inclusion and Accessibility in Metaverse's Digital Realm. Despite the fact that metaverse in education aims to foster broader participation, prioritize the needs of special learners, and ensure inclusivity [68, 71] raise concerns about educational equity and the "Metaverse divide", highlighting ethical and security challenges. Equitable access to metaverse learning requires proactive design, inclusive strategies, and enforcement of legislation to prevent exclusion and promote equity [2, 59]. There is a concern that the use of the metaverse in education could widen the "digital divide" and exclude individuals [14].

Health-Related Concerns. Addressing the challenge of cyber addiction, also known as the cyber-syndrome, is crucial [4, 6, 33, 64, 68]. Additionally, the metaverse presents challenges due to limited physical space and cybersickness caused by sensory feedback mismatches [18, 39]. Mitigating cyber-sickness is important for metaverse classroom success, considering factors like gender, gaming experience, age, and ethnic origin. Technical settings like latency, frame rates, and navigation parameters influence cybersickness occurrence [60]. Another concern related to mental health is the physical isolation of the learners [30].

Teachers training and learning. According to Zhang (2022) [68], the primary concern with the metaverse revolves around human factors, specifically achieving an education that is centered around people. They highlight issues related to individual autonomy, technical training, and challenges faced by teachers, such as skill development and increased workload.

Ethics and Morality. The metaverse's global accessibility [6, 29, 39] raises concerns about conflicting ideologies, worldviews, simulated experiments, data privacy, racial issues, religious conflicts, bullying, violence, and more. These challenges stem from virtual identities [44]. Furthermore, the use of artificial intelligence in spatial computing and metaverse data generation raises ethical and security concerns [68]. User data collection in the metaverse is essential for governance, encompassing detailed information like facial images, physical state, transactions, and consumption records [29, 69]. However, the metaverse also poses risks for socially inexperienced learners, exposing them to criminal activities facilitated by online anonymity (e.g., fraud, surveillance, data breaches).

Content Democratization and Intellectual Property Rights. The metaverse encourages user-generated content in the virtual-physical space [59] Learners and educators are encouraged to contribute educational content, supported by NFTs [75] and sustainable economic models. Privacy considerations are important to ensure appropriate content overlays and mitigate privacy risks and copyright infringements [47, 68, 71].

Identity and Social Interaction. The integration of real and virtual worlds blurs the distinction between users' "real-me" and "virtual-me" identities, causing confusion [29]. Overreliance on social connections with avatars and NPCs can hinder real-world relationship development, creating emotional and social barriers [65]. Students who lack a clear sense of identity may struggle with transitioning to the real world [29, 47]. Additionally, creating avatars in the virtual environment poses a challenge as there is no consensus on whether avatar actions originate from real individuals or if individuals assign new roles to their avatars [47, 70].

3.3 Limitations of the Study

While SLRs are a powerful tool for synthesizing existing research, when investigating a topic as complex and evolving as the metaverse in education. Therefore, there is the risk that not all papers have been identified on the subject matter as not all databases were searched. Metaverse is becoming a popular area of both applied and theoretical research more and more work is being published every day, so, it is possible that someone reading the paper when this work is published, the study may not capture the most recent developments or unpublished studies in this rapidly evolving field of the metaverse in education.

In addition, SLRs may be affected by various types of biases [23] as studies with positive results are more likely to be published than those with negative or inconclusive results. This could potentially skew the understanding of the advantages and disadvantages of the metaverse in education.

Nevertheless, this SLR has identified insights about the impact of using metaverse in education. The identification of the advantages and limitations opens new fields of research giving directions on the challenges that researchers should focus. Moreover, it provides policy makers and law makers with new insights of the impact of metaverse and the legal challenges that they would have to face.

4 Discussion

The metaverse, being the embodiment of a multi-technology convergence, where various technologies and disciplines come together to create a unified virtual space. It combines elements of virtual reality, augmented reality, artificial intelligence, blockchain, internet of things, and other emerging technologies. Each of these components plays a crucial role in shaping the Metaverse, and their details are described in further detail in [41]. This convergence enables immersive and interactive experiences, seamless communication and collaboration, and the integration of physical and digital realms. The metaverse leverages these technologies to create a dynamic and interconnected environment where users can interact, explore, and create new experiences. It represents a fusion of technologies that blur the boundaries between the physical and digital worlds, offering vast possibilities for social, economic, and educational interactions.

Till recently, the metaverse was depicted in science fiction and movies, where it was often portrayed as a fully immersive virtual reality where individuals are represented by avatars that can interact with each other and the environment. Examples include the novel "Snow Crash" by Neal Stephenson, where the term "metaverse" is described as a virtual reality-based successor to the internet, and "Ready Player One" a novel by Ernest Cline (and its subsequent film adaptation), where people escape a dystopian future by spending their lives in an expansive virtual reality universe.

In the real world, tech companies like Meta Platforms Inc.[1] Are investing heavily in the development of a metaverse. They envision it as the next evolution of the internet, a place where people can work, learn, play, and create in a more immersive and interactive way. In order to enable this sort of utopian, researchers and policy makers need to look into the advantages, concerns and challenges of metaverse similar to the ones identified in the previous section. As the development of the metaverse continues, it will be interesting to see how these issues are addressed both in reality and in speculative fiction.

Moreover, having in mind that the global market for metaverse in education is expected to rise to 79.01 Billion by 2030[2], providing a unique outlook of the future educational path while a Gartner survey identifies that by 2026, 25% of the world's population will spend at least one hour a day in metaverse for activities that include among others socializing, working or learning [58]. The Metaverse's drive for innovation and entrepreneurship are visible both in the developed and the developing countries, creating a new economy that yields to socioeconomic advantages like job creation and revenue generation [28]. Leveraging technology presents an opportunity to enhance students' overall educational experience while offering freedom and diverse educational opportunities along with gamification [51].

However, as the development of metaverse technology is still in its early stages, it's important to be mindful of issues related to digital monopoly and privacy in metaverse education. While adhering to the fundamental principles of being people-oriented and balancing virtual with reality, efforts should be made to enhance teachers' digital literacy and skills. For the new generations such as generation Z, new technologies are a natural environment for them while for the older generations, this is not the case.

Furthermore, it's crucial that governments, businesses, educational institutions, and households all play a part in advancing the application of metaverse ecological education practices. This collective effort will guarantee that the complete potential of metaverse technology is utilized in building an ecologically conscious and environmentally sensitive society [35]. The development of the metaverse should contribute to the realization of human happiness, and not only the creation of a utopian where people are escaping from their dystopian reality. Balances should be kept so that the metaverse does not replace the real world, but is a virtual digital space reflected by the real world, and cannot exist in isolation from the real world.

Even though similar studies have been performed in the past [5, 10, 31, 39, 49] those studies do not provide a comprehensive overview and analysis of the advantages and challenges of metaverse in education. Our study, provides such a comprehensive

[1] https://about.meta.com/.

[2] Metaverse in Education Market-Global Industry Assessment and Forecast Report, https://www.vantagemarketresearch.com/industry-report/metaverse-in-education-market-1515.

identification and analysis, useful for researchers looking into new research avenues. It provides points of consideration for designers and developers of metaverse environments. In addition, the challenges identified such as data security concerns and privacy, and the intellectual property rights on content are useful for regulators and policy makers when drafting the corresponding policies and law.

5 Conclusions

The integration of the metaverse in education brings numerous benefits and challenges. The metaverse provides immersive experiences, overcoming traditional limitations and enabling visualization and interaction with complex concepts. Among others, it allows for exploration, collaboration, and creative expression. However, challenges include privacy and security concerns, the need for digital literacy, access disparities, and ethical dilemmas. This study examined the benefits and challenges of integrating metaverse in education through a SLR from technological and human perspectives. The findings emphasize the significance of designing and implementing metaverse systems to enhance educational practices and improve learner satisfaction. Moreover, they provide valuable insights for researchers in addressing the various challenges associated with metaverse tools and environments in education. Our future work will focus on further investigation of the challenges identified for proposing appropriate frameworks for tackling with them with a focus on accessibility and inclusion. As the metaverse becomes more integrated into education, it will be important to ensure that it is accessible to all students, regardless of their socioeconomic status, geographical location, or physical abilities.

References

1. Abrar, M.F., et al.: Motivators for large-scale agile adoption from management perspective: a systematic literature review. IEEE Access **7**, 22660–22674 (2019)
2. AbuKhousa, E., El-Tahawy, M.S., Atif, Y.: Envisioning architecture of metaverse intensive learning experience (MiLEx): career readiness in the 21st century and collective intelligence development scenario. Future Internet **15**(2), 53 (2023). https://doi.org/10.3390/fi15020053
3. Alam, A., Mohanty, A.: Metaverse and posthuman animated avatars for teaching-learning process: interperception in virtual universe for educational transformation. In: Panda, M., et al. (eds.) ICIICC 2022. CCIS, vol. 1737, pp. 47–61. Springer, Cham (2022). https://doi.org/10.1007/978-3-031-23233-6_4
4. Al-Kfairy, M., Al-Fandi, O., Alema, M., Altaee, M.: Motivation and hurdles for the student adoption of metaverse-based classroom: a qualitative study. In: 2022 International Conference on Computer and Applications (ICCA) (2022). https://doi.org/10.1109/ICCA56443.2022.10039672
5. Camilleri, M.A.: Metaverse applications in education: a systematic review and a cost-benefit analysis. Interact. Technol. Smart Educ. (2023)
6. Chen, Z.: Exploring the application scenarios and issues facing metaverse technology in education. Interact. Learn. Environ. 1–13 (2022). https://doi.org/10.1080/10494820.2022.2133148
7. Cheng, R., Wu, N., Chen, S., Han, B.: Will metaverse be NextG internet? Vision, hype, and reality. arXiv preprint arXiv:2201.12894 (2022). https://doi.org/10.48550/arXiv.2201.12894

8. Dahan, N.A., Al-Razgan, M., Al-Laith, A., Alsoufi, M.A., Al-Asaly, M.S., Alfakih, T.: Metaverse framework: a case study on E-learning environment (ELEM). Electronics **11**(10), 1616 (2022). https://doi.org/10.3390/electronics11101616
9. De Felice, F., Petrillo, A.: Green transition: the frontier of the digicircular economy evidenced from a systematic literature review. Sustainability **13**(19), 11068 (2021)
10. De Felice, F., Petrillo, A., Iovine, G., Salzano, C., Baffo, I.: How does the metaverse shape education? A systematic literature review. Appl. Sci. **13**(9), 5682 (2023). https://doi.org/10.3390/app13095682
11. Díaz, J.E.M., Saldaña, C.A.M., Avila, C.A.R.: Virtual world as a resource for hybrid education. Int. J. Emerg. Technol. Learn. **15**(15), 94–109 (2020). https://doi.org/10.3991/ijet.v15i15.13025
12. Dreamson, N., Park, G.: Metaverse-based learning through children's school space design. Int. J. Art & Design Educ. **42**(1), 125–138 (2023). https://doi.org/10.1111/jade.12449
13. García-Holgado, A., Mena, J., García-Peñalvo, FJ., González, C.: Inclusion of gender perspective in computer engineering careers: elaboration of a questionnaire to assess the gender gap in tertiary education. In: 2018 IEEE Global Engineering Education Conference (EDUCON), pp. 1547–1554 (2018)
14. Gülen, S., Dönmez, İ, Şahin, İD.İN.: STEM education in metaverse environment: challenges and opportunities. J. STEAM Educ. **5**(2), 100–103 (2022)
15. Guo, H., Gao, W.: Metaverse-powered experiential situational english-teaching design: an emotion-based analysis method. Front. Psychol. **13**, 1–9 (2022). https://doi.org/10.3389/fpsyg.2022.859159
16. Han, X., Hu, Y., Li, Y., Tan, B., Tu, X., Jiang, Y.: Design and research of campus culture application based on ar and metaverse technology. In: 2022 International Conference on Computation, Big-Data and Engineering (ICCBE) (2023). https://doi.org/10.1109/ICCBE56101.2022.9888114
17. He, N., Ding, K., Zhang, J.-B.: Exploration and research on digital education scenarios from the perspective of metaverse. In: 2022 10th International Conference on Orange Technology (ICOT) (2022). https://doi.org/10.1109/ICOT56925.2022.10008167
18. Hedrick, E., Harper, M., Oliver, E., Hatch, D.: Teaching & learning in virtual reality: metaverse classroom exploration; virtual dimension—a primer to metaverse. In: 2022 Intermountain Engineering, Technology and Computing (IETC) (2022). https://doi.org/10.1109/IETC54973.2022.9796765
19. Higgins, J.P., Green, S. (eds.): Cochrane Handbook for Systematic Reviews of Interventions. Wiley, Hoboken (2011)
20. Ho, C.: Research on teaching of metaverse technology flipped the MICE education. In: 2022 3rd International Conference on Education, Knowledge and Information Management (ICEKIM), pp. 592–596 (2022). https://doi.org/10.1109/icekim55072.2022.00136
21. Hwang, G.-J., Chien, S.-Y.: Definition, roles, and potential research issues of the metaverse in education: an artificial intelligence perspective. Comput. Educ.: Artif. Intell. **3**, 100082 (2022). https://doi.org/10.1016/j.caeai.2022.100082
22. Jaber, T.A.: Security risks of the metaverse world. Int. J. Interact. Mob. Technol. **16**(13) (2022)
23. Jahan, N., Naveed, S., Zeshan, M., Tahir, M.A.: How to conduct a systematic review: a narrative literature review. Cureus **8**(11) (2016)
24. Jiang, C., Xu, J.: The application research of education metaverse under the framework of SWOT analysis. In: Zhang, L.J. (ed.) METAVERSE 2022. LNCS, vol. 13737, pp. 55–67. Springer, Cham (2022). https://doi.org/10.1007/978-3-031-23518-4_5
25. Jovanović, A., Milosavljević, A.: VoRtex metaverse platform for gamified collaborative learning. Electronics **11**(3) (2022). https://doi.org/10.3390/electronics11030317

26. Khansulivong, C., Wicha, S., Temdee, P.: Adaptive of new technology for agriculture online learning by metaverse: a case study in faculty of agriculture, national university of laos. In: 2022 Joint International Conference on Digital Arts, Media and Technology with ECTI Northern Section Conference on Electrical, Electronics, Computer and Telecommunications Engineering (ECTI DAMT & NCON), pp. 428–432 (2022). https://doi.org/10.1109/ECTIDAMTNCON53731.2022.9720366
27. Kitchenham, B.: Procedures for performing systematic reviews. Keele UK Keele Univ. **33**(2004), 1–26 (2004)
28. Kshetri N.: Metaverse and Developing Economies. IT Prof. **24**(4), 66–69 (2022). https://doi.org/10.1109/MITP.2022.3174744
29. Kye, B., Han, N., Kim, E., Park, Y., Jo, S.: Educational applications of metaverse: possibilities and limitations. J. Educ. Eval. Health Prof. **18**, 32 (2021). https://doi.org/10.3352/jeehp.2021.18.32
30. Lee, I., Sung, Y.M., Kim, T.: The expanding role of metaverse platform in college education. Emerg. Sci. J. **13**(10), 1037–1044 (2022). https://doi.org/10.24507/icicelb.13.10.1037
31. Lee, J.: A study on the intention and experience of using the metaverse. Eur. J. Bioethics **13**(1), 177–192 (2022). https://doi.org/10.21860/j.13.1.10
32. Li, M., Yu, Z.: A systematic review on the metaverse-based blended English learning. Front. Psychol. **13**, 1–15 (2023). https://doi.org/10.3389/fpsyg.2022.1087508
33. Lin, H., Wan, S., Gan, W., Chen, J., Chao, H.-C.: Metaverse in education: vision, opportunities, and challenges. In: IEEE International Conference on Big Data, pp. 2857–2866 (2022). https://doi.org/10.1109/bigdata55660.2022.10021004
34. Liu, F., Zhang, Y., Zhao, L., Dai, Q., Liu, X., Shi, X.: A metaverse-based student's spatiotemporal digital profile for representing learning situation. In: 2022 8th International Conference of the Immersive Learning Research Network (iLRN) (2022). https://doi.org/10.23919/iLRN55037.2022.9815985
35. Liu, X.: The application of the metaverse in ecological education. In: Zhang, L.J. (ed.) METAVERSE 2022. LNCS, vol. 13737, pp. 95–102. Springer, Cham (2022). https://doi.org/10.1007/978-3-031-23518-4_8
36. Liu, X., Fan, Z., Gu, S., Peng, S., Wang, S.: A preliminary study on education and teaching based on the concept of metaverse—take "information technology" as an example. Front. Artif. Intell. Appl. **358**, 300–306 (2022). https://doi.org/10.3233/faia220396
37. Ma, L., Yang, Z., Yang, W., Yang, H., Lao, Q.: Study on the organization and governance of bigdata for lifelong education. In: Qiu, M., Gai, K., Qiu, H. (eds.) SmartCom 2021. LNCS, vol. 13202, pp. 493–500. Springer, Cham (2022). https://doi.org/10.1007/978-3-030-97774-0_45
38. Moher, D., Liberati, A., Tetzlaff, J., Altman, D.G.: Preferred reporting items for systematic reviews and meta-analyses: the PRISMA statement (reprinted from annals of internal medicine). Phys. Ther. **89**(9), 873–880 (2009)
39. Mystakidis, S.: Metaverse. Encyclopedia. **2**, 486–497 (2022). https://doi.org/10.3390/encyclopedia2010031, https://www.researchgate.net/publication/358497370_Metaverse
40. Ng, D.T.K.: What is the metaverse? Definitions, technologies and the community of inquiry. Aust. J. Educ. Technol. **38**(4), 190–205 (2022). https://doi.org/10.14742/ajet.7945
41. Ning, H., et al.: A survey on the metaverse: the state-of-the-art, technologies, applications, and challenges. IEEE Internet Things J. (2023)
42. Onggirawan, C.A., Kho, J.M., Kartiwa, A.P., Anderies, Gunawan, A.A.S.: Systematic literature review: the adaptation of distance learning process during the COVID-19 pandemic using virtual educational spaces in metaverse. Procedia Comput. Sci. **216**, 274–283 (2023). https://doi.org/10.1016/j.procs.2022.12.137
43. Page, M.J., et al.: The PRISMA 2020 statement: an updated guideline for reporting systematic reviews. Int. J. Surg. **88**, 105906 (2021)

44. Park, S., Kim, S.: Identifying world types to deliver gameful experiences for sustainable learning in the metaverse. Sustainability **14**(3), 1261 (2022). https://doi.org/10.3390/su14031361
45. Praherdhiono, H., et al.: Synchronization of virtual and real learning patterns in e-learning systems with metaverse concept. In: 2022 8th International Conference on Education and Technology (ICET), pp. 185–189 (2022). https://doi.org/10.1109/ICET56879.2022.9990891
46. Ruwodo, V., Pinomaa, A., Vesisenaho, M., Ntinda, M., Sutinen, E.: Enhancing software engineering education in Africa through a metaversity. In: 2022 IEEE Frontiers in Education Conference (FIE) (2022). https://doi.org/10.1109/FIE56618.2022.9962729
47. Sá, M.J., Serpa, S.: Metaverse as a learning environment: some considerations. Sustainability **15**(3), 2186 (2023). https://doi.org/10.3390/su15032186
48. Sánchez-López, I., Roig-Vila, R., Pérez-Rodríguez, A.: Metaverse and education: the pioneering case of minecraft in immersive digital learning. Prof. Inform. **31**(6), 1–16 (2022). https://doi.org/10.3145/epi.2022.nov.10
49. Sarıtaş, M.T., Topraklıkoğlu, K.: Systematic literature review on the use of metaverse in education. Int. J. Technol. Educ. (IJTE) **5**(4), 586–607 (2022). https://doi.org/10.46328/ijte.319
50. Shen, T., Huang, S., Li, D., Lu, Z., Wang, F., Huang, H.: Virtual classroom: a lecturer-centered consumer-grade immersive teaching system in cyber–physical–social space. IEEE Trans. Syst. Man Cybern.: Syst. **53**(6), 3501–3513 (2022). https://doi.org/10.1109/TSMC.2022.3228270
51. Srisawat, S., Piriyasurawong, P.: Metaverse virtual learning management based on gamification techniques model to enhance total experience. Int. Educ. Stud. **15**(5), 153–163 (2022)
52. Suh, W., Ahn, S.: Utilizing the metaverse for learner-centered constructivist education in the post-pandemic era: an analysis of elementary school students. J. Intell. **10**(17), 1–15 (2022). https://doi.org/10.3390/jintelligence10010017
53. Talan, T., Kalınkara, Y.: Students' opinions about the educational use of the metaverse. Int. J. Technol. Educ. Sci. (IJTES) **6**(2), 333–346 (2022). https://doi.org/10.46328/ijtes.385
54. Themistocleous, M., Christodoulou, K., Katelaris, L.: An educational metaverse experiment: the first on-chain and in-metaverse academic course. In: Papadaki, M., Rupino da Cunha, P., Themistocleous, M., Christodoulou, K. (eds.) EMCIS 2022. LNBIP, vol. 464, pp. 678–690. Springer, Cham (2023). https://doi.org/10.1007/978-3-031-30694-5_47
55. Themistocleous, M. Cunha, P. Tabakis, E., Papadaki, M.: Towards cross-border CBDC interoperability: insights from a multivocal literature review. J. Enterp. Inf. Manage. **36**(5), 1296–1318 (2023). https://doi.org/10.1108/JEIM-11-2022-0411. 1741-0398
56. van Dinter, R., Tekinerdogan, B., Catal, C.: Automation of systematic literature reviews: a systematic literature review. Inf. Softw. Technol. **136**, 106589 (2021)
57. Virtanen, S., Räikkönen, E., Ikonen, P.: Gender-based motivational differences in technology education. Int. J. Technol. Des. Educ. **25**, 197–211 (2015)
58. Walk-Morris T.: Gartner: a quarter of consumers will use the metaverse daily by 2026 (2022). https://www.retaildive.com/news/gartner-a-quarter-of-consumers-will-use-the-metaverse-daily-by-2026/618474/#:~:text=As%20companies%20invest%20heavily%20into,or%20consume%20entertainment%20by%202026. Accessed August 2023
59. Wang, M., Yu, H., Bell, Z., Chu, X.: Constructing an edu-metaverse ecosystem: a new and innovative framework. IEEE Trans. Learn. Technol. **15**(6), 685–696 (2022). https://doi.org/10.1109/TLT.2022.3210828
60. Wang, Y., Lee, L.-H., Braud, T., Hui, P.: Reshaping post-COVID-19 teaching and learning: a blueprint of virtual-physical blended classrooms in the metaverse era. In: 2022 IEEE 42nd International Conference on Distributed Computing Systems Workshops (ICDCSW), pp. 241–247 (2022). https://doi.org/10.1109/ICDCSW56584.2022.00053

61. Watanabe, T.: Space from line: what can metaverse support in education/learning activity?. In: 2023 17th International Conference on Ubiquitous Information Management and Communication (IMCOM) (2023). https://doi.org/10.1109/imcom56909.2023.10035616
62. Williamson, B., Gulson, K.N., Perrotta, C., Witzenberger, K.: Amazon and the new global connective architectures of education governance. Harv. Educ. Rev. **92**(2), 231–256 (2022)
63. World Economic Forum, THE METAVERSE: How could the metaverse impact education? (2022). https://www.weforum.org/agenda/2022/12/metaverse-impact-education-learning/
64. Wu, F., Javed, W., Popoola, O.R., Abbasi, Q., Imran, M.: An embodied approach for teaching advanced electronics in metaverse environment. In: 2022 29th IEEE International Conference on Electronics, Circuits and Systems (ICECS) (2022). https://doi.org/10.1109/ICECS202256217.2022.9970782
65. Yang, F., Ren, L., Gu, C.: A study of college students' intention to use metaverse technology for basketball learning based on UTAUT2. Heliyon **8**(9), e10562 (2022). https://doi.org/10.1016/j.heliyon.2022.e10562
66. Yilmaz, M., O'Farrell, E., Clarke, P.: Examining the training and education potential of the metaverse: Results from an empirical study of next generation SAFe training. J. Softw.: Evol. Process e2531 (2023). https://doi.org/10.1002/smr.2531
67. Yue, K.: Breaking down the barrier between teachers and students by using metaverse technology in education: based on a survey and analysis of Shenzhen City, China. IC4E 2022: Proceedings of the 2022 13th International Conference on E-Education, E-Business, E-Management, and E-Learning, pp. 40–44 (2022). https://doi.org/10.1145/3514262.3514345
68. Zhang, X., Chen, Y., Hu, L., Wang, Y.: The metaverse in education: definition, framework, features, potential applications, challenges, and future research topics. Front. Psychol. **13** (2022). https://doi.org/10.3389/fpsyg.2022.1016300
69. Zhao, Z., Zhao, B., Wan, X.: Research on personalized learning space in educational metaverse. In: Fourth International Conference on Computer Science and Educational Informatization (CSEI 2022), pp. 1–4 (2022). https://doi.org/10.1049/icp.2022.1479
70. Zheng, W., Yan, L., Zhang, W., Ouyang, L., Wen, D.: D→K→I: data-knowledge-driven group intelligence framework for smart service in education metaverse. IEEE Trans. Systems Man Cybern.: Syst. **53**(4), 2056–2061 (2022). https://doi.org/10.1109/TSMC.2022.3228849
71. Zhong, J., Zheng, Y.: Empowering future education: learning in the edu-metaverse. In: 2022 International Symposium on Educational Technology (ISET), pp. 292–295 (2022). https://doi.org/10.1109/ISET55194.2022.00068
72. Zhou, B.:. Building a smart education ecosystem from a metaverse perspective. Mob. Inf. Syst. (2022)
73. Zonaphan, L., Northus, K., Wijaya, J., Achmad, S., Sutoyo, R.: Metaverse as a future of education: a systematic review. In: Proceedings of 2022 8th International HCI and UX Conference in Indonesia, pp. 77–81 (2022)
74. Tlili, A., et al.: Is metaverse in education a blessing or a curse: a combined content and bibliometric analysis. Smart Learn. Environ. **9**(1), 1–31 (2022). https://doi.org/10.1186/s40561-022-00205-x
75. Christodoulou, K. Katelaris, L., Themistocleous, M, Christodoulou P., Iosif, E.: NFTs and the metaverse revolution: research perspectives and open challenges. In: Lacity, M., Treiblmaier, H. (eds.) Blockchains and the Token Economy: Theory and Practice, pp. 139–178. Palgrave Macmillan, Cham (2022)
76. Themistocleous, M., Christodoulou, K., Iosif, E., Louca, S., Tseas, D.: Blockchain in academia: where do we stand and where do we go? In: Proceedings of the Fifty-third Annual Hawaii International Conference on System Sciences, (HICSS 53), 7–10 January 2020. Maui, Hawaii, USA. IEEE Computer Society, Los Alamitos (2020). https://scholarspace.manoa.hawaii.edu/handle/10125/64398

Voice Assistants - Research Landscape

Alaa Almirabi(✉), Nikolay Mehandjiev, and Panagiotis Sarantopoulos

Alliance Manchester Business School, The University of Manchester, Booth Street West, Manchester M13 0PB, UK
{alaa.almirabi,n.mehandjiev,p.sarantopoulos}@manchester.ac.uk

Abstract. AI-powered Voice Assistants (VAs) emerge as attractive facilitators of the increasing interactions between people and machines in the Metaverse. VAs are still at early stages of adoption, so this systematic literature review charts the landscape of existing VA research with a focus on their use in different context. This helps us identify important aspects that require further examination. The findings indicate that while academic interest in this novel subject is increasing, the literature regarding the factors that drive continuous use does not take into account VAs' intelligence. Indeed, users perceive such devices as collaborative sentient actors rather than as tools. Taking these perceptions into consideration can help to increase the engagement of users with VAs.

Keywords: AI · Voice assistant (VA) · Usage · Interactions

1 Introduction

A Voice Assistant (VA) is a software that operates as a personal information manager [3], providing easy-to-use voice interface to IT. VAs have gained worldwide popularity over the last few years, changing the way users complete tasks, consume content, purchase products and search for information [37]. Up to 27% of the global online population now uses voice searches (McCue, 2018), and Juniper [28] has predicted an increase of up to 1000% in-home VA usage between 2018 and 2023. The rising popularity of VA applications in users' everyday lives is closely linked to convenience [67] and facilitating the user's daily tasks [12].

Vas and voice interfaces are especially relevant in the context of Metaverse, with its intensive immersive interaction making conventional keyboard and mouse interfaces unsuitable. User engagement with VAs is a key to success, yet we need to understand obstacles and risks associated with the technology, and how it can be developed better for multiple contexts. This paper aims to provide a comprehensive landscape of different perspectives onto VA technology by answering the following research question:

What are the current VA research efforts and the main research directions?

This paper is organised as follows. The next section focuses on the systematic literature review process. The third section reports the SLR results. The fourth section discusses an agenda for research into Vas, charting the next steps to future studies in VAs, to ensure their usability for different contexts including the Metaverse.

M. Papadaki et al. (Eds.): EMCIS 2023, LNBIP 501, pp. 18–37, 2024.
https://doi.org/10.1007/978-3-031-56478-9_2

2 The Method of the Systematic Literature Review

We use the process described in the PRISMA systematic review statement [39] (shown in Fig. 1) and a thematic analysis [10] of the papers thus collected. Inclusion and exclusion criteria are presented in Table 1 below.

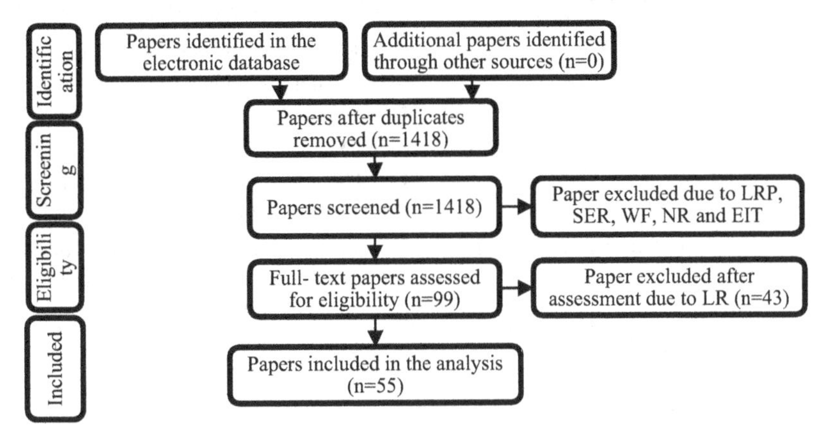

Fig. 1. The PRISMA flow chart that reports the different phases of the systematic literature review

Table 1. Inclusion and exclusion criteria

Incl/Excl.	Criteria	Principle
Exclusion	Low rank paper (LRP)	LRP: A paper from a conference which ranked B, C or lower or a journal that ranked 2 or lower
	Search engine reason (SER)	SER: A paper which has only its title, abstract, and keywords in English but not its full text
	Without full-text (WF)	WF: A paper without full text to be assessed
	Non-related (NR)	NR: A paper which is not an academic article. For example, editorial materials, conference reviews, contents, or forewords
	Loosely related (LR)	LR: A paper which does not focus on the review, survey, discussion, or problem-solving of VAs
	Exclude identified terms (EIT)	EIT: A paper does not include any of the identified terms (used in the search string) in its title, abstract or keywords

(continued)

Table 1. (*continued*)

Incl/Excl.	Criteria	Principle
Inclusion	High rank paper (HRP)	HRP: A paper which was published in a conference which ranked A* OR A, or a journal ranked 4*, 4 or 3
	Partially related (PR)	PR-1: VA is only used to support the description of some challenges, issues, or trends that a paper intends to deal with PR-2: VA is one of several objects that are reviewed or discussed
	Closely related (CR)	CR: The research efforts of a paper are explicitly and specifically dedicated to VAs
	Includes identified terms (IIT)	IIT: A paper includes at least one of the identified terms (used in the search string) in its title, abstract or keywords

2.1 Data Collection

Three electronic databases were used for the systematic search to collect academic research papers: SCOPUS, Science Direct and Web of Science. All the collected papers were published online before the end of November 2020 and were (a) published in conference proceedings or journals, and (b) written in English.

In order to collect a comprehensive set of papers, the following search was transcribbed in the syntax of each of the search engines:

`'voice assistant' and ('usage' or 'usability' or 'using' or 'artificial intelligence' or 'AI' or 'intelligent digital assistant' or 'intelligent assistant' or 'attitude' or 'motivation' or 'accept*' or 'adoption' or 'impact*' or 'factor' or 'Alexa' or 'Siri' or 'google home' or 'google now' or 'trust*' or 'satisf*').`

After removing duplicates, two steps were applied in the screening process stage (see Fig. 1). In the first step, the screening process was carried out to exclude papers with a low rank (LRP) based on the ABS 2018 list for the journal papers and the core conference portal website (http://portal.core.edu.au/conf-ranks/). To clarify, only journal papers ranked as 4*, 4 or 3 and conference papers ranked as A* or A were included. The papers without access to their full texts (WF) were removed, along with papers not written in English beyond their titles, abstracts and keywords (SER). In addition, we excluded papers which were not academic articles (NR), for example, editorials or book reviews, or did not include any of the identified terms (used in the search string) in its title, abstract or keywords (EIT).

The remaining papers went through the second stage of the screening process: each paper's title, abstract or full text (if required) was reviewed to exclude papers that did not focus primarily on VAs (LR). After that, all the qualifying papers were reviewed in-depth and then classified into one of the following sub-categories: High-rank paper (HRP), Closely related (CR) and Partially related (PR-1 and PR-2).

The following types of information were gathered regarding each included paper: titles, keywords, source database, 'conference' or 'journal'. For conference papers, the names of the conferences and their years were recorded, but for journal papers, their titles and categorisations, considering the impact factors of each paper.

2.2 Paper Categorisation

The included [53] papers were examined in terms of their focus and intent [60]. Afterwards, the included papers were categorised into four main categories based on a two-step analysis: *The first step* analysed the papers for keyword co-occurrences based on their titles, abstracts and keywords using the VOSviewer software, whilst *the second step* allocated papers to the clusters thus discovered by recording the contribution of the included papers to the relevant core concepts in each cluster.

Figure 2 presents a graphic representation of the clustering outcomes of the first step: four main clusters, their core concepts, and their connections. Careful examination of the network identifies four discrete clusters, each providing distinctive perspectives on the domain of VA technology as follows:

1) The yellow cluster, which is anchored on the core concepts of 'system', 'Google Assistant' and 'Siri', is focused on the technology itself, with a specific emphasis on voice assistant brands like Siri and Alexa.
2) The green cluster, anchored on the core concepts of 'home', 'attack', 'security' and 'accuracy', focuses on the performance, vulnerabilities and security.
3) The blue cluster, anchored on the core concepts of 'artificial intelligence' and the way it changes user perceptions of 'voice assistants' such as 'anthropomorphism', 'engagement' and 'user experience', comprises studies focused on the user experience, specifically analysing elements of engagement and overall satisfaction associated with interactions with voice assistants.
4) Finally, the red cluster, anchored on the core concepts of 'task', 'interaction' and 'efficiency', 'adoption' and 'trust', is focused on investigating the adoption, use and perceived utility of voice assistant systems, based on conventional utility- and ease-of-use- based adoption models such as TAM.

These clusters provide a suitable framework for comprehending the many facets of the ongoing research endeavour that is shaping the landscape of voice assistants.

A secondary analysis that considers the change of VA research focus over time is shown in Fig. 3. A noticeable shift in emphasis is evident, with a growing focus on the user interaction, anthropomorphism, and user engagement. This transition signifies the increasing development and sophistication of VA technology, as well as its integration with practical applications in everyday life.

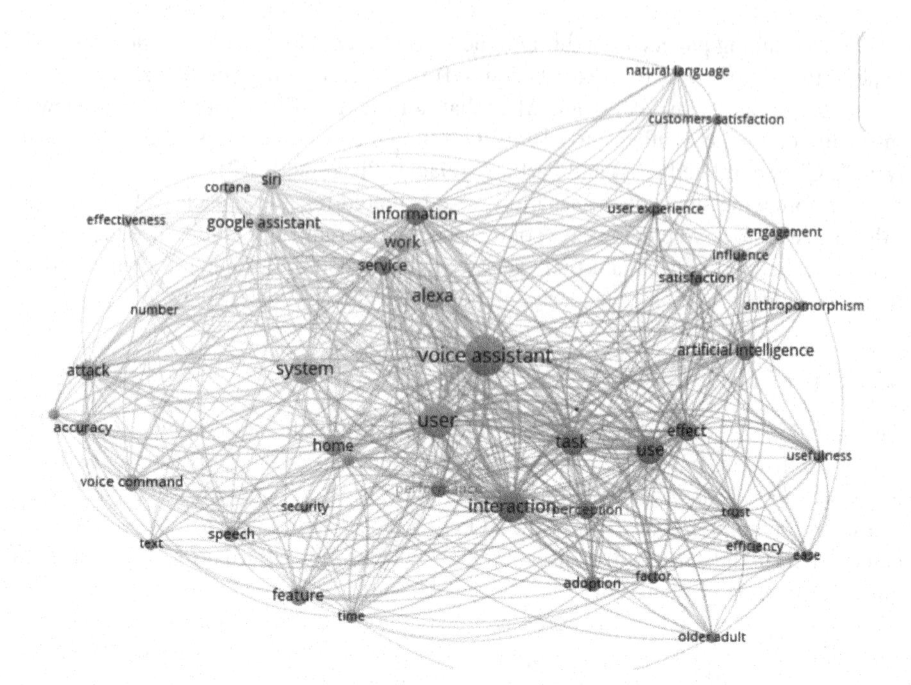

Fig. 2. VOSviewer map of the included papers

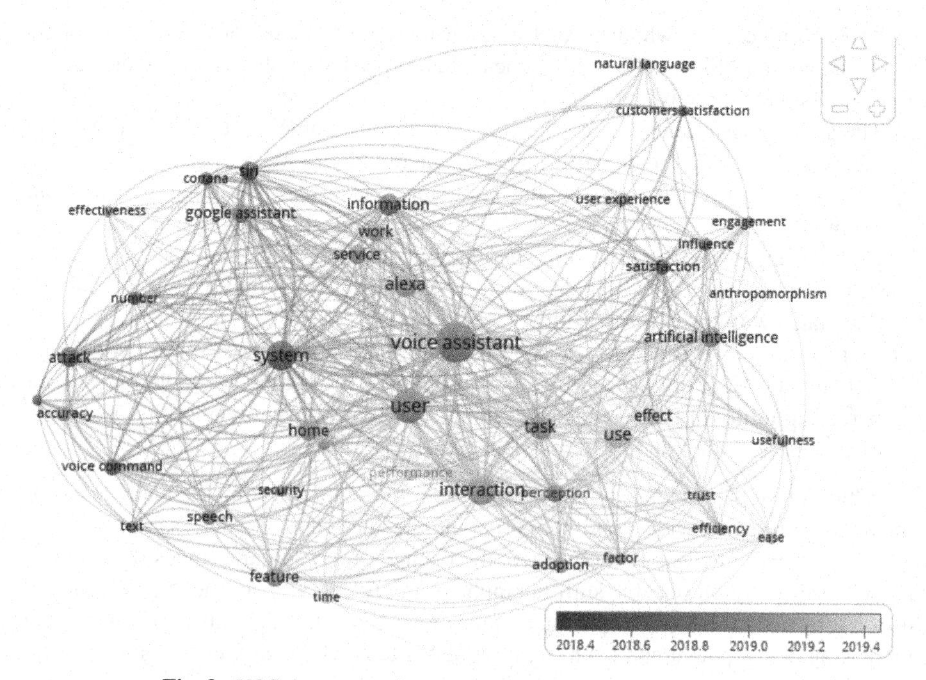

Fig. 3. VOSviewer map illustrating the shift of focus over time

3 Results: Themes of VA Research

The four clusters identified in the previous section give rise to four main themes of VA Research. In the second step, each of the selected papers was examined to establish how it relates to the four main themes, helping to identify three distinct aspects of the first theme. The resultant taxonomy is shown in Fig. 4.

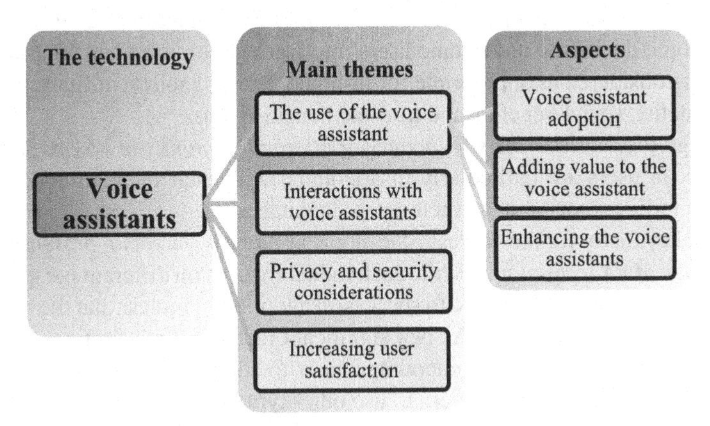

Fig. 4. The structure of the identified categories

3.1 The Use of Voice Assistants

The VAs use is researched from three different perspectives: the adoption of VAs, the value added to the users, and how VAs can be improved to encourage wider adoption.

Voice Assistant Adoption. Technology adoption is a well-researched area with widely used general models. However, VAs exhibit human-line characteristics, which makes them different from conventional technologies [61]. This is expected to impact the adoption dynamics and models, necessitating new customised adoption models for VAs [43]. McLean and Osei-Frimpong [37] confirm that AI-based VA technology has different attributes, making existing adoption models such as TAM and UTAUT unsuitable to explain users' behaviour towards the VA. Pal et al. [46] evaluate the efficacy of four widely used technology acceptance theories, namely TPB, TAM, VAM, and UTAUT, in forecasting the implementation intentions of VAs. Their results indicate that VAM outperforms TAM, UTAUT, and TPB in terms of peformance, and the lack of suitability of TAM is confirmed elsewhere [46].

Differences between voice assistants and conventional technology have been highlighted as (a) more implicit and transparent intraction with users, requiring new HCI evaluation assessment [43]; and (b) attractiveness of the devices involved and the social presence conveyed by the VAs [37]. Responding to these differences, a new adoption model is proposed [43] which combines three theoretical perspectives: the social exchange theory (SET), Human-Computer Interaction (HCI), and the Information systems success model (ISSM). This takes account of the effect of personal innovation on users' trust in their VAs to understand the quality preferences which most affect the adoption of the VA and describe the adoption of new AI-based technologies. This model concentrates on a new *system quality* construct as factor in the VA adoption, and combines it with the concepts of service quality and information quality [43].

Further gains can be made by considering the different types of VAs and a *customised model for in-home devices* [37] combines HCI attributes with Uses and Gratification theory (U>). This considers perceived privacy risks as an attitudinal dimension relevant to in-home devices to understand users' motivations for using and adopting them. Other factors considered as appropriate for in-home devices include utilitarian benefits, symbolic benefits, social benefits, and hedonic benefits [40].

Amugongo [4] confirms *the significance of interaction quality in VA adoption*, finding that African users are particularly disappointed with their interactions due to the technology's failure to understand their native languages.

A further key factor for the success of in-home systems is *the social presence created through the use of VA technology*. While each model focuses on different perspectives to identify the factors which contribute to successful adoption, it is clear that the humanlike social presence associated with VAs is a significant factor in the adoption of this new technology and the reasons which encourage users to adopt VAs in their home may be different to those which motivate them to use other types of VA technology. Fernandes and Oliveira [20] confirm that users' motivations play an essential role in adopting VAs in services. They demonstrate that relational, functional, and social factors all contribute to adoption, reveal their crossover impacts, and illuminate the moderating influence of experience and the desire for human interaction.

The practice of applying human characteristics to technological devices and the tendency to make objects 'humanlike' is known as *anthropomorphism*, and it is a distinctive facet of VAs, along with AI [25]. As a result, anthropomorphism has received particular attention as a key factor affecting the adoption of VAs. It has been studied from two primary aspects: firstly, its position in relation to digital VAs using a qualitative approach [61] and, secondly, the impact of an augmented reality visual embodiment of a VA on collective decision-making under three conditions: conducting the role alone, collaborating with a disembodied VA, and collaborating with an embodied VA [30]. In addition, Moriuchi [41] analysed three studies to ascertain the effect of anthropomorphism on users' intention to re-use a VA. The results of these studies indicated that anthropomorphism is viewed as having a positive impact on VA adoption. However, more recent research [20] concluded that its effect is not uniformly positive, but it sheds new light on the understudied customer-robot rapport-building role. *This suggests that further study may be necessary to validate these findings and contribute to a comprehensive understanding of anthropomorphism's impact on the acceptance of VAs.*

Adding Value to VA. Another significant aspect in promoting the use of new technology is its value as perceived by users. A value is defined as a "belief that a specific mode of conduct or end-state of existence is personally or socially preferable to an opposite or converse mode of conduct or end-state of existence" [50]. Factors that add value contribute to adopting the VA [43]. Advances in AI offer new ways to create value and improve user experience, notably by enabling people to interact with devices using everyday language [4]. Rzepka [52] confirms the importance of adding value by providing an in-depth analysis of means-ends value chains. These explain users' decision-making processes regarding human-computer speech interaction by focusing on identifying, defining, and understanding the user values that can be achieved using VAs. Other approaches which have been suggested to increase the VAs' value include *integrating VAs to increase functionality*. Zhao and Rau [65] propose an integrated VA that performs tasks in both personal and professional domains, combining two independent Vas and thus increasing the usefulness of the VA. They also propose a new 'synchronisation' interaction between a corporate VA and a personal VA. This enables two different VAs (a housekeeper for personal issues and a secretary for work-related issues) to share information and work together in order to complete tasks which could not be done without their collaboration. For instance, the first VA might complete the first part of the task and then wake the second VA to complete it. Taking the idea of collaboration a step further, Hettiachchi et al. [23] created Crowd Tasker (on Google Assistant), a novel system that enables crowd tasks to be performed via the VA. Using the VA, rather than a traditional online platform, offers crowd workers much greater flexibility, facilitating multitasking and allowing them to access work simply by talking to their devices, making this type of work more attractive.

Norval and Singh [45] provided another approach to adding value by *providing explanation functionality* about the Internet of Things (IoT) system embedding the VA. In order to provide meaningful answers, the VA uses recipe, log and provenance information, which enables it to track the sequence of steps involved and their inputs and outputs. This allows a user to ask about what a system should do and what a system has done. Therefore, it established the first stage in considering how VAs could help users better understand, challenge, evaluate, and accept VAs as new technology.

Another approach to making the VA more widely adopted is to *focus on specific countries, societies, or demographic groups* and consider the attributes required based on their particular needs. For example, Amugongo [4] conducted a study focused on six African countries (Namibia, Zimbabwe, South Africa, Nigeria, Zambia and Niger) to understand African users' opinions about VAs. Likewise, Hwang [27] analysed the conversational styles of some South Korean VAs (Bixby, Nugu, Kakao mini, Clova and Giga Ginie) to examine the effect of female gender stereotypes in assistants with female voices and to understand how these assistants interact with their users.

Creating VAs which are *adapted for use by older users* may also increase adoption. A study found that, over a year, most elderly participants stopped using their in-home VA (Amazon Echo), even though they had been early adopters [57]. This is due to a lack of perceived usefulness, concerns about their own technical abilities, and difficulties using the devices in shared accommodation. This suggests that current interfaces, which use the same style of auditory response in every situation, are unlikely to be widely adopted by older users unless they are adapted to meet their specific needs. Therefore, Kowalski et al. [31] established a study to explore the potential of VAs to meet the needs of older adults in the context of smart home technology. The four key needs emerged: a) an accessible design with a low barrier of entry, b) understanding technology and receiving feedback, c) control and assurance of security and d) seamless incorporation into everyday life. This issupported by Chattaraman et al. [12], who demonstrate that a one-size-fits-all design strategy fails to meet older users' needs by illustrating the degree to which the level of cognitive, functional and social support within an online shopping task depends on both the style of the agent interaction and the task-competency of the user. This indicates that personalisation would significantly increase the value of VAs and encourage their wider adoption. To illustrate this, Braun et al. [9] created four different assistant personalities ('Admirer', 'Friend', 'Butler' and 'Aunt') and compared them to a baseline (Default VA) in order to investigate the effect of personalised characters on user experience, acceptance, trust, and workload. Their findings indicate that personalisation increases confidence and likability when the VA character fits the user's personality, but mismatches can increase dissatisfaction.

Another demographic group that has received attention is *visually impaired users.* Storer et al. [56] confirm that users with visual impairments value VAs more highly than sighted users, and the technology's potential to aid people with visual impairments has increased usage within this group. As a result, Choi et al. [16] examined the perceptions of people with visual impairments in relation to the speech rate of their VAs and found that they were more satisfied with a default speed rate similar to human speech rather than at higher speeds. This study highlights both the impact of speech rate design on people with vision impairments and the potential of using speech rate as a design feature to promote wider adoption. Although VAs enhance accessibility for visually impaired users, Storer et al. [56] found that challenges can arise in mixed-visual-ability families, and these should be considered before bringing VAs into the home.

Skidmore and Moore [55] also identified problems when they examined Alexa with a view to enhancing its value in the education field, highlighting elements that hinder language learning, notably the absence of multilingual speech recognition. While their study confirmed that Alexa could support basic language learning, it requires further development to become an efficient language-learning tool. By focusing on specific countries or groups, each study provides different results based on the characteristics of their particular demographic. However, considering their findings together could prove an effective baseline for examining a new area targeting the adoption of VAs. For example, Hwang et al. [27] findings concerning the stereotypical aspects of femininity present in some South Korean VAs and Amugongo's [4] conclusions that VAs are developed without considering developing or emerging markets could provide new directions for development, adding user value (by broadening the range of languages and accents available, for example), and thus helping the wider adoption of VAs.

Moreover, Mayer, Laput and Harrison [36] offer a different value-adding approach embedded in WorldGaze, a smartphone software tool that engages the front and rear cameras to establish its physical location and follow the user's gaze. This enables the VA to provide context-specific information and respond more naturally. In a similar vein, Shen et al. [54] focus on the feasibility of determining the user's position using voice signals obtained from microphones, facilitating the addition of user localisation as a modern feature of VAs within the home. Furthermore, Zhao et al. [66] describe a reliable, low-power detector capable of operating in various scenarios to identify a novel way of interacting with VAs on smartwatches: simply by lifting your hand.

These approaches to added value have identified different factors that affect the adoption of the VA. According to Rzepka [52], five factors maximise users' overall value in using Vas: *convenience, efficiency, enjoyment, cognitive effort and ease of use.* The last three factors are related to existing constructs of "hedonic motivation" and "effort expectancy" in UTAUT2 theory [59] and also "perceived ease of use" in TAM [17]. However, Zhao and Rau [65] conclude that integrating VAs' functionality is the best-preferred type for users in terms of perceived usefulness, satisfaction, mental workload, and performance. Indeed, integrated VAs cover both work and personal domains, meeting a wider variety of user needs, thereby enhancing the user experience.

The benefits that users expect most from their VA use are *efficiency and convenience* [52]. By looking at the integrated VAs as an example, Zhao and Rau [65] confirm these are superior in terms of a) the required instructions (fewer), b) perceived usefulness (higher), c) completed tasks' quality (higher); d) satisfaction (higher) and e) mental workload required (less). This means an increase in the Vas' efficiency [52]. Furthermore, increasing the efficiency and entertainment value offered by VAs can increase users' satisfaction, thereby enhancing adoption [65]. The outcomes of applying these different approaches show the importance of certain common factors and their effect on the adoption of the VA from many perspectives.

As Rzepka [52] concluded, an integrated and comprehensive perspective on the values of using VAs can further understand when and why speech interaction is preferred over other modes of interaction with information systems. As Amugongo [4] suggests, a key concept in promoting the adoption of the VA in different regions is the need to meet local users' specific needs by resolving issues which negatively affect their user experience, thus enabling users to gain more value from using the devices. In addition, Norval and Singh [45] argued that value could be added by offering users a new way to understand the IoT ecosystems around them in terms of why specific functionality occurs and how the entities involved interact; this is achieved by providing sources of information, logging previous behaviours and actions and documenting the sequence of steps in a particular process or system. In summary, these approaches to adding value focus on two main concepts: understanding the users' values and helping users to understand how the technology operates.

Enhancing the Voice Assistants. Improving voice assistants would help to increase their adoption as an innovative new technology. As a result, the development of VAs has attracted scholarly interest in several areas, including customer queries, interactions, and how to process more complicated utterances. Due to the adoption of message communication via VAs in recent years, there is a need for tools that speed up the prototyping of feature-rich conversation systems. Therefore, Burtsev, Seliverstov, Airapetyan, Arkhipov, Baymurzina, Bushkov and Zaynutdinov [11] created DeepPavlov, an open-sourced library optimised for conversational agent development. It places a high value on modularity, extensibility, and efficiency in order to facilitate the development of dialogue systems from scratch and when limited data is available, and supports both modular and end-to-end methods of agent development. It also enables the development of multipurpose agents with a variety of skills. This is critical in real-world application contexts since abilities may be added, improved, or deleted separately after a conversation system has been installed, and existing capabilities can be used to accelerate the creation of new services and conversational solutions.

The aim of interaction development is to achieve seamless interactions which better meet user expectations. Generally, voice-activated interactive systems such as VAs operate in two stages: converting the input voice to text using an automatic speech recogniser and extracting information from text using natural language understanding to conduct the required task. However, performance can be affected by factors such as users' accents, pronunciation and use of homophones. As a result, Bhasin, Mathur, Yenigalla, and Natarajan [8] proposed a multistage CNN architecture system which uses phoneme sequences instead of word sequences to overcome these issues. Capable of processing partially generated phoneme sequences, it attained a level of accuracy comparable to the most advanced text classification systems across various stages and with phoneme sequences of varying lengths.

In addition, VAs frequently rely on a semantic parsing component to determine which actions to perform in response to a user's instruction. Usually, rule-based or statistical slot-filling systems have been used to parse simple inquiries, and shift-reduce parsers have been developed for more complicated utterances. However, while these approaches are effective, they place certain limits on the types of queries that may be parsed. Therefore, Rongali et al. [51] develop a unified architecture for semantic parsing, useful for a variety of query types with high levels of accuracy. Gamzu et al. [21] have also addressed problems that arise in voice shopping, notably mispronounced or misunderstood queries. Voice null queries like this typically fail to elicit the required response and negatively impact users' shopping experience. As a result, Gamzu et al. [21] developed a new framework for generating possible alternative queries, ranking them for suitability, and then substituting the best alternative for the null query. Testing in commercial VAs and an e-commerce website revealed fundamental differences between null voice queries and web shopping null queries, corroborating the employment of mechanisms designed specifically for the speech domain.

3.2 Interactions with Voice Assistants

The user-VA interactions significantly affect the user experience and could play a key role in encouraging wider adoption of this technology. Therefore, the interaction perspective is a significant research area. For example, Raveh et al. [49] explore how users interact with their VAs when another speaker is present. They measure the overall accommodation between participants during these interactions and analysed factors including gender, task performance, and task order to gain insights into the participants' behaviour. Additionally, Parviainen and Søndergaard [47] examine how users' experiences and interactions with VAs can be influenced by whispering. Research suggests that the manner in which users speak to their devices may reveal more about their personality and emotional condition than they realise. Hildebrand et al. [24] examine this from a business perspective, creating a conceptual framework for correlating vocal characteristics with human experiences and emotional states depending on user interaction and considering how businesses might use voice analytics in the future.

On the domestic front, Beirl et al. [7] demonstrate how VA technology influences in-home family interactions and the temporal change of usage patterns. Aslo, Lahoual and Frejus [33] perform an in-depth investigations of household and driving environments to better understand how VAs are used and to assess the efficacy of vocal interactions in these contexts. Moreover, Kim et al. [29] demonstrate that non-verbal voice cues play a crucial role in promoting positive emotional, motivational, and cognitive outcomes, moderated by users' characteristics, such as age. Therefore, Aeschlimann et al. [1] examine children's interactions with VAs and the impact these had on their prosocial behaviour and communication. They found that the children enjoyed playing with these technologies but did not expect to cooperate with them like a human.

The results of these studies indicate that the accommodation between users is more evident when the user first interacts with the gadget alone, and the chronological order of interactions affects participants' speech behaviour [49]. However, the interpersonal relationship's key variables are connectedness and intimacy, supported by the use of nonverbal empathic signals, and this has implications for human-AI interaction research [29]. It is also significant to note that in cases where VAs proved to be a barrier to the completion of tasks in household and driving environments [33], users often forgave and accepted these flaws rather than abandoning their devices, especially among experienced users with higher levels of knowledge and technical competence.

Beirl et al. [7] conclude that interactions with VAs motivate distinct trends of family engagement, including increased teamwork to scaffold children's relationships with the devices. Children eagerly adopt and interact with VAs, yet even very young children are able to distinguish AI from human intelligence and adapt their behaviour accordingly [1]. As Aeschlimann et al. [1] demonstrate, human cooperation principles are distinct from human-computer cooperation standards; however, both the positive and negative sides of family interactions are necessary components of understanding one another and improving empathy and conversational skills.

3.3 Privacy and Security Considerations

VAs are being rapidly improved to address existing privacy and security issues, yet their open nature makes it challenging to secure them from cyber-attacks [19]. VAs continuously monitor conversations for trigger words. When activated they send the conversation to a cloud system, where they are stored, processed, and often transmitted to other service providers. However, users have no chance to control this process, such as by tracking conversation recordings, nor can they control the recording behaviour of surrounding VAs. As a result, Cheng et al. [13] propose a prototype acoustic tagging device to signal how VAs and their back-end systems should handle recordings. Tags indicate recording consent and track recording time and location. This prototype which was established based on PocketSphinx [26], enables devices to tag conversations and retrieve the tagging signal from conversations saved in the cloud back-end system. In addition, the Aretha project [53] allows users to engage in conversations about network security and privacy to reduce such risks in their homes.

However, despite such developments, user concerns about privacy persist [34]. Security concerns could impede the adoption of VA technology, and existing research is focused on addressing issues such as *leakage of recording content* without users' authorisation, *adversarial audio attacks*, *voice squatting attacks* and *voice masquerading attacks*, *hidden voice command attacks*, and *spoofing attacks* [64]. Malicious software can cause damage by impersonating users and launching 'man-in-the-middle' attacks against the VAs' functionalities, targeting the Skill/App ecosystem through interface loopholes to redirect input (users' voices) to bogus Skills/Apps, thereby hijacking the conversation [38]. VAuth [19] is the first system to provide usable continuous VA authentication. It achieves a high matching accuracy, with a false positive rate of less than 0.1%. In addition, Gao et al. [22] propose a framework imposing privacy/security perimeter around a VA, where it jams the VA's microphones to block unauthorised voice commands, through a scheme similar to two-factor authentication, allowing access only when the authorised user is close to the VA.

Relatively little is known in the literature about defending against audio adversarial attacks [63]. As a result, audio compression and code modulation is proposed for proactive defence mechanisms against audio adversarial attacks. Moreover, Kwak et al. [32] explore the possibility of using text-converted voice command analysis to support security-critical command processes (e.g. making payments, checking emails, and unlocking devices) by identifying simple user commands, detecting suspicious commands, and maintaining high detection accuracy. They develop a globally trained model, open to all users (including new users) and gradually switched to a user-specific model tailored to a target user's speech patterns. When an optimal threshold value was used, the model detected 95.7% of attacks and achieved an equal error rate of 3.4%. However, some of the most common attacks on current voice interface systems are "hidden voice commands". In these cases, the attack is embedded in the user's voice command through synthetically rendered sounds in order to trick the speech recognition process into executing malicious commands without the user noticing [62]. Therefore, Wang et al. [62] propose a system that detects attacks through low-cost motion sensors, focusing on the unique audio vibration signatures of the issued voice commands.

Additionally, to avoid spoofing attacks launched by adversaries who record users' voice commands and then replay them, Ahmed et al. [2] provide a solution that analyses the spectral power patterns of voice signals using a classification model with 97 features. It has high reported detection rates of the following types of attack: inaudible voice command, hidden voice command, voice synthesis, combining replay attacks with live-human voices and equalisation manipulation attacks [2].

Moreover, Vaidya and Sherr [58] introduce new techniques that seek to prevent malicious technologies from falsifying or misdirecting users' speech. These work by sanitising voice inputs and removing distinctive voice characteristics, leaving only the information needed for the cloud-based services to recognise the speech. Zhang et al. [64] focus on voice masquerading threats, and propose a precise context-sensitive detector. Despite such developments, these attacks are likely to increase [64]. Given the fact that some users would never use a smart speaker because of privacy concerns and risk perceptions [52], integrating solutions in VAs is very important.

Furthermore, in terms of users' privacy with regard to health-related information, Cho [15] observe the effect of the modality (voice or text) and device types (mobile VA or in-home VA) on users' perceptions, specifically in terms of retrieving sensitive health information from their VAs. Voice interaction enhances the perceived social presence of VAs only if the users require less sensitive health information, and fewer privacy concerns were raised in these cases. Generally, only 'low-sensitivity' users felt positive about their voice interactions, and the type of device used had no significant effect on users' attitudes toward the VA, regardless of the sensitivity level of the health information being sought or the level of users' privacy concerns.

3.4 Increasing User Satisfaction

User satisfaction has been studied from a number of different perspectives. For example, Ernst and Malzahn [18] propose a model based on the Expectation Confirmation Theory to explain consumers' post-purchase behaviour, evaluating the role of Past Technology Category Satisfaction on people's use behaviour. Meanwhile, Lee et al. [34] test the effect of post-adoption use of Alexa on group dynamics among users sharing the device in order to understand the factors which drive satisfaction, the relationship between satisfaction and continuance usage, and the role of habituation as a mediator between them. In addition, Baier, Rese and Röglinger [6] used the Kano approach and the Segmented Kano perspective to explore conversational user interfaces' use in online shopping and their influence on customer satisfaction. They also investigate which are the most attractive use cases for different customer segments.

Another perspective that could affect user satisfaction is that of speech interaction systems, an area which is becoming more important due to their increasing popularity. Ning *et al.* [44] focus on the role of satisfactory VA responses for use intention. They have created a benchmark data set using the Sogou VA (a Chinese brand), which uses a real-world speech interaction system to predict intention prominence. Focusing on modality contribution analysis and multitask learning effects showed a 7% increase in understanding user intention, opening up new ways to increase user satisfaction.

Ernst and Malzahn [18] demonstate that use behaviour is indirectly influenced by satisfaction with similar technologies in the past (through Perceived Usefulness and Perceived Enjoyment). Regular users familiar with the VAs exhibit greater enthusiasm and less rejection when deciding which VA use cases are most attractive [6]. Hedonic Motivation, Compatibility, and Perceived Security are shown to explain over 60% of the variance of the satisfaction of post-adoption VA usage [34]. The same study shows that satisfaction is a predictor of continuance intent in most systems, yet the key drivers of satisfaction differ depending on the system's value propositions and the context in which the system operates [34].

To sum up, all these studies aimed to increase user satisfaction; therefore, it would be valuable to assess the percentage increase in satisfaction concerning each approach by examining the findings and applying them in specific case studies using different types of VAs (mobile and in-home VAs) and different brands (Siri, Google and Alexa for example) to ensure that similar results can be generated for all the VAs.

4 Discussion and Conclusion

This paper provides a systematic review and analysis of studies related to VAs in order to provide insights into the main directions of current research and thus to identify gaps in existing knowledge and thus motivate further work. It begins by providing an overview of VAs drawn from papers selected from prominent journals and conference proceedings concerned with voice-assistant-related content, identified via the Chartered Association of Business Schools (CABS) 2018 list and the Core Conference Portal website (http://portal.core.edu.au/conf-ranks/), and illustrating the VAs' significance. Following this, particular consideration is given to four particular aspects which emerged from the included papers: firstly, the use of the VA, including VA adoption, adding value to the

VA and enhancing the VAs; secondly, interactions with VAs; thirdly, privacy and security issues; and finally, user satisfaction.

The analysis of the papers thus organised established that despite the increasing academic interest in this area, the factors that drive continuous use and engagement require further work. Research on adopting digital technology, ranging from chatbots to humanoid robots, is sparse and fragmented. Additionally, contemporary research studies from a consumer perspective (e.g. [42]) often rely on existing acceptance mechanisms such as UTAUT and TAM, and this could be constrained as the usefulness of these frameworks is highly context-dependent. Furthermore, they are mostly concerned with functional characteristics without fully encapsulating automated technologies' extended dimensions and unique, empathic characteristics [35], with antropomorphism especially important in immersive contexts such as the Metaverse.

As Chi, Denton and Gursoy [14] note, while recent research indicates that a growing number of consumers are willing to embrace VA technology, resistance to change remains substantial and forms a fundamental impediment to adoption. One strategy for increasing user acceptance is to incorporate human characteristics into the design of the VA in order to maximise user familiarity and satisfaction by suggesting a human-like counterpart [48]. In respect of this, Araujo [5] concluded that critical distinctions should be made between anthropomorphic signals provided by embodied agents like robots (such as voice, body movements, facial expressions, dialogue) and disembodied agents like chatbots (e.g. dialogue, personality, voice, name). Fernandes and Oliveira [20] confirmed that further investigation of the importance of perceived humanness, rapport and interactivity could be a fruitful area for future studies. We believe this is especially important area for Metaverse research.

The effect of anthropomorphism on VAs has become a controversial topic. Based on the analysis presented here, no study has yet explored the role of anthropomorphism in relation to automated VAs and examined the most influential humanlike characteristics using a quantitative approach. More specifically, no one has yet provided an understanding of how anthropomorphism could affect the acceptance of the VA based on quantitative results and which humanlike characteristics could enhance their adoption. Addressing this gap will contribute to solving some of the current issues associated with adopting this technology by understanding what can be done to make users more willing to accept VAs and less resistant to change. This leads to the formulation of the following research question: To what extent could anthropomorphism play a role in enhancing VAs' adoption? Two main objectives to answer the research question were then identified: the first to explore anthropomorphism's significance in relation to VAs, and the second to examine the most powerful humanlike characteristics in different use contexts, e.g. in-home versus the Metaverse. The research will lead to a more comprehensive view of the adoption of VAs and offer managerial advice on integrating that technology successfully.

The limitation of this study is that all the papers were collected from only two multi-disciplinary databases (Web of Science and Science Direct) and one peer-reviewed literature database (Scopus). Another is that the search criteria were restricted to English language publications, which means that papers reporting VA research in other languages were excluded. From a completeness point of view, including more databases and more languages would give a more comprehensive picture; however, these restrictions were required to ensure the systematic literature review conducted here was feasible. The search of the three electronic databases yielded 1547 papers, of which [53] were included. Considering that voice assistance is a new topic in the field, and given the aim of identifying a specific knowledge gap, [53] papers would be a logical and reasonable number of papers whose data and findings could be relied on without doubts about the accuracy of their results or their reliability.

This study contributes to the literature on artificial intelligence-based technology usage, specifically VAs, by explaining the current research efforts and the main research directions to identify some clear potential avenues for future studies in using VAs in novel contexts such as the Metaverse. Despite the limitations set out above, this paper has provided an analysis of the current status of VA research and the key approaches adopted. The next stage of this research project will focus on addressing the gap in current knowledge identified here.

References

1. Aeschlimann, S., Bleiker, M., Wechner, M., Gampe, A.: Communicative and social consequences of interactions with voice assistants. Comput. Hum. Behav. **112**, 106466 (2020)
2. Ahmed, M., Kwak, I., Huh, J., Kim, I., Oh, T., Kim, H.: Void: a fast and light voice liveness detection system. In 29th USENIX Security Symposium (USENIX Security 20), pp. 2685–2702 (2020)
3. Almirabi, A., Chesney, T.: The effectiveness of the interaction and trust on the intelligent digital assistants usage. In: Proceedings of 140th IASTEM International Conference, London, UK (2018)
4. Amugongo, L.M.: Understanding what Africans say. In: Extended Abstracts of the 2018 CHI Conference on Human Factors in Computing Systems, pp. 1–6 (2018)
5. Araujo, T.: Living up to the chatbot hype: the influence of anthropomorphic design cues and communicative agency framing on conversational agent and company perceptions. Comput. Hum. Behav. **85**, 183–189 (2018)
6. Baier, D., Rese, A., Röglinger, M., Baier, D., Rese, A., Röglinger, M.: Conversational user interfaces for online shops? A categorization of use cases. In: ICIS (2018)
7. Beirl, D., Rogers, Y., Yuill, N.: Using voice assistant skills in family life. In: Computer-Supported Collaborative Learning Conference, CSCL, vol. 1, pp. 96–103. International Society of the Learning Sciences, Inc. (2019)
8. Bhasin, A., Mathur, G., Yenigalla, P., Natarajan, B.: Phoneme based domain prediction for language model adaptation. In: 2020 International Joint Conference on Neural Networks (IJCNN), pp. 1–6. IEEE (2020)
9. Braun, M., Mainz, A., Chadowitz, R., Pfleging, B., Alt, F.: At your service: designing voice assistant personalities to improve automotive user interfaces. In: Proceedings of the 2019 CHI Conference on Human Factors in Computing Systems, pp. 1–11 (2019)
10. Braun, V., Clarke, V.: Using thematic analysis in psychology. Qual. Res. Psychol. **3**(2), 77–101 (2006)
11. Burtsev, M., et al.: Deeppavlov: open-source library for dialogue systems. In: Proceedings of ACL 2018, System Demonstrations, pp. 122–127 (2018)
12. Chattaraman, V., Kwon, W., Gilbert, J.E., Ross, K.: Should AI-based, conversational digital assistants employ social-or task-oriented interaction style? A task-competency and reciprocity perspective for older adults. Comput. Hum. Behav. **90**, 315–330 (2019)

13. Cheng, P., Bagci, I.E., Yan, J., Roedig, U.: Smart speaker privacy control-acoustic tagging for personal voice assistants. In: 2019 IEEE Security&Privacy Workshops (SPW), pp. 144–149. IEEE (2019)
14. Chi, O.H., Denton, G., Gursoy, D.: Artificially intelligent device use in service delivery: a systematic review, synthesis, and research agenda. J. Hosp. Market. Manag. **29**(7), 757–786 (2020)
15. Cho, E.: Hey Google, can I ask you something in private?. In: Proceedings of the 2019 CHI Conference on Human Factors in Computing Systems, pp. 1–9 (2019)
16. Choi, D., Kwak, D., Cho, M., Lee, S.: "Nobody speaks that fast!" An empirical study of speech rate in conversational agents for people with vision impairments. In: Proceedings of the 2020 CHI Conference on Human Factors in Computing Systems, pp. 1–13 (2020)
17. Davis, F.D., Bagozzi, R., Warshaw, P.: User acceptance of computer technology: a comparison of two theoretical models. Manage. Sci. **35**(8), 982–1003 (1989)
18. Ernst, C. P. H., Malzahn, B.: If at first you don't succeed, try, try again'might not always make sense: on the influence of past technology category satisfaction on technology usage. In: Proceedings of AMCIS 2019 (2018)
19. Feng, H., Fawaz, K., Shin, K.G.: Continuous authentication for voice assistants. In: Proceedings of the 23rd Annual International Conference on Mobile Computing and Networking (2017)
20. Fernandes, T., Oliveira, E.: Understanding consumers' acceptance of automated technologies in service encounters: Drivers of digital voice assistants adoption. J. Bus. Res. **122** (2021)
21. Gamzu, I., Haikin, M., Halabi, N.: Query rewriting for voice shopping null queries. In: Proceedings of the 43rd International ACM SIGIR Conference on Research and Development in Information Retrieval, pp. 1369–1378 (2020)
22. Gao, C., Chandrasekaran, V., Fawaz, K., Banerjee, S.: Traversing the quagmire that is privacy in your smart home. In: Proceedings of the 2018 Workshop on IoT Security and Privacy (2018)
23. Hettiachchi, D., et al.: "Hi! I am the crowd tasker" crowdsourcing through digital voice assistants. In: Proceedings of the 2020 CHI Conference on Human Factors in Computing Systems, pp. 1–14 (2020)
24. Hildebrand, C., Efthymiou, F., Busquet, F., Hampton, W.H., Hoffman, D.L., Novak, T.P.: Voice analytics in business research: conceptual foundations, acoustic feature extraction, and applications. J. Bus. Res. **121**, 364–374 (2020)
25. Horstmann, A.C., Bock, N., Linhuber, E., Szczuka, J.M., Straßmann, C., Krämer, N.C.: Do a robot's social skills and its objection discourage interactants from switching the robot off? PLoS ONE **13**(7), e0201581 (2018)
26. Huggins-Daines, D., Kumar, M., Chan, A., Black, A.W., Ravishankar, M., Rudnicky, A.I.: Pocketsphinx: a free, real-time continuous speech recognition system for hand-held devices. In: 2006 IEEE International Conference on Acoustics Speech and Signal Processing Proceedings, vol. 1, p. I (2006)
27. Hwang, G., Lee, J., Oh, C.Y., Lee, J.: It sounds like a woman: exploring gender stereotypes in South Korean voice assistants. In: Extended Abstracts of the 2019 CHI Conference on Human Factors in Computing Systems, pp. 1–6 (2019)
28. Juniper Research. Voice Assistants used in smart homes to grow 1000%, reaching 275 million by 2023, as Alexa leads the way (2018). https://www.juniperresearch.com/press/press-releases/voice-assistants-used-in-smart-homes
29. Kim, J., Kim, W., Nam, J., Song, H.: " I can feel your empathic voice": effects of nonverbal vocal cues in voice user interface. In: Extended Abstracts of the 2020 CHI Conference on Human Factors in Computing Systems, pp. 1–8 (2020)

30. Kim, K., de Melo, C.M., Norouzi, N., Bruder, G., Welch, G.F.: Reducing task load with an embodied intelligent virtual assistant for improved performance in collaborative decision making. In: 2020 IEEE Conference on Virtual Reality and 3D User Interfaces (VR), pp. 529–538. IEEE (2020)
31. Kowalski, J., et al.: Older adults and voice interaction: a pilot study with Google Home. In: Extended Abstracts of the 2019 CHI Conference on Human Factors in Computing Systems (2019)
32. Kwak, I.Y., Huh, J.H., Han, S.T., Kim, I., Yoon, J.: Voice presentation attack detection through text-converted voice command analysis. In: Proceedings of the 2019 CHI Conference on Human Factors in Computing Systems, pp. 1–12 (2019)
33. Lahoual, D., Frejus, M.: When users assist the voice assistants: from supervision to failure resolution. In Extended Abstracts of the 2019 CHI Conference on Human Factors in Computing Systems, pp. 1–8 (2019)
34. Lee, K., Lee, K.Y., Sheehan, L.: Hey Alexa! A magic spell of social glue?: sharing a smart voice assistant speaker and its impact on users' perception of group harmony. Inf. Syst. Front. **22**, 563–583 (2020)
35. Lin, Q., Sun, X., Chen, X., Shi, S.: Effect of pretreatment on microstructure and mechanical properties of Nafion™ XL composite membrane. Fuel Cells **19**(5), 530–538 (2019)
36. Mayer, S., Laput, G., Harrison, C.: Enhancing mobile voice assistants with worldgaze. In: Proceedings of the 2020 CHI Conference on Human Factors in Computing Systems, pp. 1–10 (2020)
37. McLean, G., Osei-Frimpong, K.: Hey Alexa… examine the variables influencing the use of AI in-home voice assistants. Comput. Hum. Behav. **99**, 28–37 (2019)
38. Mitev, R., Miettinen, M., Sadeghi, A.R.: Alexa lied to me: skill-based man-in-the-middle attacks on virtual assistants. In: Proceedings of the 2019 ACM Asia Conference on Computer and Communications Security, pp. 465–478 (2019)
39. Moher, D., Liberati, A., Tetzlaff, J., Altman, D.G., Prisma Group: Preferred reporting items for systematic reviews and meta-analyses: the PRISMA statement. Int. J. Surg. **8**(5), 336–341 (2010)
40. Moriuchi, E.: Okay, Google!: An empirical study on voice assistants on consumer engagement and loyalty. Psychol. Mark. **36**(5), 489–501 (2019)
41. Moriuchi, E.: An empirical study on anthropomorphism and engagement with disembodied AIs and consumers' re-use behavior. Psychol. Mark. **38**(1), 21–42 (2021)
42. Moussawi, S., Koufaris, M., Benbunan-Fich, R.: How perceptions of intelligence and anthropomorphism affect adoption of personal intelligent agents. Electron. Mark. **31**, 343–364 (2021)
43. Nasirian, F., Ahmadian, M., Lee, O.K.D.: AI-based voice assistant systems: evaluating from the interaction and trust perspectives (2017)
44. Ning, Y., et al.: Multi-task deep learning for user intention understanding in speech interaction systems. In: Proceedings of the AAAI Conference on Artificial Intelligence, vol. 31, no. 1 (2017)
45. Norval, C., Singh, J.: Explaining automated environments: Interrogating scripts, logs, and provenance using voice-assistants. In: Adjunct Proceedings of the 2019 ACM International Joint Conference on Pervasive and Ubiquitous Computing and Proceedings of the 2019 ACM International Symposium on Wearable Computers, pp. 332–335 (2019)
46. Pal, D., Arpnikanondt, C., Funilkul, S., Chutimaskul, W.: The adoption analysis of voice-based smart IoT products. IEEE Internet Things J. **7**(11), 10852–10867 (2020)
47. Parviainen, E., Søndergaard, M.L.J.: Experiential qualities of whispering with voice assistants. In: Proceedings of the 2020 CHI Conference on Human Factors in Computing Systems, pp. 1–13 (2020)

48. Purington, A., Taft, J.G., Sannon, S., Bazarova, N.N., Taylor, S.H.: "Alexa is my new BFF" social roles, user satisfaction, and personification of the Amazon Echo. In: Proceedings of the 2017 CHI Conference Extended Abstracts on Human Factors in Computing Systems, pp. 2853–2859 (2017)
49. Raveh, E., Siegert, I., Steiner, I., Gessinger, I., Möbius, B.: Three's a crowd? Effects of a second human on vocal accommodation with a voice assistant. In: INTERSPEECH, pp. 4005–4009 (2019)
50. Rokeach, M.: The Nature of Human Values. Free Press (1973)
51. Rongali, S., Soldaini, L., Monti, E., Hamza, W.: Don't parse, generate! A sequence to sequence architecture for task-oriented semantic parsing. In: Proceedings of the Web Conference 2020, pp. 2962–2968 (2020)
52. Rzepka, C.: Examining the use of voice assistants: a value-focused thinking approach. In: Twenty-fifth Americas Conference on Information Systems, Cancun (2019)
53. Seymour, W.: Privacy therapy with aretha: what if your firewall could talk?. In: Extended Abstracts of the 2019 CHI Conference on Human Factors in Computing Systems, pp. 1–6 (2019)
54. Shen, S., Chen, D., Wei, Y.L., Yang, Z., Choudhury, R.R.: Voice localization using nearby wall reflections. In: Proceedings of the 26th Annual International Conference on Mobile Computing and Networking, pp. 1–14 (2020)
55. Skidmore, L., Moore, R.K.: Using Alexa for flashcard-based learning. In: Proceedings of Interspeech 2019, pp. 1846–1850. ISCA (2019)
56. Storer, K.M., Judge, T.K., Branham, S.M.: "All in the same boat": tradeoffs of voice assistant ownership for mixed-visual-ability families. In: Proceedings of the 2020 CHI Conference on Human Factors in Computing Systems, pp. 1–14 (2020)
57. Trajkova, M., Martin-Hammond, A.: "Alexa is a toy": exploring older adults' reasons for using, limiting, and abandoning echo. In: Proceedings of the 2020 CHI Conference on Human Factors in Computing Systems, pp. 1–13 (2020)
58. Vaidya, T., Sherr, M.: You talk too much: limiting privacy exposure via voice input. In: 2019 IEEE Security and Privacy Workshops (SPW), pp. 84–91. IEEE (2019)
59. Venkatesh, V., Thong, J. Y., Xu, X.: Consumer acceptance and use of information technology: extending the unified theory of acceptance and use of technology. MIS Q. 157–178 (2012)
60. Vessey, I., Ramesh, V., Glass, R.L.: Research in information systems: an empirical study of diversity in the discipline and its journals. J. Manage. Inf. Syst. **19**(2), 129–174 (2002)
61. Wagner, K., Schramm-Klein, H.: Alexa, Are You Human? Investigating Anthropomorphism of Digital Voice Assistants-A Qualitative Approach. In: ICIS (2019)
62. Wang, C., Anand, S.A., Liu, J., Walker, P., Chen, Y., Saxena, N.: Defeating hidden audio channel attacks on voice assistants via audio-induced surface vibrations. In: Proceedings of the 35th Annual Computer Security Applications Conference, pp. 42–56 (2019)
63. Zhang, J., Zhang, B., Zhang, B.: Defending adversarial attacks on cloud-aided automatic speech recognition systems. In: Proceedings of the Seventh International Workshop on Security in Cloud Computing, pp. 23–31 (2019)
64. Zhang, N., Mi, X., Feng, X., Wang, X., Tian, Y., Qian, F.: Dangerous skills: understanding and mitigating security risks of voice-controlled third-party functions on virtual personal assistant systems. In: 2019 IEEE Symposium on Security and Privacy (SP), pp. 1381–1396. IEEE (2019)
65. Zhao, J., Rau, P.L.P.: Merging and synchronizing corporate and personal voice agents: comparison of voice agents acting as a secretary and a housekeeper. Comput. Hum. Behav. **108**, 106334 (2020)

66. Zhao, S., et al.: Raise to speak: an accurate, low-power detector for activating voice assistants on smartwatches. In: Proceedings of the 25th ACM SIGKDD International Conference on Knowledge Discovery & Data Mining, pp. 2736–2744 (2019)
67. Zhou, S., Jia, J., Wang, Q., Dong, Y., Yin, Y., Lei, K.: Inferring emotion from conversational voice data: a semi-supervised multi-path generative neural network approach. In: Proceedings of the AAAI Conference on Artificial Intelligence, vol. 32, no. 1 (2018)

Blockchain Technology and Applications

Web Mining for Estimating Regulatory Blockchain Readiness

Andreas Vlachos, Elias Iosif, and Klitos Christodoulou(✉)

Department of Digital Innovation, School of Business, Institute for the Future (IFF), University of Nicosia, Nicosia, Cyprus
{vlachos.a,iosif.e,christodoulou.kl}@unic.ac.cy
https://www.unic.ac.cy/iff/

Abstract. The regulatory landscape for cryptocurrencies and blockchain tokens is a critical factor shaping business decisions and opportunities. This study proposes a computational model that leverages Web mining through search engines to quantitatively estimate the regulatory readiness of countries in relation to cryptocurrencies. The model's performance is validated through experimental trials supplemented with un- supervised clustering techniques for deeper analysis of the derived estimations. The findings demonstrate the effectiveness of the model in assessing regulatory tendencies and offer valuable insights for policymakers and industry stakeholders. This algorithmic approach presents an algorithmic methodology over manual regulatory assessments and subjective methods, showcasing its potential to guide regulatory frameworks in the rapidly evolving space of cryptocurrencies.

Keywords: Blockchain Regulation · Cryptocurrency Blockchain Business Intelligence · Web Mining · Regulatory · Readiness · Policy Guidance

1 Introduction

The transformative potential of blockchain technologies is well acknowledged [3, 4, 10, 25]. Indicative studies about the current challenges, as well as the future opportunities are presented in [6, 17, 22, 28, 29]. Although the underlying technology has shown its potential in various industries, the regulatory treatment of cryptocurrencies (and, in general, blockchain tokens) has become a critical concern subject to the intricacies of national jurisdictions [30]. In certain nations like the United States (USA), the legal treatment of cryptocurrencies (aka cryptoassets) varies significantly across diverse intranational jurisdictions [26].

In contrast to the regulatory landscape observed in jurisdictions like the United States, the European Union (EU) has taken a significant step towards enhancing the regulatory clarity and oversight of cryptocurrencies. On April 20, 2023, the European Parliament adopted the Markets in Crypto-Assets Regulation (MiCA), a regulatory framework proposed to address the challenges posed by the rapid growth of cryptocurrencies at the same time protecting investors, introduce security measures and anti-money laundering rules [11].

M. Papadaki et al. (Eds.): EMCIS 2023, LNBIP 501, pp. 41–51, 2024.
https://doi.org/10.1007/978-3-031-56478-9_3

Such diversification in the regulatory treatment for cryptocurrencies and blockchain technology in general, highlight the importance for stakeholders within the cryptocurrency industry to thoroughly assess the regulatory landscape of specific regions before embarking on pivotal decisions related to potential business ventures, encompassing activities such as establishing blockchain companies and making investment commitments.

To this date, attempts mainly involve manual and subjective assessment of the regulatory landscape. This assessment is a time-consuming process requiring very specific expertise. Even if those conditions are met, the subjective and manual character of it may lead to erroneous estimations affecting decision making procedures.

The present work suggests that the proper assessment of cryptocurrency regulation shall be based on objective information and be able to scale into a wide range of assessed countries. The acquisition of such information should be conducted in an algorithmic manner. In this work, we propose an algorithmic model based on web-harvested information for assessing cryptocurrency regulation. This is done in order to measure the aforementioned regulatory tendency for each country. The proposed algorithm estimates a numerical score for each country, which represents the degree of "non-hostile" local cryptocurrency regulations. The underlying methodology is based on Web-harvested information retrieved from Web search engines, and extends metrics used in the areas of information retrieval and natural language processing.

By bridging the gap between the increasing regulatory demands and the need for efficient assessment methods, our work contributes to shaping informed decision-making processes for industry stakeholders, policymakers, and regulators in an era marked by both advancements in cryptocurrency regulation and the continuous evolution of the global regulatory landscape [12].

2 Related Work

A favorable regulatory landscape surrounding the treatment of cryptocurrencies might be able to boost the adoption by local industries and governments, thereby attracting regional and foreign investments. Regulation is currently the biggest obstacle to widespread cryptocurrency adoption for investment funds [8]. Prior attempts to assess the regulatory landscape of emerging technologies have been based on surveys and assessment of legal systems, documents and articles by panels of experts. Examples of such assessments have been implemented for areas like autonomous vehicles by the Autonomous Vehicles Readiness Index [20], digital transformation by the Network Readiness Index [2] and automation by the Automation Readiness Index [23]. To the best of our knowledge, no end–to–end algorithmic framework was employed for the development of the aforementioned indices. We believe that creating an algorithmic and automated method of assessing the legal environment of a specific sector, will be an appropriate tool to be adopted globally for an unlimited number of countries and industries.

The same observation holds for the specific case cryptocurrencies and, in general, blockchain tokens. There is limited work done to date, with the aim to provide an assessment of the legal environment surrounding Bitcoin and other cryptocurrencies.

Even though some documents have been published which provide a glimpse of the regulatory areas in some countries (e.g., see [9, 13, 21, 24]) an algorithmic model of global assessment, as mentioned before, is profoundly lacking. Those studies do not exhibit a clear quantitative character (e.g., score-based rankings) which could significantly reduce the assessment time.

An industry-oriented study has been recently published as an initial attempt to translate the cryptocurrency regulatory environment of 249 countries into a numerical score [7]. This study was based on five indicators summarized as follows:

1. **Legality of cryptocurrencies.** This indicator examines whether cryptocurrencies are considered legal or are banned by local governments. The governments which have not banned the use of cryptocurrencies, are considered to be within a legal grey area which is outlined as "dangerous".
2. **Initial Coin Offerings (ICOs) restrictions.** The assessment of local re- strictions of ICOs, including complete bans and hostile regulations, which could affect investors' ability to invest and raise obstacles that developers will need to overcome in order to develop their projects. An example of this indicator is the assessment of the jurisdictions in the USA, which classifies coin offerings under securities laws, and have forced a number of blockchain projects to reconsider the location of their company's registration.
3. **ICOs locations.** The consideration of the number of ICOs which have officially been registered at a given nation.
4. **Exchanges locations.** The availability of exchanges in countries is important for traders and potential investors. It is still essential for cryptocurrency users to be able to easily convert crypto–to–fiat and fiat-to-crypto, especially when certain exchanges have faced hostile regulations in specific regions of the world. Note that it is unclear whether this indicator assesses the number of the official registration of exchanges per country or the ability of local users per country to register for any exchange worldwide.
5. **User opinions.** This represents a subjective indicator, whereas website users are encouraged to vote for the countries they believe they have the most friendly cryptocurrency regulations.

The aforementioned work can be regarded as a first attempt of providing a numerical assessment of cryptocurrency regulations worldwide. However, it comes with a number of inefficiencies. The procedure of obtaining the scores is highly unclear, as the formula used is not published. Also, there is no explanation of how the score of each indicator was calculated. Those observations imply a manual (and not fully transparent) methodology. We believe that a proper algorithmic and automated procedure, driven by web-harvested data can be used and, at some extent, automate the manual approach. The proposed algorithmic approach is formalized and experimentally validated in the sections that follow.

3 Model: Regulatory Stance Hypothesis

In this section, we present a computational model used for estimating the regulatory stance of a given country with respect to cryptocurrencies. That is, the degree of "non-hostile" local cryptocurrency regulations as mentioned in Sect. 1. This estimation is conducted based on lexical information harvested from the world wide web using search engines. The regulatory stance for the countries of interest is assumed to be reflected in the web documents indexed by the used search engines. Those documents are meant to thematically cover all the indicators summarized in Sect. 2 (and, of course, more). Next, the following aspects of the proposed model are described:

- the underlying hypothesis,
- the model parameters also including the needed web search queries, and
- a metric along with the respective query complexity.

Regulatory Stance Hypothesis Based on Lexical Co-occurrence. Consider a country c. The co-occurrence of positive/negative regulation-related cues with c's references within a coherent linguistic environment, implies c's tendency to- wards a positive/negative stance. For example, the positive stance of a particular country with regards to cryptocurrency regulation is expected to be observable (i.e., expressed) in articles that are publicly available in the World Wide Web. For short, the above hypothesis is also referred to as the *Regulatory Stance Hypothesis (RSH).*

The foundations of RSH lie in a variant of the widely-used *Distributional Hypothesis of Meaning (DHM)* [14] which has been applied in the area of lexical semantics (and, in general, in the areas of natural language processing and information retrieval) for estimating the semantic similarity between words, as well as multi-word terms. DHM is the core of distributional semantic models (DSMs) suggesting that the similarity of context implies similarity of meaning [19]. A variation of DHM considers the co-occurrence of words for deriving association measurements which can quantify the semantic similarity of the corresponding words, for example see [1] and [18]. One of the earliest works on word co-occurrence and associations was proposed in [27]. Such semantic models are of great applicability also in the broader area of semantic web [5].

The RSH-based model proposed in the present work constitutes a new approach. Motivated by the hypothesis that the world wide web can be regarded as the largest resource of lexical information, the regulatory stance hypothesis is implemented as a "contrast measurement" between positive and relative lexical cues. Specifically, the "contrast" aspect is what makes the present approach different compared to previous DSM-based models utilized for semantic similarity estimates. Unlike the case of semantic similarity for which the notion of "contrast measurement" cannot be defined, this type of measurement is instrumental for the case of regulatory stance. Another difference is that DHM and DSMs deal with pairs of words, while the proposed RSH-based model takes as input a single argument (i.e., the country of interest – see below).

Given a country c, the output of the model is a numerical score, R_c, which quantifies c's tendency towards a positive/negative stance. Following the proposed RSH, the R_c score is computed according to the co-occurrence of c's references with positive and negative lexical cues. The linguistic environment where the co-occurrence is considered is the textual context of web documents. Assume two sets, namely, P and N, which contain lexical cues that imply positive and negative stance respectively. The model parameters needed for defining R_c are:

- p_c: Total number of web documents in which word c co-occurs with positive cues.
- n_c: Total number of web documents in which word c co-occurs with negative cues.
- t_c: Total number of web documents in which c is mentioned.

Following the terminology of web search [18], the values above are also referred to as "number of results", "number of hits". For a country c, the R_c score is computed as follows:

$$R_c = \frac{p_c - n_c}{\max\{p_c, n_c\}} \frac{p_c + n_c}{t_c} \quad (1)$$

for pc > 0 and/or nc > 0. If any of those two constraints holds, then it can be inferred that also tc > 0. It is observed that Rc is expressed as the product of two factors:

(i) $\frac{p_c - n_c}{\max\{p_c, n_c\}}$

and

(ii) $\frac{p_c + n_c}{t_c}$

The first factor ranges within the $(-1, 1)$ interval and it implements the aforementioned contrast by taking into account the difference between pc and nc.

The max operator that appears in the denominator is employed for normalization purposes. If $p_c \gg n_c$, then $R_c \approx 1$, while $R_c \approx -1$ when $n_c \gg p_c$. If $p_c \approx n_c$, then $R_c \approx 0$. The normalized difference between p_c and n_c is weighted by the second factor that takes values in the (0, 1) interval. The underlying idea is that the "positive vs. negative" signal –as expressed by the first factor– should be weighted analogously to number of web documents exhibiting regulation-related topics. The sum of p_c and n_c in the enumerator suggests that the polarity is irrelevant for this factor. This sum is normalized by the overall presence of c in the world wide web (or in a subset of the world wide web being thematically filtered according to the proper query formulation, as explained in the following paragraphs) as quantified by the t_c value. Overall, a positive R_c score can be interpreted as a tendency towards a positive stance. Similarly, a negative R_c score can be considered as a tendency towards a negative stance. Also, as R_c approaches zero, and regardless of its sign, the stance becomes more neutral. Absolute neutrality can be assumed for $R_c = 0$, which can take place when $p_c = n_c$ (given that $p_c > 0$ and $n_c > 0$).

In order to retrieve the values of the aforementioned model parameters, a series of web search queries are required as follows:

- $Q_{c,p}$: query(ies) for retrieving p_c
- $Q_{c,n}$: query(ies) for retrieving n_c
- $Q_{c,t}$: query(ies) for retrieving t_c

The above represent three query types rather than three queries[1], i.e., for each case one or more queries can be formulated. Each query can take the form of a text string,

[1] Having only three queries is a special case that is explained at the end of this section.

which is the typical data type passed to web search APIs. Regarding the first two query types, $Q_{c,p}$ and $Q_{c,n}$, the query can be the concatenation of three lexical fields. The first field deals with the country itself, where lexical variants can be also used, e.g., ("USA" "United States of America").

In this example, note the incorporation of variants in a single query via the use of the disjunctive " " operator[2]. Let C denote the set of c's lexical variants. The second field is a set of lexical entries representing positive or negative lexical cues. As mentioned, P and N denote those sets for the case of positive and negative cues, respectively. Each set can consist of one or more single- or multi-word entries. Consider the following example for the case of P, ("favors' "strongly supports"). In a similar way, we can have the following query fragment for the case of N, ("bans" "strictly prohibits"). The third field can be a set of lexical entries that can be used for thematically restricting the overall scope of the web search. For example, the fragment ("cryptocurrencies" "cryptos") can be appended in the query. A related example, even from a different domain, is the "apple fruit" search query that can be used for retrieving results where the word "apple" is used according to the sense of "fruit" (and filtering out results that may refer to the homonymous technology company). Such entries are also referred to as "pragmatic constraints". Let S denote the set of pragmatic constraints. Regarding the third query type, $Q_{c,t}$, a similar methodology can used. Specifically, the query is simpler compared to the case of $Q_{c,p}$ and $Q_{c,n}$ because the fragments that correspond to P and N are not needed. Thus, for the case of $Q_{c,t}$ only the lexical entries included in C and S are considered.

Based on the above definitions and discussion, it is clear that query complexity is a key characteristic of the proposed model. For the case of a single country c, this simply refers to the number of queries needed for estimating the R_c score. There are various ways that can be followed for the formulation of queries. This variety can be characterized by two ends as follows. For each of the aforementioned set (P, N, C, and S), all of its lexical entries are combined in a single query fragment using disjunctive operators. According to (1), this requires three queries per country, while the query complexity, O_l, is constant and denoted as follows

$$O_l(1) \tag{2}$$

Alternatively, one may formulate a separate query fragment for each entry of the aforementioned sets. This approach significantly increases the query complexity, O_h, as

$$\mathrm{O}_h((|P||C||S|) + (|N||C||S|) + (|C||S|)) \tag{3}$$

which can be factorized as

$$\mathrm{O}_h((|P|+|N|+1)|C||S|). \tag{4}$$

This factorization directly reveals a cubic degree of complexity. For the sake of clarity, it should be noted that "|.|" stands for the cardinality of the respective set, which is irrelevant to the "" search operator also used in this paper.

[2] Assuming that such operators are supported by the utilized search engine. The "|" operator is also referred to as "OR".

It is obvious that there is a significant difference between the two ends, i.e., constant vs. cubic query complexity. In general, many query formulation strategies can be devised at any point between the two ends. From a modeling perspective, the aforementioned complexities represent the lowest ($O_l(.)$) and the highest ($_h(.)$) bounds, respectively, as far as (1) is concerned for estimating R_c. The advantage of employing a complex query formulation strategy is the semantic control gained by reducing the use of search operators like the disjunctive " " operator. The application of the proposed model for a list of countries exhibits linear query complexity, with respect to the size of the list, since countries are independent.

4 Pilot Experiment

In this section, we present the results of a pilot experiment. To justify the effectiveness of the proposed algorithmic methodology a pilot study is introduced for validating the regulatory stance hypothesis along with the computational model. For this purpose, a validation dataset comprising of 15 countries has been compiled by a team of three domain experts. The chosen countries represent a diverse spectrum of regulatory stances, encompassing both positive and negative regulatory environments.

Experimental Setup. The experimental setup proposed for this study follows a structured approach:

- **Web search.** Google's Programmable Search Engine was utilized[3]. Specifically, this engine was configured in order to take into account the entire world wide web (as indexed by Google and offered via the used service), while the supported language was set to English.
- **Query formulation.** Regarding C, only a single lexicalization was used the country names included in Table 1. For the case of pragmatic constraints (members of set S), the following were used: "cryptocurrencies", and "bitcoin". Also, the hq parameter[4] of the search engine was set to "central bank security exchange commission". In order to achieve constant query complex ity, for the case of positive and negative lexical cues (members of P and N) the following cues were used: "allows" and "bans" in combination with the " " search operator (i.e., " allows" and " bans"). This operators enables the consideration of synonyms.

Results and Insights. The outcome of the pilot experiment is presented in Table 1, where each country in the validation dataset is associated with three distinct scores: the final score, R_c, computed according to (1), as well as, the values of the *first* and *second* factor of (1) as explained in Sect. 3.

It is observed that the highest R_c scores are obtained for Gibraltar (0.325) and China (0.202). Nepal and Algeria are associated with the lowest R_c scores, 0.031 and 0.035, respectively. The role of the weighting implemented by the second factor of (1) is clearly

[3] https://developers.google.com/customsearch.

[4] More information about the supported parameters can be found in the documentation: https://developers.google.com/customsearch/v1/reference/rest/v1/cse/list.

Table 1. Pilot experiment: results.

Country (c)	$\frac{p_c - n_c}{\max\{p_c, n_c\}}$	$\frac{p_c + n_c}{t_c}$	Rc
Algeria	0.114	0.305	0.035
China	0.673	0.300	**0.202**
Cyprus	0.277	0.334	0.093
Egypt	0.313	0.198	0.062
France	0.720	0.180	0.130
Gibraltar	0.544	0.598	**0.325**
Greece	0.487	0.225	0.110
Hong Kong	0.497	0.264	0.131
Malta	0.386	0.255	0.098
Morocco	0.211	0.216	0.046
Nepal	0.127	0.247	0.031
Singapore	0.590	0.212	0.125
South Africa	0.496	0.179	0.089
Switzerland	0.637	0.166	0.106
Taiwan	0.326	0.335	0.109

demonstrated for the case of France. Without this weighting, France would be ranked as the top country according to R_c (in contrast to the 4_{th} position just after Hong Kong). The correctness of the aforementioned scores and observations were independently justified by the three researchers. Of course, this should be further validated by the utilization of a benchmark[5]. Currently, and even in the absence of such a benchmark, the three human evaluations provided sufficient indications that the proposed model performs well (at least at the two ends of the ranked list).

Table 2. Pilot experiment: clustering of countries based on Rc.

Cluster ID	Countries
1	China, Gibraltar
2	France, Cyprus, Greece, Hong Kong, Malta, Singapore, South Africa, Switzerland, Taiwan
3	Algeria, Egypt, Morocco, Nepal

Another (subjective) observation is the coarse formulation of three categories taking into account the computed R_c scores, namely, "high R_c", "mid R_c", and "low R_c". In

[5] This is part of our ongoing work.

order to computationally investigate this observation, the k–means clustering algorithm was applied (with $k = 3$) over the R_c scores listed in Table 1. The identified clusters are shown in Table 2. Based on the cluster memberships, we can observe that the two R_c thresholds used for identifying the three clusters are 0.202 (China) and 0.062 (Egypt). Specifically, the following centroid values were computed and used by the clustering algorithm, 0.264, 0.110, and 0.044. The middle cluster (ID 2) is the most populous cluster, which justifies the intuition that the "high vs. mid" and "mid vs. low" boundaries are blurred. This can be further explored by applying k–means (i) only over the members of the middle cluster, and/or (ii) hierarchical clustering. The latter does not requires the a–priory setting of k, which is a desirable configuration. The automatically computed results of Table 2 are in agreement with the aforementioned findings suggesting that our model provided reasonable results at least for the two ends (i.e., "high R_c" and "low R_c").

Summary and Further Exploration. The aforementioned findings provide preliminary support for the effectiveness of the model across disparate regulatory stances. In conclusion, while the pilot experiment exhibited promising results, further validation against a benchmark is required. Nevertheless, the experiment highlighted promising results indicating the model's proficiency in differentiating between regulatory dispositions and its potential to offer valuable insights into diverse regulatory landscapes.

5 Conclusions

In this work, we have proposed a computational model for estimating the regulatory stance of countries with respect to cryptocurrencies. The key idea is the regulatory stance hypothesis suggesting that the co-occurrence of positive/ negative regulation-related cues within web documents can be exploited for inferring the tendency towards a positive or negative stance. The major finding of this work was the experimental justification of the above hypothesis using web search engines. In this framework, a metric was proposed for yielding a single score quantifying the overall regulatory stance. The basic operation is the "contrast" of search results corresponding to positive and negative lexical cues. A straight- forward calculation of this contrast is the normalized difference of the respective numbers of hits. The proposed model has a series of advantageous technical characteristics as follows: (i) it is fully automatic saving a significant amount of time when compared to a manual process which demands expert knowledge, (ii) under the proper configuration, constant query complexity can be reached, (iii) given a set of countries, full parallelization is enabled since there are no dependencies between the countries, i.e., one thread per country. The only requirements are the initial query development and the employment of a proper web search API. Regarding the latter, the utilization of advanced search features is recommended as they provide more expressive power to the query developer. This can lower the query complexity (and, thus, the execution time). Currently, we are working on the development of formal evaluation dataset having larger country coverage compared to the present pilot study, while we are investigating techniques from the area of automatic grammar induction (e.g., see [16]) in order to automate the formulation of web search queries. Also, the output of the proposed model fits the broader algorithmic framework developed by authors that deals with the robust (i.e., missing information resilience) estimation of blockchain readiness scores [15]. One of our immediate goals is the integration of the two models.

References

1. Bollegala, D., Matsuo, Y., Ishizuka, M.: Measuring semantic similarity between words using web search engines. In: Proceedings of the 16th International World Wide Web Conference (WWW2007) (2007)
2. Bratt, M., Dutta, S., Lanvin, B., Rossini, C.: The network readiness index 2019: towards a future-ready society. In: Dutta, S., Lanvin, B. (eds.) The Network Read ness Index 2019: Towards a Future-Ready Society, pp. 1–308. Portulans Institute, Washington D.C. (2019)
3. Carayannis, E.G., Christodoulou, K., Christodoulou, P., Chatzichristofis, S.A., Zinonos, Z.: Known unknowns in an era of technological and viral disruptions—implications for theory, policy, and practice. J. Knowl. Econ. 1–24 (2021)
4. Karamitsos, I., Papadaki, M., Themistocleous, M.: Blockchain as a service (BCaaS): a value modeling approach in the education business model. J. Softw. Eng. Appl. **5**, 165–182 (2022)
5. Christodoulou, K., Paton, N.W., Fernandes, A.A.A.: Structure inference for linked data sources using clustering. In: Hameurlain, A., Küng, J., Wagner, R., Bianchini, D., De Antonellis, V., De Virgilio, R. (eds.) Transactions on Large-Scale Data- and Knowledge-Centered Systems XIX. LNCS, vol. 8990, pp. 1–25. Springer, Heidelberg (2015). https://doi.org/10.1007/978-3-662-46562-2_1
6. Cunha, P.R., Soja, P., Themistocleous, M., 2021, "Blockchain for development: a guiding framework", Information Technology for Development, 27(3): 417–438, https://www.tandfonline.com/doi/full/https://doi.org/10.1080/02681102.2021.1935453
7. Cointobuy.io: Cryptocurrency Regulation Analysis (2020). https://cointobuy.io/countries. Accessed 23 Mar 2021
8. Crypto Fund Research: Crypto Fund Report (2021). https://cryptofundresearch.com/. Accessed 23 Mar 2021
9. Cumming, D.J., Johan, S., Pant, A.: Regulation of the crypto-economy: managing risks, challenges, and regulatory uncertainty. J. Risk Financ. Manag. **12**(3), 126 (2019)
10. Kapassa, E., Themistocleous, M.: Blockchain technology applied in iov demand response management: a systematic literature review. Future Internet **14**(5), 1–19 (2022)
11. European Parliament and Council: Markets in crypto-assets regulation (MiCA) (2023). https://eur-lex.europa.eu/eli/reg/2023/1114/oj. Accessed 23 July 2023
12. Fenwick, M., Kaal, W.A., Vermeulen, E.P.: Regulation tomorrow: what happens when technology is faster than the law. Am. U. Bus. L. Rev. **6**, 561 (2016)
13. Global Legal Insights: Blockchain Laws and Regulations (2021). https://www.globallegalinsights.com/practice-areas/blockchain-laws-and-regulations. Accessed 23 Mar 2021
14. Harris, Z.S.: Distributional structure. Word **10**(2–3), 146–162 (1954)
15. Iosif, E., Christodoulou, K., Vlachos, A.: A robust blockchain readiness index model. arXiv preprint arXiv:2101.09162 (2021)
16. Iosif, E., et al.: Speech understanding for spoken dialogue systems: from corpus harvesting to grammar rule induction. Comput. Speech Lang. **47**, 272–297 (2018)
17. Iosif, E., Louca, S.: Blockchains and data management: current landscape and challenges. In: Proceedings of the 12th Mediterranean Conference on Information Systems (Research–in–progress Papers) (2018)
18. Iosif, E., Potamianos, A.: Unsupervised semantic similarity computation between terms using web documents. IEEE Trans. Knowl. Data Eng. **22**(11), 1637–1647 (2009)

19. Iosif, E., Potamianos, A.: Similarity computation using semantic networks created from web-harvested data. Nat. Lang. Eng. **21**(1), 49 (2015)
20. KPMG: 2020 Autonomous Vehicles Readiness Index (2020). https://kpmg.com/xx/en/home/insights/2020/06/autonomous-vehicles-readiness-index.html. Accessed 23 Mar 2021
21. Library Law of Congress: Regulation of Cryptocurrency around the World (2020). https://www.loc.gov/law/help/cryptocurrency/index.php. Accessed 23 Mar 2021
22. Makridakis, S., Christodoulou, K.: Blockchain: current challenges and future prospects/applications. Future Internet **11**(12), 258 (2019)
23. The Economist: Who is Ready for the Coming Wave of Automation? (2018). https://www.automationreadiness.eiu.com/static/download/PDF.pdf. Accessed 23 Mar 2021
24. Yeoh, P.: Regulatory issues in blockchain technology. J. Financ. Regulation Compliance (2017)
25. Themistocleous, M., Cunha, P., Tabakis, E., Papadaki, M.: Towards cross-border CBDC interoperability: insights from a multivocal literature review. J. Enterp. Inf. Manage. **36**(5), 1296–13182023. https://doi.org/10.1108/JEIM-11-2022-0411. 1741-0398
26. Chohan, U.: Oversight and regulation of cryptocurrencies: BitLicense. SSRN (2018)
27. Church, K., Hanks, P.: Word association norms, mutual information, and lexicography. Comput. Linguist. **16**(1), 22–29 (1990)
28. Kapassa, E., Themistocleous, M., Christodoulou, K., Iosif E.: Blockchain application in internet of vehicles: challenges, contributions and current limitations. Future Internet **13**(12), 313 (2021). https://www.mdpi.com/1999-5903/13/12/313
29. Christodoulou, K., Katelaris, L., Themistocleous, M, Christodoulou P., Iosif E.: NFTs and the metaverse revolution: research perspectives and open challenges. In: Lacity, M., Treiblmaier, H. (eds.) Blockchains and the Token Economy: Theory and Practice, pp. 139–178. Palgrave Macmillan, Cham (2022)
30. Drozd, O., Lazur, Y., Serbin, R.: Theoretical and legal perspective on certain types of legal liability in cryptocurrency relations. Baltic J. Econ. Stud. **3**(5), 221–228 (2017)

Reviewing the Role of Secret Sharing Schemes in Electronic Payment Protocols

Rym Kalai[1](✉), Wafa Neji[2], and Narjes Ben Rajeb[3]

[1] LIPSIC Laboratory, Faculty of Sciences of Tunis, University of Tunis El Manar, Tunis, Tunisia
rym.kalai@gmail.com

[2] LIPSIC Laboratory, The General Directorate of Technological Studies, Higher Institute of Technological Studies of Beja, Tunis, Tunisia

[3] LIPSIC Laboratory, National Institute of Applied Sciences and Technology, University of Carthage, Carthage, Tunisia

Abstract. Since e-Cash and electronic transactions consist of digital data, there are risks that this data can be easily falsified, compromising the security and efficiency of the entire electronic payment protocol. To ensure the security of the protocol, these data must be protected. This is why researchers have focused on several cryptographic studies including secret sharing. Secret sharing techniques are used to divide a secret into multiple shares and to distribute them among a group of participants. Only when a sufficient number of shares are combined together, the original secret can be reconstructed. Moreover, no information about the secret can be deducted from the shares by a smaller group.

In this paper, we investigate the different phases of an electronic payment protocol where secret sharing can be used and more particularly distributed key generation and distributed multikey generation protocols. This paper also highlights the importance of using secret sharing in enhancing the security and confidentiality of electronic payment systems, as well as the ability to prevent unauthorized access to sensitive information. The trust is shared among multiple parties.

Keyword: Secret Sharing (SS) · Distributed Key Generation (DKG) · Distributed Multi-Key Generation (DMKG) · electronic payment protocols

1 Introduction

Secret sharing is a cryptographic technique that enables one party to split a secret into multiple shares and to distribute the shares among n participants. Individual shares are of no use on their own. Only a subgroup of these participants can combine their shares together to reconstruct the secret [7]. The secret sharing scheme ensures that an adversary in control of t out of n participants will learn no information about the secret. However, traditional secret sharing schemes are often insufficient, particularly for applications in which the set of participants might change over the time [3]. Distributed key generation and distributed multi key generation protocols are cryptographic techniques based on secret sharing. These techniques allow a group of participants to cooperate together

M. Papadaki et al. (Eds.): EMCIS 2023, LNBIP 501, pp. 52–60, 2024.
https://doi.org/10.1007/978-3-031-56478-9_4

to generate a shared secret key without revealing any information about the key to outsiders. The main contribution of these cryptographic techniques is the cooperation and the involvement of the participants. By dividing sensitive keys into multiple shares and distributing them among different parties, these protocols can reduce the risk of key compromise and unauthorized access. This can help ensure the integrity, confidentiality, and privacy of electronic payment transactions [4].

In this article, we present the role of secret sharing as well as DKG and DMKG protocols in electronic payment protocols. We first provide an overview of the theoretical foundations of these techniques and discuss their advantages in the context of electronic payment protocols. We then examine the different steps of an electronic payment protocol where secret sharing protocols can be applied, including authorization request, authentication, encryption and trans- mission transaction data, confirmation and data tracing. After that, we discuss their advantages and limitations in the context of electronic payment protocols. Finally, we present some of the current challenges of these techniques and suggest future directions for research.

2 Theoretical Foundations of Secret Sharing, DKG and DMKG

In group communications such as electronic payment, there is a real need to hide secret data such as passwords, encryption key, etc. The members of the group must be identified and must share one or more secret information such as a common group confidentiality key via group key management protocols for example. Many group key management schemes hide the secret key in a single secret location. This solution is not reliable. Another solution is to make multiple copies of the key in different places. But it is still not secure enough. That is where the idea of using a secret sharing scheme came from.

The secret sharing was introduced by Shamir [12] and Blakley [1] in 1979. This contribution has had a major impact on research in the field of cryptography. The idea is to have a trusted party that holds a secret s, then splits it into multiple shares $s_1,\ldots,$ s_n and then distribute them to a group of n participants. As illustrated in Fig. 1, in a $(t,$ $n)$ threshold secret sharing scheme, only a subgroup of t participants can cooperate to reconstruct the secret s [16]. In secret sharing, a dealer knows the secret it shares. The problem is that the trusted party is the unique party that have all the privileges and that holds the secret s. The system security can be compromised by this trusted party which is the point of attack: a dishonest participant can own the secret or distribute and broadcast invalid information by only attacking this trusted entity.

A solution to this problem was presented by Pedersen [10] in 1991. A distributed key generation protocol without a trusted party. The power is decentralized: No party holds the secret but a group of participants that cooperate to generate and share this information in a completely distributed way such that any subset of size greater than a threshold can reveal or use the shared secret, while smaller subsets do not have any knowledge about it. The most important aspect is that there is no dealer or trusted party [15]. The trust is shared among multiple parties. The DKG protocol is equivalent to having n parallel executions of a secret sharing scheme [5]. In each execution, we have a different dealer from the group of n participants. However, in the DKG protocol, all participants share a unique secret information. In many situations, there is a need to handle and manage multiple secrets at the same time.

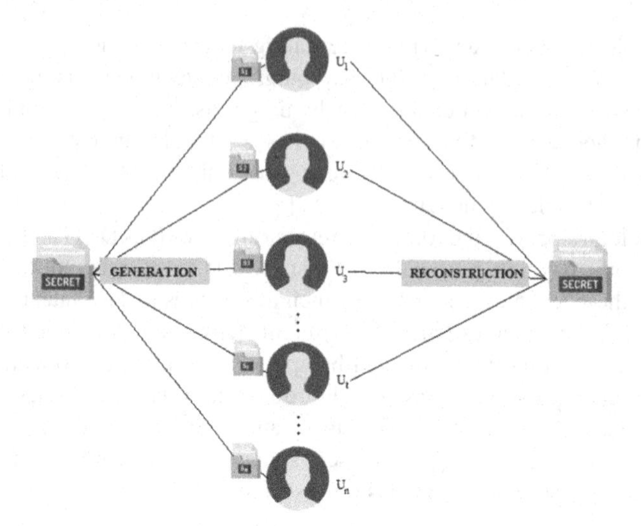

Fig. 1. A (t, n) threshold secret sharing scheme illustration.

The solution was introduced in 2020 by Tianjun Ma et al. [6]. A distributed multi key generation protocol that is equivalent to having *n* parallel executions of a multi secret sharing scheme [4]. In each execution, we have a different dealer from the group of *n* participants. The participants cooperate together to manage simultaneously *l* different secrets.

3 Application of Secret Sharing in Electronic Payment Protocol

Whenever we have secret, confidential or critical information to handle, we can use secret sharing techniques. It is especially helpful when the honesty of all participants is not required. Just a predefined order subgroup of honest participants can accomplish the need. Their cooperation is the key to the success of the protocol [11]. The DKG and DMKG protocols are the solution whenever we want to decentralize power and not to rely on a single trusted authority. The trust is distributed between multiple parties. We choose DKG or DMKG depending on the context: there is one or more secrets to manage at a time.

Thus, the secret sharing techniques can be applied in multiple phases of the electronic payment protocols.

Authorization Request. The user submits a payment request to a payment processor or bank for a given transaction. The payment processor or bank then verifies the information provided by the user to con rm that funds are available and authorize the transaction.

Secret sharing can be used in this phase to protect sensitive information such as credit card information. Indeed, it can be used to split an encryption key into several parts and distribute them among different parties involved in the transaction, enhancing the security of the transmission of sensitive information.

Authentication. The user is authenticated by the payment processor or bank to con rm his identity and ensure that the transaction is being carried out by an authorized person. This data should not be stored in a database or anywhere else.

Secret sharing can be used to manage authentication keys and reinforce authentication security. For instance, in [9], a secret sharing scheme is used to split the authentication key into multiple parts and distribute them among different parties implied in the transaction. The key can be reconstructed only when a sufficient number t of shares are combined together. Any $(t\ 1)$ shares or less cannot reveal anything about the authentication key, making it harder for an attacker to compromise the authentication and to usurp the user's identity.

Encryption and Transmission of Transaction Data. Transaction data such as the transaction amount and recipient information are encrypted and securely transmitted between the parties implicated in the transaction. The encryption key must be protected and not be reachable.

Secret sharing can be used to split the encryption key into several parts and distribute them among the different participants, improving the security of the transmission of transaction data. The involved parties can then use their shares of the encryption key to encrypt and decrypt the transaction data.

Transaction Confirmation. The payment processor or bank con rms the transaction and transfers funds from the user to the recipient. Secret sharing can be used to manage the encryption and validation keys necessary for trans- action confirmation.

In fact, it can be used to split a validation key into multiple parts and distribute them among different parties implied in the transaction, strengthening the security of transaction confirmation which also avoids participants from denying their payment obligations.

Tracing Data. In some electronic payment protocols with privacy protection, such as Blockchain applications, we are sometimes faced with the need to data tracing. This is to prevent dishonest users from committing crimes without being identified [4] and to prohibit any participant from denying his involvement or actions related to the payment transactions.Secret sharing can be used to manage the data tracing: The traceable key is divided and distributed among a group of participant. The cooperation of the group permit to reconstruct the traceable key and to allow the tracing process. To prevent a single entity from abusing his right of data tracing without having any restrictions or any control measures, a distributed secret sharing technique is necessary [4].

An example of this technique is used in [6]. A regulator need to trace user's identities of Blockchain data. In order to be allowed to do this, it must receive the committee's consent. The committee is formed by a group of participants which cooperate together by applying a distributed multi key generation technique to generate a traceable key. This key is sent to the regulator to authorize tracing.

4 Impact of Secret Sharing in Electronic Payment Protocols

In this section, we discuss the advantages and limitations of using secret sharing techniques in the context of electronic payment protocols.

4.1 Advantages

Secret sharing schemes offer many benefits in electronic payment protocols. Also, several secrecy properties are satisfied thanks to the use of secret sharing.

Threshold Access. Secret sharing allows threshold access to the secret. A predefined number of shares, known as the threshold, is required to reconstruct the secret. As long as the threshold is not reached, the secret remains secure. This provides an additional layer of protection, as it requires cooperation and compromise of multiple participants to reconstruct the secret.

Confidentiality. Secret sharing ensures the confidentiality of sensitive information in electronic payment protocols. In fact, secret information such as payment credentials or encryption keys are divided into multiple shares and distributed among multiple participants. Even if one share is compromised, it is useless and does not reveal any meaningful information about the original secret. In addition, only when a sufficient number of shares are combined, the original secret can be reconstructed. This prevents unauthorized participants or parties from gaining access to the complete secret: Only authorized participants can reconstruct the original secret, while the others are unaware of its value.

Privacy Preservation and Anonymity. Applying secret sharing techniques in electronic payment protocols helps to achieve a main goal which is assuring privacy. Secret information like user identities or payment details are splitted into multiple parts and distributed among different participants involved in the protocol. This prevent any participant or any other party from having complete knowledge of the transaction or the identities involved. In addition, from a received share, no participant is able to deduct any information about the payment transaction or to link the payment information to a particular party [2]. Thus, the anonymity of the payer and payee can be preserved.

Integrity. Secret sharing can guaranty integrity in electronic payment protocols. Any attempt to falsify the secret or alter its value can be detected. In fact, participants can verify the integrity of the secret by comparing their shares with those of others.

Availability. Secret sharing can ensure the availability of the secret in electronic payment protocols. Even if some shares are compromised or are unavailable, as long as the required threshold of shares is reached, the secret can still be reconstructed and used for payment transactions.

Non-repudiation. Secret sharing can support non-repudiation in electronic payment protocols. By involving multiple parties in the secret sharing process, it becomes di cult for any participant to deny their involvement or actions related to the secret. This helps establish accountability and prevents participants from denying their payment obligations.

Scalability and Flexibility. Secret sharing, specifically DKG and DMKG o er scalability and flexibility in electronic payment protocols. They can accommodate a varying number of participants and adapt to changing requirements. As the number of participants increases, the secret can be split into more shares, enabling a larger number of participants to be accommodated without scarifying security or performance [13, 14, 17]. This makes them ideal for both small-scale and large-scale payment systems.

Trust and Transparency. In traditional payment systems, they used to have a unique trusted party that can misbehave or misuse or manipulate the information. It can also be easily attacked and compromised. It can lead to a complete security breach. In DKG and DMKG protocols, the information and the responsibility are distributed among multiple parties. It ensures that no single party has complete control over the keys or sensitive information, reducing the risk of a single point of failure and making it di cult for attackers to compromise the system. It helps building trust among participants and ensures a fair and transparent system.

Computational Security. Secret sharing relies on computational security assumptions to ensure the secrecy properties. The security of secret sharing schemes is based on mathematical problems that are computationally hard to solve, such as factoring large numbers or solving discrete logarithm problems. These assumptions make it computationally infeasible for an attacker to reconstruct the secret without having a sufficient number of shares.

Resilience to Attacks. In electronic payment protocols, there is always a risk of a participant being compromised or becoming malicious. Secret sharing, DKG and DMKG protocols enhance the resilience of electronic payment protocols against various attacks.

First, Secret sharing hides the original secret by dividing it among multiple participants. Each participant holds a share of the secret, but he do not have access to the complete secret on his own. It becomes harder for an attacker to compromise the entire secret. Even if some shares are compromised or intercepted, the secret remains secure as long as a sufficient number of honest participants is reached: The threshold requirement ensures that the system can still function securely.

Second, DKG and DMKG protocols mitigate the risk of participants being compromised or becoming malicious. By involving multiple parties in the key generation process and splitting the secrets, it becomes harder for any malicious participant that tries to manipulate or misuse the information, to gain complete control over the system.

By employing secret sharing in electronic payment protocols we ensures that the secrecy properties, such as confidentiality, integrity, availability, etc. are satisfied. This helps to protect sensitive information such as credentials, to preserve privacy of electronic payment transactions and to assure the flexibility and the transparency of the entire electronic payment process.

4.2 Limitations

Despite that secret sharing technique adds an extra layer of security to electronic payment protocols, it still have some limitations.

- High Communication Secret sharing schemes often require high communication between multiple parties to distribute and reconstruct shares. This communication overhead can introduce delays and increase the bandwidth requirements of the payment protocol. In cases with limited network resources or high latency, this can be a limitation.

- Increased Complexity Implementing secret sharing schemes in electronic payment protocols can introduce additional complexity.

First, this includes securely generating, storing, and distributing the shares, ensuring proper access control, and reconstructing secret information. Especially in DKG and DMKG protocols, managing and protecting the keys can be complex and can requires additional computational resources and it may increase the overall system complexity which can impact the efficiency and performance of the payment protocol.

Second, as the number of participants in the payment protocol increases, the complexity and the communication overhead of secret sharing schemes can become more challenging to manage. Scaling the scheme to accommodate a large number of participants may require additional resources and careful design considerations.

Finally, secret sharing schemes can have some difficulties adapting to constantly evolving payment systems, as adding new participants or changing sharing thresh- olds can require complex adjustments and may slow down transactions.

Trusted Parties. Secret sharing schemes rely on the cooperation and involvement of multiple trusted parties to generate, distribute, store and reconstruct secrets. If any of these parties are compromised or act maliciously, it can under- mine the security of the scheme. Insiders are legitimate participants who have successfully passed the registration step. Since they can bene t from all the information shared on the network between all the participants, it is easier to perform various attacks with a high impact [8]. Ensuring the trustworthiness of all involved parties can be challenging which may introduce potential vulnerabilities.

Threshold Access. The security of a secret sharing scheme relies on the discretion and cooperation of a predefined number of honest participants. The set threshold is only a security assumption and therefore cannot be guaranteed. Beyond this threshold, if a certain number of participants are compromised or malicious, the security of the entire system may be compromised.

Before implementing secret sharing schemes in electronic payment protocol, it is important to carefully evaluate these limitations, to consider the specific requirements and constraints and to carefully select security and key management mechanisms.

5 Other Cryptographic Schemes Based on Secret Sharing

In our researches about techniques based on secret sharing and applied in electronic payment systems, we were focused on distributed key generation and distributed multi key generation protocols. However, there are other cryptographic schemes based on secret sharing like Visual Cryptographic Schemes.

Visual cryptography is a secret sharing scheme basically for images and built on a visual (t, n) threshold secret sharing scheme. Indeed, a secret image is hidden within n transparencies. Only a sufficient number t of transparencies permit to reconstruct the secret image before payment can be performed.

To be more precise, a secret image s is encoded and divided into n shares. Each share is sent to a different participant. The secret image s can be reconstructed by a subgroup of t participants that pulls the shares together. No information can be obtained by any subgroup of less than t participants [9].

This scheme is specially used in the authentication phase or confirmation phase. It can be used alone or combined with other secret sharing techniques such as DKG and DMKG to get the maximum benefits of these different techniques at the same time.

6 Conclusion

In this paper, we presented several secret sharing techniques specifically DKG and DMKG protocols and their use in improving the security of electronic payment protocols as well as the ability to prevent unauthorized access to sensitive information.

Despite the advantages of using secret sharing techniques, some participants can cheat, exchange invalid shares and aim to eliminate honest participants which presents a considerable attack. These cheaters must be detected by the protocol without revealing any information about their shares or those of honest participants. Moreover, each participant should be able to complain against any other suspicious participant [4]. Thus, in addition to secret sharing techniques, a com- plaints strategy must be defined to overcome this problem.

References

1. Blakley, G.R.: Safeguarding cryptographic keys. In: International Workshop on Managing Requirements Knowledge, p. 313. IEEE Computer Society (1979)
2. Dehez-Clementi, M., Lacan, J., Deneuville, J.C., Asghar, H., Kaafar, D.: A blockchain-enabled anonymous-yet-traceable distributed key generation. In: 2021 IEEE International Conference on Blockchain (Blockchain), p. 257–265. IEEE (2021)
3. Goyal, V., Kothapalli, A., Masserova, E., Parno, B., Song, Y.: Storing and retrieving secrets on a blockchain. In: Hanaoka, G., Shikata, J., Watanabe, Y. (eds.) PKC 2022. LNCS, vol. 13177, pp. 252–282. Springer, Cham (2022). https://doi.org/10.1007/978-3-030-97121-2_10
4. Kalai, R., Neji, W., Ben Rajeb, N.: A distributed multi-key generation protocol with a new complaint management strategy. In: Papadaki, M., Rupino da Cunha, P., Themistocleous, M., Christodoulou, K. (eds.) EMCIS 2022. LNBIP, vol. 464, pp. 150–164. Springer, Cham (2022). https://doi.org/10.1007/978-3-031-30694-5_12
5. Kim, J., Kim, P., Choi, D., Lee, Y.: A study on the interoperability technology of digital identification based on WACI protocol with multiparty distributed signature. Sensors **23**(8), 4061 (2023)
6. Ma, T., Xu, H., Li, P.: A blockchain traceable scheme with oversight function. In: Meng, W., Gollmann, D., Jensen, C.D., Zhou, J. (eds.) ICICS 2020. LNCS, vol. 12282, pp. 164–182. Springer, Cham (2020). https://doi.org/10.1007/978-3-030-61078-4_10
7. Neji, W., Blibech, K., Ben Rajeb, N.: Distributed key generation protocol with a new complaint management strategy. Secur. Commun. Netw. **9**(17), 4585–4595 (2016)
8. Noh, J., Kwon, Y., Son, J., Cho, S.: Blockchain-based one-time authentication for secure v2x communication against insiders and authority compromise attacks. IEEE Internet Things J. **10**(7), 6235–6248 (2022)
9. Ogheneruemu, A.S., Taiye, A.O.: Electronic payment system using visual cryptographic scheme

10. Pedersen, T.P.: A threshold cryptosystem without a trusted party. In: Davies, D.W. (ed.) EUROCRYPT 1991. LNCS, vol. 547, pp. 522–526. Springer, Cham (1991). https://doi.org/10.1007/3-540-46416-6_47
11. Shalini, I., Sathyanarayana, S., et al.: A comparative analysis of secret sharing schemes with special reference to e-commerce applications. In: 2015 International Conference on Emerging Research in Electronics, Computer Science and Technology (ICERECT), pp. 17–22. IEEE (2015)
12. Shamir, A.: How to share a secret. Commun. ACM **22**(11), 612–613 (1979)
13. Touloupou, M., Themistocleous, M, Iosif E., Christodoulou, K.: A systematic literature review towards a blockchain benchmarking framework. IEEE Access **10**, 7630–7644 (2022). https://ieeexplore.ieee.org/abstract/document/9813702/authors#authors
14. Touloupou, M., Christodoulou, K., Inglezakis, A., Iosif, E., Themistocleous, M.: Benchmarking blockchains: the case of XRP ledger and beyond. In: Proceedings of the Fifty-third Annual Hawaii International Conference on System Sciences, (HICSS 55), 4–7 January 2022, Maui, Hawaii, USA. IEEE Computer Society, Los Alamitos (2022). https://scholarspace.manoa.hawaii.edu/bitstream/10125/80070/0586.pdf
15. Voudouris, A., Politis, I., Xenakis, C.: Secret sharing a key in a distributed way, Lagrange vs Newton. In: Proceedings of the 17th International Conference on Availability, Reliability and Security, pp. 1–7 (2022)
16. Zhou, X.: Threshold cryptosystem based fair online e-cash. In: 2008 Second International Symposium on Intelligent Information Technology Application, vol. 3, pp. 692–696. IEEE (2008)
17. Christodoulou, K, Iosif, E., Inglezakis, A., Themistocleous, M.: Consensus crash testing: exploring ripple's decentralization degree in adversarial environments. Future Internet **12**(3), 53 (2020). https://www.mdpi.com/1999-5903/12/3/53/htm

Decentralization of DAOs: A Fundamental Analysis

Stamatis Papangelou, Klitos Christodoulou, and Marinos Themistocleous(✉)

Department of Digital Innovation, School of Business, Institute For the Future (IFF), University of Nicosia, Nicosia, Cyprus
{Papangelou.m,Christodoulou.kl,Themistocleous.m}@unic.ac.cy

Abstract. This paper addresses a significant gap in the current understanding of decentralization within Decentralized Autonomous Organizations (DAOs). While many theoretical discussions generalize governance models, there is a lack of in-depth analysis of the unique governance structures and their inherent decentralization mechanisms in individual DAOs. To bridge this gap, the paper presents two primary contributions. First, it offers a foundational analysis of potential points of failure in decentralization within DAOs. Second, the study introduces a "decentralization layer system", a new framework aimed at guiding future quantitative analyses of decentralization. This system is based on two pillars: a critical evaluation of governance structures and a stratification approach to identify distinct layers within these structures. The paper further demonstrates the application of this framework by analyzing the governance models of several DAOs, including Uniswap, Compound, and ApeCoin. Through these analyses, the paper provides insights into the strengths and vulnerabilities of each DAO's decentralization mechanisms, offering a comprehensive perspective on DAO governance.

Keywords: DeFi · DAOs · Governance · Decentralization

1 Introduction

Blockchain technology has transformed business innovation by offering transparency and security and promoting innovative decentralized commerce and decision-making models [23–26]. The rise of Decentralized Autonomous Organizations, or DAOs [27], signals a revolutionary paradigm shift in organizational structure and operational dynamics, marking a pivotal moment in the evolution of businesses. At their core, DAOs are digital entities that exist and function on blockchain platforms, relying on predefined rules encoded as smart contracts [20]. Unlike traditional organizations, which are centralized and often subject to hierarchical decision-making processes, DAOs operate in a decentralized manner, ensuring that power and control are distributed among their members [21]. The concept of decentralization is fundamental to DAOs and is explored in collective decision making: In a DAO, decisions are typically made through a consensus mechanism, where members vote on proposals based on their stake or reputation. This ensures that no single entity has undue influence over the organization's direction.

M. Papadaki et al. (Eds.): EMCIS 2023, LNBIP 501, pp. 61–74, 2024.
https://doi.org/10.1007/978-3-031-56478-9_5

The DAO decentralization is based on these four main pillars. Firstly, transparency, all transactions and decisions within a DAO are recorded on the blockchain, providing a transparent and immutable record [22]. This openness fosters trust among members and reduces the chances of fraudulent activities [11]. Second, is autonomy, DAOs operate autonomously, without the need for intermediaries or centralized authorities. Once the rules are set in the form of smart contracts [13], the DAO follows them automatically, reducing the potential for human error or bias. Third, global collaboration, DAOs enable people from all over the world to collaborate without the constraints of geographical boundaries or centralized control [6]. This fosters innovation and allows for a diverse set of perspectives to influence the organization's direction [5]. Fourth, is resilience, Due to their decentralized nature, DAOs are less vulnerable to attacks, censorship, or shutdowns. Even if one part of the network is compromised, the DAO can continue to function normally [15].

However, due to the early stages of development of Decentralized Finance (DeFi) and their DAOs, there are some pitfalls to their current decentralization status. Many of the DeFi protocols, such as Automated Market Makers, are indeed operating automatically in an on-chain smart decentralized environment. However, even with their decentralized operation, it does not necessarily mean that their governance power is equally distributed [17]. The relevant work reveals a notable gap in the quantitative assessment of decentralization within these entities. Specifically, there is a lack of in-depth discussion and analysis of the unique governance structures underpinning each DAO and it's decentralization. Most theoretical papers tend to generalize governance models without delving into the intricate mechanisms of decentralization inherent to each DAO's governance.

This research investigates various DAOs with the ambition to observe and visualize the governance process of these protocols and the decentralization mechanics in them. The overall goal is to propose a new methodology that will take into account fundamental aspects of DAOs, on which quantitative decentralization analysis can be then applied. Addressing this void, our contributions are twofold. Firstly, the article provides a fundamental analysis of the potential points of failure in decentralization within DAOs. Secondly, this study introduces a "decentralization layer system", a novel framework designed to be instrumental in future quantitative analyses of decentralization. This framework not only fills the existing research gap but also offers a robust foundation for future explorations in the realm of DAO governance and decentralization metrics.

The remainder of this paper is organized as follows: Sect. 2 presents and critically evaluates the normative literature, while Sect. 3 introduces the proposed framework for analyzing fundamental DAO decentralization. The paper concludes with a summary and outlines directions for future work.

2 Relevant Work

The existing literature on the decentralization of DeFi governance primarily centers on the distribution of governance tokens by leading DeFi platforms. The study by [17] delves into the token distribution of platforms like IDEX, MakerDAO, Compound, Curve, and Uniswap. Their methodology examines the contrast between the token distribution of all holders versus the top 20 holders. Their findings indicate a significant concentration of voting power across these platforms, but they don't offer concrete mathematical evidence to support their conclusions.

Jensen et al., [12] also investigate the decentralization levels among DeFi platforms, employing specific decentralization metrics. This research elaborates on the governance structures of Balancer, Compound, Uniswap, and Yearn Finance, sourcing data from Defipulse. The study reveals pronounced centralization in these applications, with Gini Coefficient values [4] ranging from 0.82 to 0.98 and Nakamoto Coefficients [14] between 82 and 9. Notably, Compound emerges as the most centralized of the lot. While this research offers valuable insights into applying decentralization metrics in DeFi and potential wallet sampling exclusions, it lacks a temporal perspective and doesn't address vote delegation.

Additionally, Barbereau et al., [1] build upon prior research by introducing novel decentralization metrics suitable for DAOs. It assesses the governance decentralization of platforms like Uniswap, Maker, ShushiSwap, UMA, and Yearn Finance. This study also incorporates a time factor by sampling data at six-month intervals, and the results consistently highlight the pronounced centralization in DeFi governance.

A particularly pertinent study akin to this paper is by Fritsch [10]. The authors undertake an empirical analysis of the governance systems of three notable Decentralized Autonomous Organizations on the Ethereum blockchain: Compound, Uniswap, and ENS. Utilizing a comprehensive dataset encompassing governance token holders, delegates, proposals, and votes, they scrutinize the distribution and utilization of voting power. Furthermore, they assess the decentralization levels and examine the voting patterns of various delegate categories.

Another study by Sharma et al., [16] highlights the significance of DAOs in understanding algorithmic authority, governance in decentralized systems, and potential pitfalls in incentive design. Interviews with DAO experts revealed key metrics for de centralization, such as token distribution, voter participation, and geographical distribution. Empirical analyses showed a trend of voting pattern polarization among large token holders and a correlation between a proposal author's token balance and the proposal's success. The research also found that DAOs allowing governance tokens to be purchased from secondary markets tend to become less decentralized over time. The study further explored DAOs' autonomy, revealing that many off-chain DAOs rely on Multisig wallets, which can compromise autonomy. The paper suggests design implications, emphasizing the need for a clear separation between voting power and monetary value, AI-based intervention protocols for informed voting, on-chain proposal pipelines for enhanced autonomy, and automated tracking tools for proposal transparency.

There are also other two studies that focus on the fundamental aspects of DAO governance models. Ding et al., [8] offers an in-depth analysis of the various voting schemes employed in DAO governance, highlighting their distinct features and drawing comparisons between them. The study introduces a novel hypothetical voting mechanism tailored for decentralized and permissionless DAO governance. This proposed mechanism, enriched with incentive designs, promises enhanced efficiency over existing models and can be modified for permissioned settings. After reviewing and comparing several prominent voting schemes, it was observed that each has its advantages and drawbacks. These schemes have been developed and adopted by notable DAO service platforms like Aragon, Moloch, DAOstack, and SubDAO10. The evolution of DAOs should not be static; instead, they should continuously adapt by integrating effective designs. While the proposed scheme offers improvements, it is not without potential challenges during its implementation. The study underscores the broader implications of these voting mechanisms, noting their application in societal governance, such as elections and corporate decision-making.

Another paper from Ding [7] offers a systematic review of DAOs, shedding light on their characteristics, categorizations, and real-world applications like Maker DAO. The study delves into DAO governance, addressing challenges like proposal spamming and emphasizing the importance of incentive design for effective voting. Traditional dispute resolution mechanisms are found to be ill-suited for DAOs, with decentralized methods showing promise. The research not only provides valuable insights for those interested in DAOs but also highlights numerous areas for future exploration, from the role of delegation in DAO governance to the harmonization of legal responsibilities with on-chain governance [27].

3 Fundamental DAO Decentralization Analysis Framework

This framework is built on two foundational pillars. The first involves dissecting the governance structure to critically assess the inherent level of decentralization in each component. The default decentralization system, termed as the "theoretical level of decentralization", can be exemplified by token distribution. A fair and equal distribution across the community implies a low likelihood of centralization, ensuring higher decentralization. Conversely, if governance decisions are monopolized by a single entity, it introduces a point of failure, indicating reduced decentralization. For the purposes of this paper, we classify decentralization into three tiers: "green" signifies optimal decentralization, "purple" indicates moderate decentralization, often due to potential barriers or thresholds, and "red" denotes potential centralization, as seen when a specific group dominates a governance step. The second pillar focuses on stratification. Recognizing distinct layers is crucial to differentiate the multifaceted subsystems within the governance structure. These layers follow a hierarchical pattern, with each representing a pivotal decision-making process within a DAO. Centralization in one layer can cascade and influence subsequent layers. The decentralization layer system offers a comprehensive approach to analyzing a DAO's decentralization. By evaluating each layer's decentralization metrics, a holistic view of a DAO's decentralization can be derived. While this method doesn't yield definitive results, it establishes a theoretical foundation for subsequent quantitative analyses.

For the purpose of this paper, a fundamental decentralization analysis will be conducted in order to showcase this methodology. Uniswap, Compound, ApeCoin, and Synthetix DAOs shall be discussed. This set of DAO consists of platforms that have different functions (decentralized exchange, NFTs etc.) and they have different governance models. The differentiation of the governance structure provides a good example of the utility of the proposed framework.

3.1 Uniswap

Uniswap is a DEX (Decentralized Exchange) that stores the most capital in the market [2].Uniswap gives the ability to exchange any token with another in an on-chain process without the need for intermediates and offers the chance to participate in the market-making process through liquidity providing DEXs are one of the most common types of DeFi swap platforms from the ecosystem. It's DAO governance model includes the proposal and vote system for its governance token. UNI is the underlying governance token that was introduced in September of 2020 with total ever supply of 1 Billion of UNI. According to Uniswap's website [19], the initial allocation of UNI tokens were 21.51% to the developing team,

17.8% to the investors, 0.69% to advisors and 60% allocated in the community. With the 10-year supply projections is allocating 67.19% total UNI supply to the community which results with greater distribution of voting power.

The governance model of Uniswap starts with the UNI Token holders. These holders are eligible to join the governance structure of Uniswap and every token represents one vote. Holders must delegate or self-delegate their tokens in order to be included in the governance process. Users can then vote or propose (if they have delegated votes over 2.5 million). In order to propose a user must post their ideas in Uniswap's governance forum and pass through to steps. Step number one is the temperature check (a poll that requires more than 25 thousand votes to produce a result) and the consensus check (a poll that requires more than 50 thousand votes to produce a result). After these two steps, this proposal idea can be officially proposed (in a form of executable code) for implementation in the form of an on-chain voting procedure. If the vote is successfully activated and positively voted, then it stays in the Timeclock smart contract for at lead 2 days before is executed.

The identifiable sub-systems in the form of layers are presented in Fig. 1. The weakness of Layers 0, 1 and 2 is the danger of an address (or a group of addresses) controlling most of the tokens. If the protocol distributes the token unfairly then potentially users will be able to influence the entire governance system. Additionally, a 51% attack on the UNI holders would be possible if a user ordains the 51% of the circulating tokens from the market. But even if the UNI holder distribution is completely decentralized don't result in a complete decentralization in the delegated sub-system. A user that doesn't hold any UNI tokes can be delegated votes, thus, if a group of UNI token holders were to create an alliance, they can delegate all their tokens to one address to take decisions for them. Then for a user to submit a proposal, the user needs to have at least 2,5 million votes to his address. If a user was to buy UNI and self-delegate the votes to himself they need almost 15.4 Million USD (March 2023 prices) in order to be eligible to propose. This process puts a restriction on users and makes the proposing system less decentralized. The final layer is the system of voting which includes how users vote and how the potential voting power is actually used in existing proposals.

3.2 Compound

Compound Finance [9] is a lending and borrowing platform which has similar attributes and governance model to Uniswap. Through Ethereum based smart contracts, Compound

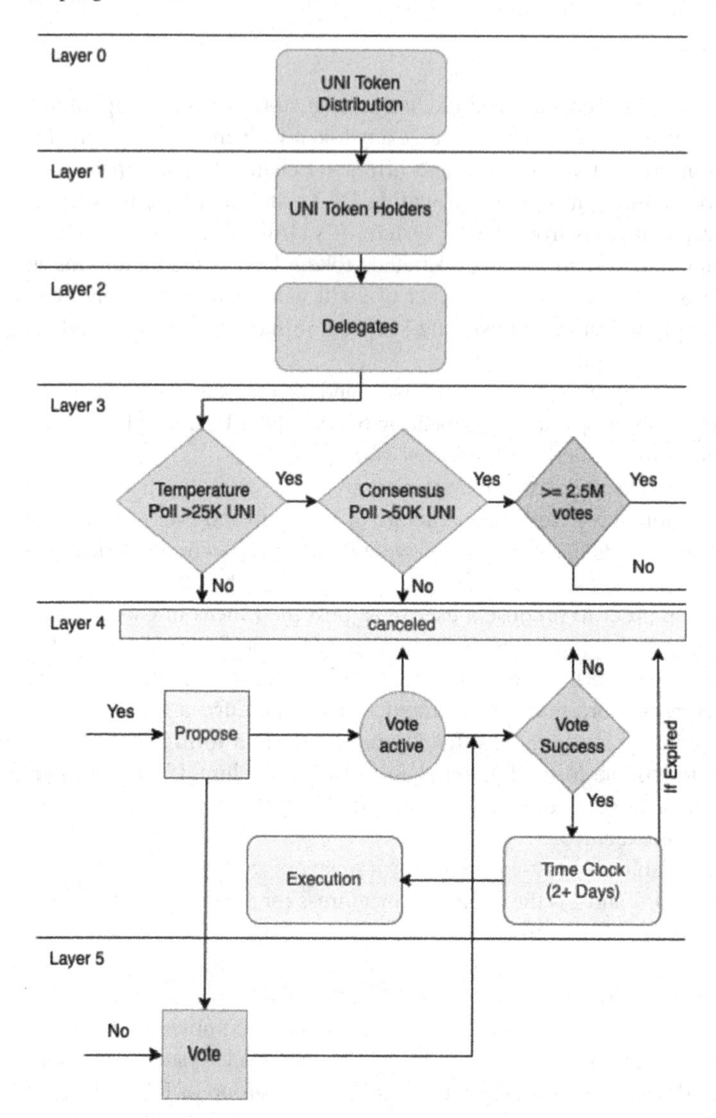

Fig. 1. Uniswap DAO governance structure

allows borrowers and lenders to provide or receive loans via crypto assets which are locked in the protocol. The interest rates are connected to the supply and demand of the underlying cryptocurrency and they are updated with each mined block. The lender can repay anytime the loan and the collateralized assets can be withdrawn at any time. The government token users must delegate or self-delegate their governance tokens, or they can transfer them to others users in order to give more voting power to another account (Governance as a Service). Compound also has a one-to-one token and vote ratio with a minimum of 25,000 COMP stake in order to create a governance proposal [9]. Additionally, any address can lock 100COMP in order to create an Autonomous

Proposal which can be converted to a governance proposal after being delegated the 25,000 necessary COMP delegated tokens. Copmound's governance token COMP was introduced in April 2020 with a total ever supply of 10 million COMP.

The governance process of Compound Finance consists of 4 subsystems with all the layers summarized in Fig. 2. The first system is the distribution of the Governance token. This distribution is of the highest importance due to the fact that this layer can influence all the other governance layers. The second layer is the distribution of the COMP token delegates and then two parallel subsystems, the voter impact and proposal impact.

The point of failure for the first sub-system is the possibility of a central address controlling the 51% majority of the COMP tokens. Centralization on this level can potentially have a great impact on the whole governance system of Compound; the reason being that if a centralized malicious holder owns the majority of COMP tokens can then have the majority of delegations, thus the votes also. Theoretically, this malicious holder will be able to submit any proposal of his liking and influence the result with his votes, but in practice, every proposal must pass through a 2-day review period on which all members of the compound community and users who actually have positions in the protocol can review the proposal, thus introducing a centralized review layer in the proposals. At the time of writing, the COMP circulation is 7,267,152 out of the 10 million with the price of 53$ which results that an attacker (or group of attackers) to obtain the 51% vote majority needs 196,431,118 US Dollars' worth of funds. Additionally, the same governance influence can be observed in the case of a centralization of the delegates occurs. In the delegates sub-system, an attacker can only hold the 51% of the voting power if it can collect delegations for the majority COMP holders or form a cartel from other delegated addresses.

3.3 ApeCoin

ApeCoin (APE) is an ERC-20 token on the Ethereum blockchain [3], initially airdropped to Bored Ape Yacht Club (BAYC) and Mutant Ape Yacht Club (MAYC) NFT holders, this makes layer 0 and 1 [6]. Developed by Yuga Labs and governed by the ApeDAO, APE serves as both a governance and utility token, facilitating voting on system improvements and acting as currency in web3 projects. Its launch saw significant adoption, notably in the purchase of virtual land in the Otherside metaverse. With a capped supply of 1 billion tokens and a diverse distribution strategy, ApeCoin aims to be a pivotal asset in the web3 domain.

ApeCoin DAO is a decentralized governance framework that oversees the Ecosystem Fund. The DAO operates through a proposal process, determining how the Ecosystem Fund will be allocated by the APE Foundation to foster a diverse and self-sustaining ecosystem. Membership in the ApeCoin DAO is exclusive to ApeCoin holders which consists of layer 1. This membership grants them the privilege to participate in idea submissions, commentary, proposal sub missions (layer 2), and voting (layer 3). This DAO is using Snapshot for voting. Proposals that have passed the necessary approval processes are made available for voting here. Again, wallet authentication is required for ApeCoin holders to vote. ApeCoin DAO has adopted a proposal process inspired by Ethereum's EIP system. Proposals are categorized into firstly, "Core" which includes decisions related to the Ecosystem Fund Allocation and Branding. Then "Process" proposals that suggest changes to existing processes or implementations. Additionally, "Informational" proposals that provide general guidelines or information to the community. Each proposal undergoes several phases, from idea submission in Discourse to a final implementation phase. This process ensures that each proposal is thoroughly vetted, analyzed, and moderated before being put up for a community vote. The proposing process has 9 steps which also seen in Fig. 3.

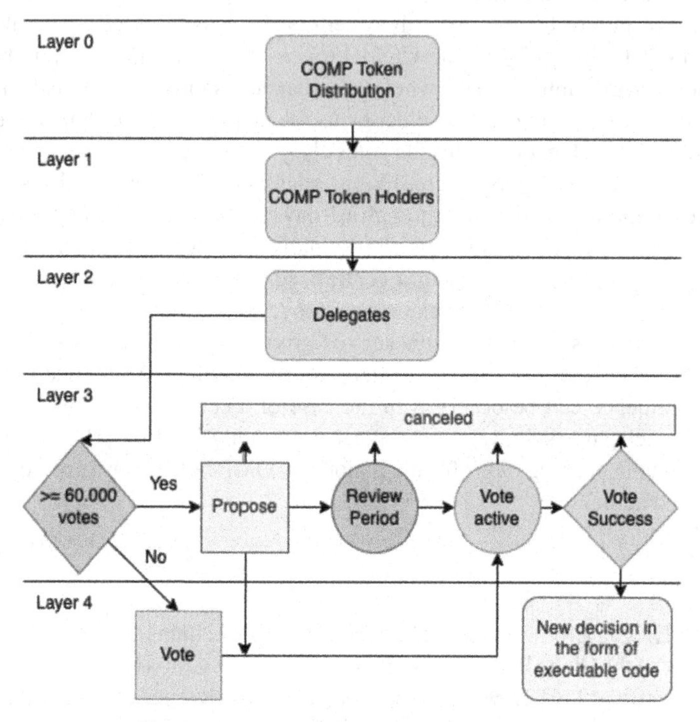

Fig. 2. Compound DAO governance structure

1. AIP Idea: Ideas are posted on Discourse, awaiting moderator approval based on DAO guidelines. Authors receive feedback for a week and can't modify the original post but can comment.
2. AIP draft: After feedback, authors refine their idea using a provided tem plate. If aligned with DAO interests, the idea gets a unique identifier.
3. AIP analysis report: A team from the APE Foundation reviews the preliminary idea, producing an evaluation report covering all aspects, including costs and legal factors.
4. AIP moderation: Moderators review the combined preliminary and report. Approved ideas move to the next stage; declined ones can be resubmitted.
5. Post-moderation tagging: Endorsed ideas are labeled "Straight to Vote" or "Needs Administrative Review"(complex implications).
6. Administrative review: Complex ideas are evaluated by the Board for clarity or actions. They're then labeled "Ready for Voting" or need further clarity/actions.

7. Live AIP: Approved drafts become Live AIPs on Snapshot, released on Thursdays at 9PM ET. They remain open for voting until the following Wednesday at 9PM ET. Only moderators can post AIPs on Snapshot after confirming their approval process.
8. Final AIP: Post-voting, AIPs with no votes or ties are tagged "Stalled" and can be resubmitted. Others are categorized as Rejected or Accepted. Rejected AIPs can be resubmitted after moderator consultation. Accepted ones proceed to implementation.
9. Implementation: Accepted AIPs are executed based on their outlined steps. The APE Foundation's project management team oversees this, ensuring proper execution without necessarily implementing it themselves.

The DAO employs a consensus mechanism to ensure that voting is fair, transparent, and cost-effective. Snapshot is the chosen tool for voting due to its gas-free nature, transparency, and inclusivity for all ApeCoin holders. Voting delegation is also possible, allowing DAO members to delegate their votes to trusted experts or other members. The voting process goes as follows.

1. Moderators upload AIPs to Snapshot after verifying their adherence to the approval process. AIPs are updated weekly, every Thursday at 9PM ET, marking the start of a six-day voting period.
2. DAO members cast their votes on Snapshot. Each ApeCoin is equivalent to a single vote. Fractional tokens are rounded down for voting (e.g., 100.1 or 100.9 tokens both equate to 100 votes). Voters can choose between "In favor" (supporting the AIP as presented) or "Against" (opposing the AIP in its current form, suggesting potential revisions).
3. Each weekly batch of proposals remains open for voting for six days, con cluding at 9PM ET the subsequent Wednesday.
4. AIPs with a majority of "In favor" votes proceed to the implementation phase. Rejected AIPs can be reconsidered if the author liaises with a moderator for resubmission. AIPs with no votes or a tie by the closing time are labeled "Stalled" and are also eligible for resubmission.
5. DAO members can assign their voting rights to another trusted DAO member. This delegation allows members with tokens still in the initial lock-up phase to participate in voting.

Potential Centralization Points (Points of Failure) based on the analysis is firstly the requirement for moderator confirmation at multiple stages of the proposal process can be a centralization point. If moderators act maliciously or are compromised, they can influence which proposals get visibility and progress to the next stages. Also, the Board has the authority to halt or send a pending AIP to vote. This power can be misused if the Board becomes centralized or acts against the community's interests. Although the coin distribution and delegation is decentralized but as with other cases it can lead to centralization if a significant number of members delegate their votes to a single entity, giving that entity disproportionate influence.

3.4 Synthetix

Synthetic assets, commonly referred to as synths, mimic the value of other assets, allowing traders to access securities or commodities they might not typically trade [18]. This

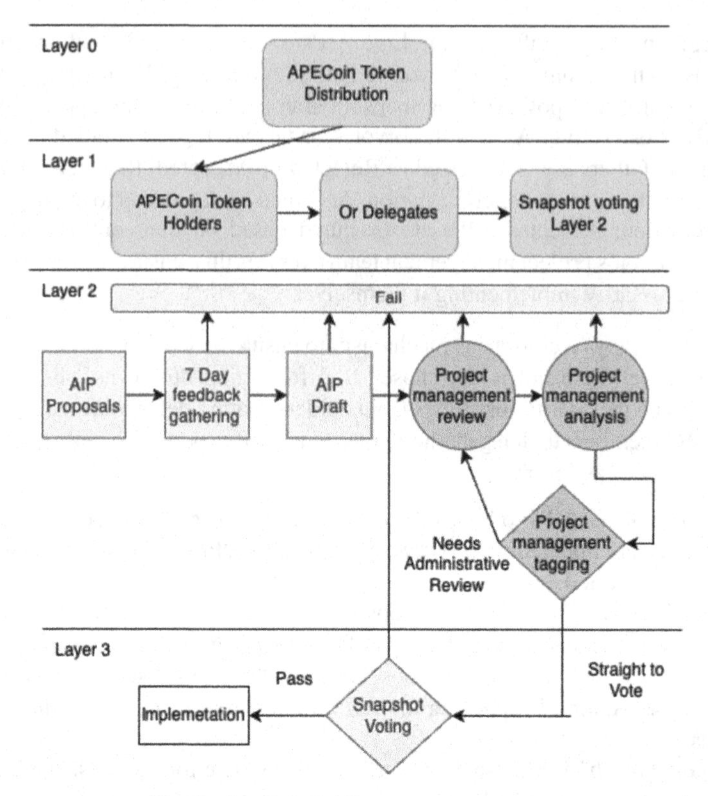

Fig. 3. ApeCoin DAO governance structure

concept mirrors traditional finance instruments like futures, which derive value from real-world assets. The Synthetix protocol facilitates the issuance of synths, such as sBTC and sUSD, using overcollateralization with the Synthetix Network Token (SNX). SNX also plays a role in staking rewards and decentralized governance. While Synthetix supports crypto derivatives markets, it primarily offers the infrastructure and liquidity for these platforms, branding itself as "the derivatives liquidity protocol".

The broader Synthetix community, especially the SNX stakers, forms the backbone of the governance process. This DAO process has five observable layers as seen in Fig. 4 By staking their SNX, they gain voting power for electing the various councils, influencing the direction of the protocol. Community engagement platforms, such as Discord and Twitter, serve as hubs for discussion, debate, and dissemination of information. This layer can be thought of as theoretically decentralized. The staking mechanism, while aiming to decentralize power, could inadvertently centralize decision-making if a few entities accumulate a significant portion of SNX.

Layer 3 of it's DAO consists of Synthetix's governance is further structured through various councils and DAOs. The Spartan Council stands as the primary governing body, overseeing SIP/SCCP processes, while other councils like the Treasury, Grants, and Ambassador Councils have specialized roles. Each council is elected by Synthetix Stakers, ensuring a democratic process. The election process, though democratic, can be influenced if a minority of stakers holds a majority of SNX. Such a scenario could lead to skewed election outcomes favoring specific agendas. The Spartan Council's control over the security bonds of all pDAO members through their Multi-Sig Gnosis Safe is another centralization concern. This power to slash a member's bond if deemed necessary places significant authority in the hands of the council. The Treasury Council's control over financial resources is another potential point of failure, especially if not managed with utmost transparency.

The foundational layer (layer 4) of Synthetix's governance is the creation and management of proposals. Two primary artifacts drive changes within the protocol: the Synthetix Improvement Proposals (SIPs) and the Synthetix Con figuration Change Proposals (SCCPs). SIPs are comprehensive documents that describe proposed alterations to the core Synthetix Protocol. In contrast, SCCPs are more specific, suggesting modifications to the system's existing parameters. The process of SIP authorship, although open to all, is ultimately under the discretion of the Spartan Council for approval. This

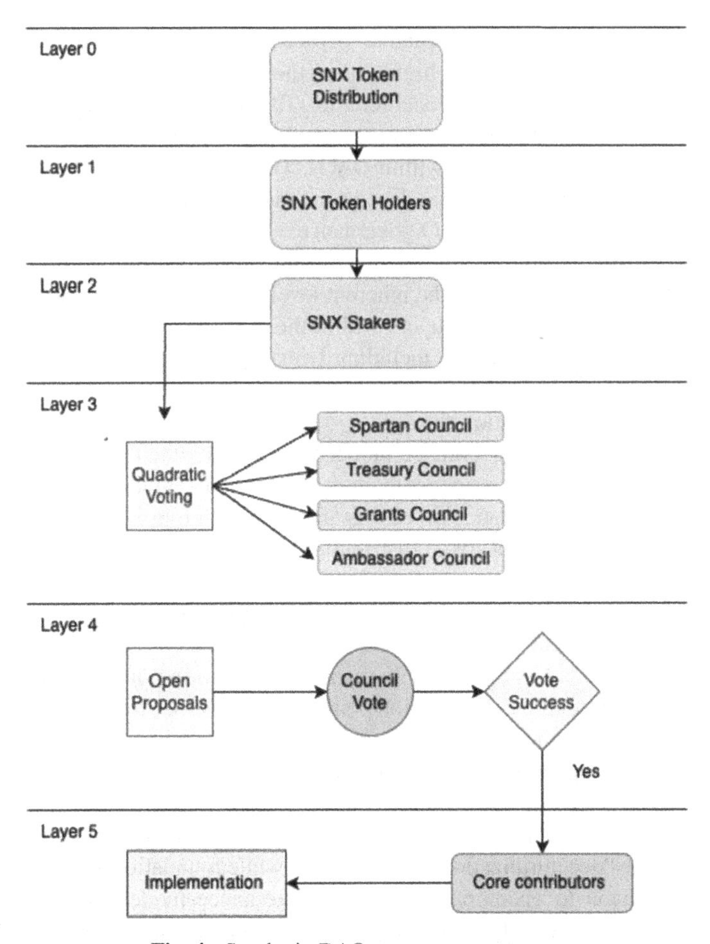

Fig. 4. Synthetix DAO governance structure

centralized decision-making can lead to potential biases or overlook community-driven proposals. Furthermore, the reliance of SCCPs on previously deployed SIPs means that the integrity and accuracy of original SIPs are paramount. Any oversight or error in the foundational SIP could propagate through subsequent SCCPs.

Once proposals are approved, they move to the implementation phase (layer 3). Core Contributors play a pivotal role here, translating approved SIPs into actionable changes. The Protocol DAO (pDAO) further refines and implements these technical changes, ensuring they align with the Spartan Council's vision. The nomination and election of Core Contributors could become a closed-loop system, where existing contributors favor certain nominations, leading to a lack of diversity in thought and approach. The pDAO, given its significant role in actualizing changes, could become a centralization risk if its actions diverge from the broader community's vision.

4 Conclusion and Future Work

The paper delves into the intricate nature of decentralization within Decentralized Autonomous Organizations (DAOs), highlighting the significant gaps in the quantitative assessment of decentralization in current literature. The study's primary contributions are the introduction of a "decentralization layer system" and a detailed analysis of potential points of failure in decentralization within DAOs. This novel framework offers a comprehensive approach to evaluating the decentralization metrics of a DAO, providing a foundation for future research in DAO governance.

The paper further introduces a fundamental DAO decentralization analysis framework built on two pillars: assessing the inherent level of decentralization in governance structures and recognizing distinct layers within the governance structure. The framework is applied to prominent DAOs, including Uniswap, Compound, and ApeCoin, to demonstrate its utility. Each DAO's governance structure is dissected, revealing potential points of failure and centralization risks.

For instance, Uniswap's governance model is explored which consists of five layers, with the framework suggesting that there is not a fundamental centralization point. Similarly, Compound's governance process is analyzed with four layers of governance, highlighting the significance of token distribution and the potential impact of centralization in some aspects of the proposal dimension. ApeCoin's governance framework, on the other hand, is detailed through its proposal process, emphasizing the importance of community involvement and transparent voting mechanisms. Apecoin consists of the fewest governance layers (3) and with many centralization points in the proposing process and various potential centralization points in the governance proposal layer. Furthermore, Synthetix provides a very good case of an alternative governance model. Al though inside the five-layer process, there are clear points with concentrated power other processes are more decentralized than the equivalent parts in other governance models.

In considering potential future work in the realm of DAO governance, several avenues emerge. Firstly, the "decentralization layer system", while foundational, presents opportunities for expansion to encompass emerging governance models in the fast-paced blockchain domain. There's also a pressing need to formulate quantitative metrics that objectively measure decentralization in each of the layers and decision points, facilitating nuanced comparisons across diverse DAOs. Delving deeper, comprehensive case studies on an even broader range of DAOs, particularly those with distinct governance structures, can shed light on the tangible challenges and strategies in decentralization. The security dimensions of decentralization, especially the balance between decentralization and susceptibility to threats like 51% attacks, warrant in-depth exploration. Lastly, a holistic examination of the economic constructs that underlie DAOs, from token allocation to incentive schemes, can provide insights into the economic determinants of decentralization.

References

1. Barbereau, T., Smethurst, R., Papageorgiou, O., Rieger, A., Fridgen, G.: DeFi, not so decentralized: the measured distribution of voting rights. In: Proceedings of the 55th Hawaii International Conference on System Sciences (2022)
2. Adams, H., Zinsmeister, N., Salem, M., Keefer, R., Robinson, D.: Uniswap v3 core. Technical report, Uniswap (2021)
3. ApeCoin: Apecoin website (2023). https://apecoin.com/
4. Ceriani, L., Verme, P.: The origins of the gini index: extracts from variabilita emutabilita (1912) by corradogini. J. Econ. Inequality **10**, 421–443 (2012)
5. Chen, W., Botchie, D., Braganza, A., Han, H.: A transaction cost perspective on blockchain governance in global value chains. Strateg. Chang. **31**(1), 75–87 (2022)
6. Christodoulou, K., Katelaris, L., Themistocleous, M., Christodoulou, P., Iosif, E.: NFTs and the metaverse revolution: research perspectives and open challenges. In: Lacity, M.C., Treiblmaier, H. (eds) Blockchains and the Token Economy. Technology, Work and Globalization, pp. 139–178. Springer, Heidelberg (2022). https://doi.org/10.1007/978-3-030-95108-5_6
7. Ding, Q., Liebau, D., Wang, Z., Xu, W.: A survey on decentralized autonomous organizations (DAOs) and their governance. Available at SSRN 4378966 (2023)
8. Ding, Q., Xu, W., Wang, Z., Lee, D.K.C.: Voting schemes in DAO governance. Annu. Rev. Fintech (2023, forthcoming)
9. Finance, C.: Compound finance website (2023). https://compound.finance/
10. Fritsch, R., Mu¨ller, M., Wattenhofer, R.: Analyzing voting power in decentralized governance: who controls daos? arXiv preprint arXiv:2204.01176 (2022)
11. Hellani, H., Sliman, L., Samhat, A.E., Exposito, E.: On blockchain integration with supply chain: overview on data transparency. Logistics **5**(3), 46 (2021)
12. Jensen, J.R., von Wachter, V., Ross, O.: How decentralized is the governance of blockchain-based finance: empirical evidence from four governance token distributions. arXiv preprint arXiv:2102.10096 (2021)
13. Karamitsos, I., Papadaki, M., Al Barghuthi, N.B.: Design of the blockchain smart contract: a use case for real estate. J. Inf. Secur. **9**(3), 177–190 (2018)
14. Kusmierz, B., Overko, R.: How centralized is decentralized? Comparison of wealth distribution in coins and tokens. In: 2022 IEEE International Conference on Omni layer Intelligent Systems (COINS), pp. 1–6. IEEE (2022)
15. Rmit, K.N., Zargham, M.: The ethnography of a 'decentralized autonomous organization' (DAO): de-mystifying algorithmic systems. In: Ethnographic Praxis in Industry Conference Proceedings, vol. 2022, pp. 74–97. Wiley Online Library (2022)
16. Sharma, T., et al.: Unpacking how decentralized autonomous organizations (DAOs) work in practice. arXiv preprint arXiv:2304.09822 (2023)

17. Stroponiati, K., Abugov, I., Varelas, Y., Stroponiatis, K., Jurgelevicience, M., Savanth, Y.: Decentralized governance in DeFi: examples and pitfalls (2020)
18. Synthetix: Synthetix website (2023). https://synthetix.io/
19. Uniswap: Uniswap website (2023). https://uniswap.org/
20. Ding, W.W., et al.: Parallel governance for decentralized autonomous organizations enabled by blockchain and smart contracts. In: 2021 IEEE 1st International Conference on Digital Twins and Parallel Intelligence (DTPI). IEEE (2021)
21. Liu, L., et al.: From technology to society: an overview of blockchain-based DAO. IEEE Open J. Comput. Soc. **2**, 204–215 (2021)
22. El Faqir, Y., Arroyo, J., Hassan, S.: An overview of decentralized autonomous organizations on the blockchain. In: Proceedings of the 16th International Symposium on Open Collaboration (2020)
23. Karamitsos, I., Papadaki, M., Themistocleous, M.: Blockchain as a service (BCaaS): a value modeling approach in the education business model. J. Softw. Eng. Appl. **5**, 165–182 (2022)
24. Cunha, P.R., Soja, P., Themistocleous, M.: Blockchain for development: a guiding framework.. Inf. Technol. Dev. **27**(3), 417–438 (2021.) https://www.tandfonline.com/doi/full/10.1080/02681102.2021.1935453
25. Kapassa, E., Themistocleous, M.: Blockchain technology applied in IoV demand response management: a systematic literature review. Future Internet **14**(5), 1–19 (2022)
26. Christodoulou, K, Iosif, E., Inglezakis, A., Themistocleous, M.: Consensus crash testing: exploring ripple's decentralization degree in adversarial environments. Future Internet **12**(3), 53 (2020) https://www.mdpi.com/1999-5903/12/3/53/htm
27. Papangelou, S., Christodoulou, K., Michoulis, G.: Exploring decentralized governance: a framework applied to compound finance. In: Pardalos, P., Kotsireas, I., Knottenbelt, W.J., Leonardos, S. (eds.) MARBLE 2023. Lecture Notes in Operations Research, pp. 152–168. Springer, Cham (2023). https://doi.org/10.1007/978-3-031-48731-6_9

Blockchain-Powered NFTs: A Paradigm Shift in Carbon Credit Transactions for Traceability, Transparency, and Accountability

Abhirup Khanna[1] and Piyush Maheshwari[2(✉)]

[1] University of Petroleum and Energy Studies, Dehradun, India
akhanna@ddn.upes.ac.in
[2] The British University in Dubai (BUiD), Dubai, United Arab Emirates
piyush.maheshwari@buid.ac.ae

Abstract. The adoption of carbon credit systems has emerged as a pivotal strategy in addressing climate change and promoting sustainable environmental practices. This research paper delves into the multifaceted landscape of carbon credit systems, specifically focusing on the challenges inherent in their design and implementation. We investigate the innovative use of Non-Fungible Tokens (NFTs) as a transformative approach to carbon credits, exploring the intricacies of NFT structures tailored for carbon credit trading. Furthermore, the paper presents a comprehensive examination of the system architecture underpinning a blockchain-enabled NFT trading platform for carbon credits. The architectural framework's design and components are meticulously scrutinized to ensure transparency, security, and efficiency in the trading ecosystem. The research also dissects the intricate functionalities that empower the blockchain-NFT trading system, facilitating seamless transactions and traceability. A pivotal facet of this study is a compelling case study spotlighting the application of a Blockchain-NFT trading system for Agricultural Carbon Credits within a rural farming community. The case study highlights the tangible benefits reaped by farmers through the innovative system, shedding light on how the technology incentivizes sustainable farming practices and offers an additional revenue stream.

Keywords: Blockchain · NFT · Carbon credits · Trading platform · SDG 13

1 Introduction

Carbon offset projects are initiatives designed to reduce or remove greenhouse gas emissions from the atmosphere in order to mitigate climate change. These projects are critical components of efforts to combat global warming and achieve carbon neutrality. They play a significant role in addressing the environmental challenges posed by the excessive release of carbon dioxide (CO2) and other greenhouse gases into the atmosphere [9, 10]. Carbon offset projects directly contribute towards reduction in the emission of greenhouse gases. They focus on activities such as reforestation, afforestation, renewable

M. Papadaki et al. (Eds.): EMCIS 2023, LNBIP 501, pp. 75–87, 2024.
https://doi.org/10.1007/978-3-031-56478-9_6

energy generation, and improved land management practices that either prevent emissions or capture and store CO2 from the atmosphere. By mitigating the effects of climate change, these projects help to limit temperature rise, reduce the frequency and severity of extreme weather events, and protect vulnerable ecosystems and communities. They promote the transition to more sustainable and environmentally friendly practices. For example, renewable energy projects replace fossil fuels, reducing reliance on carbon-intensive energy sources. These projects often involve international cooperation and partnerships. They provide a mechanism for individuals, businesses, and governments to take responsibility for their carbon emissions on a global scale, regardless of where emissions occur. Investment in carbon offset projects fosters innovation in clean technologies and sustainable practices. It encourages the development of green industries and creates economic opportunities.

Carbon offsetting allows individuals and organizations to achieve carbon neutrality by balancing their carbon emissions with an equivalent reduction in emissions elsewhere, effectively "neutralizing" their carbon footprint. Many carbon offset projects have positive social and economic impacts on local communities. They create jobs, improve access to clean energy, and enhance overall quality of life. The Paris Agreement, adopted in 2015 at the United Nations Framework Convention on Climate Change (UNFCCC), primarily focuses on global efforts to combat climate change and limit global warming to well below 2 °C above pre-industrial levels, with an aspiration to limit it to 1.5 °C. While the Paris Agreement does not explicitly detail carbon offset projects, it does make several references to related concepts and mechanisms. Article 6 includes provisions for countries to voluntarily engage in emissions trading and cooperation, essentially enabling carbon offset projects. Carbon credits, also known as carbon offsets or emission reduction units (ERUs), represent a specific quantity of emissions reduced, removed, or avoided by a carbon offset project. Carbon credits, typically quantified in metric tons of carbon dioxide equivalent (CO2e), equate one unit to a corresponding reduction or elimination of one metric ton of CO2e emissions. When a carbon offset initiative effectively curtails emissions, it garners carbon credits subject to issuance by an accredited certification entity, ultimately rendering them tradeable within carbon markets. These credits frequently find acquisition by a diverse spectrum of entities, spanning individuals, corporations, and governmental bodies, as a means to offset their individual carbon footprints while concurrently endorsing endeavors dedicated to emissions reduction. The valuation of these carbon credits primarily adheres to market forces, resulting in notable fluctuations in their pricing. The carbon credit market operates through an intricate interplay of supply and demand dynamics. In the realm of carbon credit trading, blockchain technology emerges as a pivotal innovation, serving as a conduit for transparent, secure, and efficient transactions concerning carbon credits and emissions allowances. The fundamental tenet of blockchain's utility resides in its ability to furnish an immutable and fully transparent ledger, permanently recording all transactions. The immutable nature of the data within the blockchain reinforces trust within the trading ecosystem, as once information is inscribed, it cannot be altered or eradicated without securing a consensus from the network's participants. Blockchain further augments security via the deployment of cryptographic techniques to safeguard both data and transactions, fortifying the system's resilience against hacking and fraudulent activities. Each transaction undergoes meticulous verification and subsequent incorporation into the blockchain, underpinned by a consensus mechanism that bolsters security measures. Traditional trading systems often involve a multiplicity of intermediaries, encompassing banks, clearinghouses, and brokers. Blockchain introduces a paradigm shift by enabling peer-to-peer (P2P) transactions, thereby diminishing the necessity for intermediary intervention. This not only culminates in cost reduction but also streamlines the entire trading process, a particularly advantageous feature in trading ecosystems characterized by exorbitant fees and overhead expenses [1–3]. Moreover, the versatility of blockchain extends to asset tokenization, encompassing an array of assets spanning carbon credits, equities, real estate, and more. These digital tokens represent ownership rights or stakes in assets, simplifying the process of asset trading and partitioning into smaller, more manageable units, thus rendering them more accessible and fluid within the market [4–7].

The rest of the paper is categorized as follows: Sect. 2 elucidates the research methodology. Section 3 presents the detailed literature review of blockchain technology in hospitality industry. Real world case studies discussing implementation of blockchain technology in hospitality industry are discussed in Sect. 4.

2 Challenges of Carbon Credit Systems

Traditional carbon credit pricing systems have faced several challenges and problems, which have prompted the exploration of alternative approaches like blockchain-based systems. The following is a list of common problems related to traditional carbon credit pricing systems:

- Lack of Transparency: Traditional systems often lack transparency in pricing mechanisms, making it difficult for participants to understand how prices are determined.
- Volatility: Prices for carbon credits in traditional markets can be highly volatile, which can deter long-term investments in emission reduction projects.
- Lack of Standardization: Different regions and countries have their own standards and methodologies for carbon credit pricing, leading to a lack of standardization.
- Regulatory Uncertainty: Frequent changes in carbon pricing regulations and policies can create uncertainty for market participants, making it challenging to plan long-term investments.
- Fraud and Double Counting: Traditional systems are vulnerable to fraud and double counting of emissions reductions thereby undermining the integrity of the market.
- High Transaction Costs: Trading carbon credits in traditional systems often involves significant transaction costs, including fees charged by brokers and other mediators.
- Limited Accessibility: Traditional carbon markets may be inaccessible to smaller players in developing countries due to high entry barriers and administrative burdens.
- Lack of Accountability: In some cases, traditional systems have struggled to hold participants accountable for their emissions and offset commitments.

- Complex Compliance Reporting: Compliance with carbon pricing regulations often requires extensive and complex reporting, which can be burdensome for companies.
- Inequity and Leakage: Traditional systems may inadvertently incentivize carbon leakage, where emissions reductions in one area lead to increased emissions in another, especially in industries with high mobility.
- Ineffectiveness in Achieving Emission Reduction Goals: Traditional carbon pricing systems have not been effective in achieving the necessary emissions reductions to combat climate change.

3 Non-Fungible Tokens as Carbon Credits

Non-Fungible Tokens (NFTs) represent an innovative manifestation of digital assets, serving as incontrovertible indicators of ownership or proof of the unparalleled authenticity inherent in a singular item or piece of digital content, all firmly ensconced within the expansive domain of blockchain technology. Unlike their fungible cryptocurrency counterparts, typified by the likes of Bitcoin or Ethereum, which readily permit one-to-one exchanges, NFTs fundamentally diverge by eschewing interchangeability and embracing a distinct intrinsic worth and individual identity. This distinctive quality of NFTs is harnessed to infuse scarcity, provenance, and verifiable ownership into the digital or even physical assets they encapsulate within the digital realm, thereby catalyzing a transformative ripple effect across multifarious sectors and domains. Across the multifaceted landscape of applications, NFTs have unveiled a mosaic of innovative avenues, each rapidly unfurling and burgeoning in its own unique way. Within the realm of art, NFTs have orchestrated a metamorphic shift in the modus operandi through which digital artists monetize their creations, bequeathing them the capacity to vend singular digital masterpieces fortified with verifiable ownership and irrefutable provenance, all the while ushering in an epochal revenue stream via royalties accruing from secondary sales. In a harmonious resonance with the symphonies of the music world, NFTs have ventured forth, affording musicians the capacity to unfurl limited edition opuses or harmonious compositions as NFTs, bestowing upon their ardent enthusiasts an exclusive passageway into the realms of exclusive access and proprietary entitlements. The arena of gaming, too, has witnessed the transformative incursion of NFTs, wherein they function as the embodiments of in-game accoutrements and personae, furnishing players with the ability to procure, vend, and barter sundry in-game paraphernalia across diverse gaming domains and ecosystems. Beyond the expanse of entertainment, NFTs have embarked upon journeys transcending into the intricacies of real estate, where the hallowed scrolls of property deeds and the hallowed annals of land titles undergo digitization and secure sanctification upon the blockchain's immutable ledger. Meanwhile, their multifaceted utility extends into the realm of supply chain management, wielding their potential in the authentication of luxury commodities, as well as in the domain of collectables, where they take root as indispensable conduits for trading unique items replete with historical, sentimental, or even monetary value, such as sports memorabilia and trading cards. As the metamorphosis unfolds, NFTs are poised to persist as transformative agents, continuing to reshape our fundamental conceptions of ownership and intrinsic value within the digital epoch, thereby proffering a trove of enthralling opportunities for creators, collectors, and enterprises alike.

In parallel, NFTs have found themselves amidst the fertile terrain of carbon credits and environmental conservation, transcending boundaries to inscribe their unique digital imprints upon the landscape of climate action. As unparalleled digital assets, perpetuated by the unwavering underpinning of blockchain technology, NFTs stand poised to revolutionize the very essence of carbon credit genesis, trade, and stewardship. Through the judicious implementation of NFTs, carbon credits can be seamlessly tokenized, thus affording the carbon market an unprecedented trifecta of transparency, traceability, and legitimacy. One of the foremost applications of NFTs within the realm of carbon credits materializes through their indispensable role in the validation and authentication of emissions abatement endeavors. When a carbon offset initiative, such as reforestation or renewable energy deployment, culminates in the tangible realization of emissions reductions, these quantified reductions can be painstakingly transmuted into NFTs, each meticulously emblematic of a precise quantum of duly authenticated emissions reductions. The immutable nature of NFT ownership and the transparency intrinsic to blockchain technology converge to eliminate the looming specter of double counting and fraudulent activities, thereby elevating the credence and integrity underpinning the issuance of carbon credits. Furthermore, the formidable potential of NFTs extends to the orchestration of a comprehensive lifecycle narrative for each carbon credit, from the nascent moment of emissions abatement, through rigorous verification, to the ultimate act of retirement, wherein carbon credits assume their hallowed role in offsetting carbon footprints. This meticulous traceability and documentation, etched indelibly upon the blockchain, ensures that carbon credits, bearing the imprimatur of authenticity and compliance with regulatory benchmarks, become veritable magnets for prospective buyers, gracing both the realms of voluntary and mandated carbon markets with their enhanced allure. In a remarkable synergy with the aspirations of individuals and organizations to actively engage in carbon offsetting endeavors, NFTs, standing as tangible embodiments of carbon credits, serve as conduits to establish a tangible nexus with environmental initiatives. This connection, fortified by transparency and a profound sense of ownership, carries within it the potential to galvanize and incentivize a more active participation in the exalted realm of emissions reduction projects, offering not just a glimpse but a pivotal role in the journey towards a more sustainable, carbon-conscious world.

4 Carbon Credit Trading System

4.1 NFT Structure

When using Non-Fungible Tokens (NFTs) for trading carbon credits, several parameters and metadata can be associated with each NFT to provide comprehensive information about the carbon credits being represented. These parameters help ensure transparency, traceability, and accountability in carbon credit trading. Here's a list of parameters that can be part of an NFT for trading carbon credits:

- Project Name: The name of the emissions reduction project.
- Project Description: A detailed description of the project, its goals, and methodologies.
- Verification Agency: The organization responsible for independently verifying the emissions reductions.

- Verification Report: A link to or a reference to the official verification report.
- Carbon Offset Type: The specific type of emissions reductions (e.g., reforestation, renewable energy, methane capture).
- Carbon Reduction Amount: The quantity of carbon dioxide equivalent (CO2e) emissions reduced or removed.
- Vintage Year: The year in which the emissions reductions were achieved.
- Baseline Emissions: The baseline emissions against which reductions are measured.
- Project Location: The geographic coordinates or location of the emissions reduction project.
- NFT Token ID: A unique identifier for the NFT representing the carbon credits.
- Total Supply: The total quantity of carbon credits represented by the NFT.
- Issuer Name: The entity responsible for issuing the carbon credits.
- Current Owner: The current owner of the NFT, which may change as the NFT is traded.
- Compliance Standard: The environmental standards or protocols followed (e.g., Kyoto Protocol, Verified Carbon Standard).
- Compliance Status: Whether the carbon credits comply with applicable regulations and standards.
- Co-Benefits: Any additional environmental, social, or economic benefits generated by the project.
- Sustainable Development Goals (SDGs): Indication of how the project aligns with specific SDGs.
- Creation Date: The date when the carbon credits were generated.
- Transaction History: A record of all past transfers and transactions involving the NFT.
- Blockchain Network: The blockchain platform on which the NFT is issued and traded (e.g., Ethereum, Binance Smart Chain).
- Smart Contract Address: The address of the smart contract governing the NFT.
- Terms of Use: Any licensing or usage rights associated with the carbon credits represented by the NFT.
- Additionality: Information on how the project demonstrates "additionality" by achieving emissions reductions beyond business-as-usual.
- Permanence Measures: Actions taken to ensure the permanence of emissions reductions (e.g., long-term forest protection).
- Market Segment: Whether the NFT is part of the voluntary or compliance carbon market.
- Pricing Information: Current and historical pricing data for the carbon credits.

4.2 System Functionalities

Blockchain-based carbon trading systems leverage blockchain technology to facilitate transparent, secure, and efficient trading of carbon credits and emissions allowances. These systems are designed to address the challenges associated with traditional carbon markets, such as transparency, traceability, and trust issues. Below, we describe in detail the key components and functionalities of blockchain-based carbon trading systems:

Distributed Ledger:

- Blockchain is a distributed ledger that records transactions across a network of nodes (computers) in a secure and immutable manner. All participants in the network have access to the same ledger, creating transparency.

Key Components:

- Smart Contracts: Smart contracts are self-executing agreements with predefined rules and conditions. In a blockchain-based carbon trading system, smart contracts enable autonomic execution of transactions, including the issuance, transfer, and retirement of carbon credits.
- Tokens: Tokens represent carbon credits. Each token corresponds to a specific amount of emissions reduction or allowance.
- Consensus Mechanism: Blockchain networks use a consensus mechanism (e.g., Proof of Work or Proof of Stake) to validate and add new transactions to the ledger.
- Decentralization: Blockchain works in a decentralized manner with no central authority overseeing the transactions thereby reducing the risk of fraud or manipulation.

Carbon Credit Lifecycle:

- Creation: Carbon credits are created through certified carbon offset projects, such as reforestation, renewable energy, or methane capture.
- Verification: Independent third parties verify the emissions reductions achieved by the project and issue carbon credits.
- Tokenization: Carbon credits are tokenized into digital tokens on the blockchain, each representing a specific quantity of emissions reductions.
- Trading: Participants in the blockchain network can buy, sell, or transfer carbon credits using the digital tokens.
- Retirement: When a carbon credit is used to offset emissions, it is retired, ensuring it cannot be double-counted.

Market Types:

- Voluntary Markets: These markets cater to individuals and organizations voluntarily seeking to offset their emissions to reduce their carbon footprint.
- Compliance Markets: Compliance markets are regulated and require entities to meet emissions reduction targets or purchase allowances to comply with government regulations.

Integration with IoT and Data Sources:

- Some blockchain-based systems integrate with the Internet of Things (IoT) and environmental data sources to automatically verify emissions reductions in real time. This enhances trust in the system.

4.3 System Architecture

The system architecture for a Blockchain-Based NFT Trading System for carbon credits is a complex and multifaceted structure that involves various components to ensure the integrity, security, and functionality of the platform. Figure 1 illustrates the system architecture.

1. **Blockchain Network**: An Ethereum-based consortium blockchain network is a private or semi-private blockchain network built on the Ethereum blockchain platform but with restricted access and control over participation. This type of network is typically used by a group of known and trusted entities, such as businesses, organizations, or government agencies, to collaborate on specific projects or applications that require blockchain technology. Unlike the public Ethereum blockchain, where anyone can participate, a consortium blockchain restricts access to a predefined group of participants. Participants are typically known and trusted entities within the consortium. Consortium blockchains can choose from various consensus mechanisms, such as Proof of Authority (PoA) or Proof of Stake (PoS), to validate and confirm transactions. These mechanisms are more efficient and cost-effective than Proof of Work (PoW), which is used in the public Ethereum network.
2. **User Interface (UI)**: Website or Mobile App
3. **Identity and Access Management (IAM)**: Secure user authentication and authorization mechanisms to control access to the platform. Validate the identity of participants, especially project developers and verifiers.
4. **NFT Tokenization**:
    - NFT Minting Service: A service for minting NFTs representing carbon credits based on project data and verification reports.
    - Metadata Management: Store and manage metadata associated with each NFT, including project details and emissions reduction data.
5. **Carbon Credit Verification**: Connect the platform with independent verification agencies to validate emissions reductions.
6. **Marketplace and Trading Engine**:
    - NFT Marketplace: A marketplace where users can buy, sell, and trade NFTs representing carbon credits.
    - Trading Engine: A robust trading engine to facilitate NFT transactions, including bid-ask matching and settlement.
7. **Payment Integration**: Enable users to connect digital wallets (e.g., Metamask) for making payments and receiving NFTs.
8. **Data Integration**: Ability to integrate environmental data sources and IoT devices for real-time emissions tracking and verification. Incorporation of external data sources, such as market data and climate information, to support carbon credit pricing and decision-making.
9. **Security and Privacy**: Use of encryption techniques to protect sensitive data, including user information and verification reports. Implementation of security audits, monitoring, and incident response mechanisms.
10. **Reporting and Analytics**:

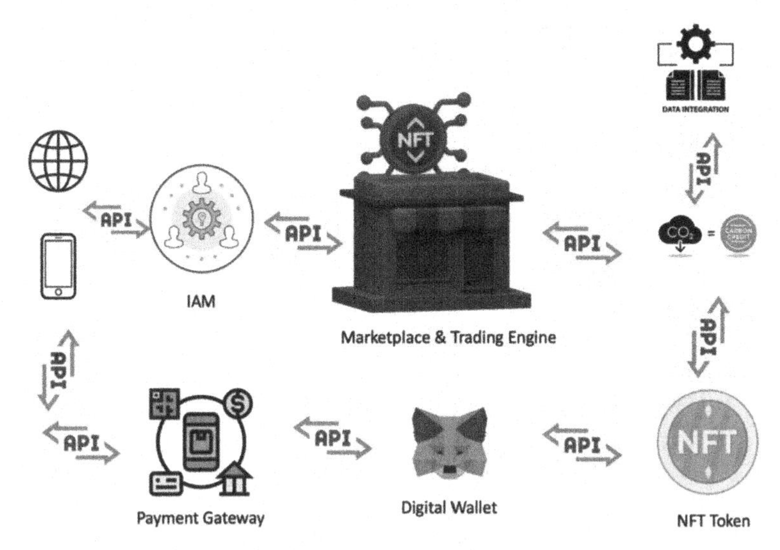

Fig. 1. NFT based Carbon Credit Trading Platform System Architecture

- Reporting Tools: Provide users with access to analytics and reporting tools to track their carbon credit holdings and trading history.
- Carbon Footprint Calculator: Offer tools to calculate and visualize the carbon footprint of users and their offsetting activities.

11. **Scalability and Performance Optimization:**

   - Load Balancing: Use load balancing techniques to distribute traffic evenly across servers and ensure system scalability.
   - Performance Optimization: Optimize the system's performance to handle a large volume of transactions and users.

12. **APIs and Integrations:** Third-Party Integrations: Support integration with third-party services, including payment gateways and verification agencies.

4.4 Smart Contract

```
// Import required libraries
import "@openzeppelin/contracts/token/ERC721/ERC721.sol";
import "@openzeppelin/contracts/access/Ownable.sol";

// Define the CarbonCredit smart contract that inherits ERC721 and Ownable
contract CarbonCredit is ERC721, Ownable {
  // Struct to represent a carbon credit
  struct CarbonCredit {
    uint256 projectId;
    uint256 vintageYear;
    uint256 emissionsReductionAmount;
    string projectMetadataURI;
    bool isRetired;
  }

  // Array to store all carbon credits
  CarbonCredit[] public carbonCredits;

  // Mapping from token ID to carbon credit data
  mapping(uint256 => CarbonCredit) public carbonCreditData;

  // Event to log the issuance of a new carbon credit
  event CarbonCreditIssued(uint256 tokenId, uint256 projectId, uint256 vintag-
eYear, uint256 emissionsReductionAmount);

  // Constructor to initialize the contract
  constructor() ERC721("CarbonCreditNFT", "CCNFT") {}

  // Function to issue a new carbon credit as an NFT
  function issueCarbonCredit(
    uint256 projectId,
    uint256 vintageYear,
    uint256 emissionsReductionAmount,
    string memory projectMetadataURI
  ) external onlyOwner {
    // Create a new carbon credit
    CarbonCredit memory newCarbonCredit = CarbonCredit({
      projectId: projectId,
      vintageYear: vintageYear,
      emissionsReductionAmount: emissionsReductionAmount,
      projectMetadataURI: projectMetadataURI,
      isRetired: false
    });

    // Mint a new NFT representing the carbon credit
    uint256 tokenId = carbonCredits.length;
    _mint(msg.sender, tokenId);
    carbonCredits.push(newCarbonCredit);
    carbonCreditData[tokenId] = newCarbonCredit;

    // Emit an event to log the issuance
    emit CarbonCreditIssued(tokenId, projectId, vintageYear, emissionsReduc-
tionAmount);
  }

  // Function to retire a carbon credit NFT
  function retireCarbonCredit(uint256 tokenId) external onlyOwner {
    require(!_exists(tokenId), "Carbon credit does not exist");
    carbonCreditData[tokenId].isRetired = true;
  }
}
```

5 Case Study

Title: Empowering Farmers with Blockchain-NFT Trading System for Agricultural Carbon Credits.

Agriculture is both a major contributor to greenhouse gas emissions and an industry susceptible to the impacts of climate change. In an effort to mitigate these issues and promote sustainable farming practices, a pioneering Blockchain-NFT Trading System for Agricultural Carbon Credits were introduced in a rural farming community. This case study explores how the proposed system benefits local farmers and the environment.

Location: A rural farming community in Uttar Pradesh, India.

Solution: A blockchain-NFT trading system was designed to incentivize farmers to adopt sustainable practices and trade their resulting carbon credits as NFTs on a blockchain platform.

Implementation:

- Carbon Credit Generation: Farmers adopted sustainable practices such as reduced tillage, cover cropping, and organic farming, which led to measurable reductions in carbon emissions and enhanced soil carbon sequestration.
- Verification: Independent environmental agencies were engaged to verify and validate the emissions reductions and soil carbon sequestration achieved by participating farmers.
- Tokenization: Verified carbon credits were tokenized as NFTs where each NFT represents a specific quantity of carbon emissions reduced or sequestered by a farmer's activities.
- Trading Platform: A user-friendly blockchain-NFT trading platform was deployed that allowed farmers to list their carbon credits as NFTs for sale or trade with other participants.

Benefits for Farmers:

- Additional Revenue Stream: Farmers received payments for their carbon credits thereby providing them with an additional source of income beyond traditional farming activities.
- Financial Incentive for Sustainable Practices: The trading system encouraged farmers to implement sustainable agricultural practices that reduced their carbon emissions, benefiting the environment.
- Access to Global Markets: Farmers gained access to a global marketplace, allowing them to connect with buyers, including corporations and individuals seeking to offset their carbon footprints from all over the world.
- Transparent and Secure Transactions: Blockchain technology ensured transparency and security in all transactions, reducing the risk of fraud.
- Environmental Stewardship: Farmers became active contributors to environmental stewardship, aligning their practices with broader sustainability goals.

Environmental Impact:

- Carbon Emissions Reduction: The implementation of sustainable farming practices led to a significant reduction in carbon emissions from agricultural activities.

- Soil Carbon Sequestration: Enhanced soil carbon sequestration contributed to improved soil health, increased crop yields, and better resilience to climate change impacts

6 Conclusion

In this paper, we embarked on a journey into the world of carbon credit systems, innovation, and sustainability. The objectives of this study revolved around dissecting the challenges inherent in traditional carbon credit systems, exploring the myriad benefits offered by Non-Fungible Tokens (NFTs) as a novel approach to carbon credits, unravelling the intricate structure of NFTs designed for trading carbon credits, and delving into the system architecture underpinning a blockchain-enabled NFT trading platform for carbon credits. Our examination revealed that the challenges associated with traditional carbon credit systems, such as transparency, verification, and inclusivity, could be significantly addressed through the transformative potential of NFTs. NFTs, with their unique and immutable digital representations, offer a game-changing approach to carbon credit creation and trading. Their divisibility, ownership traceability, and programmability not only streamline the entire process but also open up new avenues for engagement and participation. The description of the NFT structure specifically tailored for carbon credits provided insights into the essential metadata, verifiable attributes, and provenance tracking necessary for ensuring the integrity of carbon credit transactions. By merging the capabilities of NFTs with blockchain technology, a robust and transparent system architecture was conceptualized, which not only addresses current carbon credit challenges but also lays the foundation for a sustainable and scalable trading ecosystem. Furthermore, the integration of smart contracts into the trading process augments efficiency and trust by automating transactions and ensuring adherence to predefined rules and standards. The culmination of our research is exemplified in the illuminating case study centered on the use of a Blockchain-NFT trading system for Agricultural Carbon Credits. This case study underscores the tangible benefits accrued by farmers, who not only gain additional revenue streams but also become champions of environmental sustainability. The application of technology has not only mitigated carbon emissions but also fostered resilience in the agricultural sector, emphasizing the real-world impact of our exploration.

References

1. Khanna, A., Jain, S., Burgio, A., Bolshev, V., Panchenko, V.: Blockchain-enabled supply chain platform for Indian dairy industry: safety and traceability. Foods **11**(17), 2716 (2022)
2. Khanna, A., Sah, A., Bolshev, V., Burgio, A., Panchenko, V., Jasiński, M.: Blockchain–cloud integration: a survey. Sensors **22**(14), 5238 (2022)
3. Choudhury, T., Khanna, A., Toe, T.T., Khurana, M., Nhu, N.G. (Eds.): Blockchain Applications in IoT Ecosystem. Springer, Cham, Switzerland (2021). https://doi.org/10.1007/978-3-030-65691-1
4. Lehmann, J., et al.: Biochar in climate change mitigation. Nat. Geosci. **14**(12), 883–892 (2021)

5. Kim, S.K., Huh, J.H.: Blockchain of carbon trading for UN sustainable development goals. Sustainability **12**(10), 4021 (2020)
6. Jiang, T., Song, J., Yu, Y.: The influencing factors of carbon trading companies applying blockchain technology: evidence from eight carbon trading pilots in China. Environ. Sci. Pollut. Res. 1–13 (2022)
7. Elmay, F.K., Salah, K., Jayaraman, R., Omar, I.A.: Using NFTs and blockchain for traceability and auctioning of shipping containers and cargo in maritime industry. IEEE Access **10**, 124507–124522 (2022)
8. Van Wassenaer, L., van Hilten, M., Van Ingen, E., Van Asseldonk, M.: Applying blockchain for climate action in agriculture: state of play and outlook. Food & Agriculture Org. (2021)
9. Paustian, K., et al.: Quantifying carbon for agricultural soil management: from the current status toward a global soil information system. Carbon Manag. **10**(6), 567–587 (2019)
10. Kumar, S., Jain, S., Lamba, B.Y., Kumar, P.: Epigrammatic status and perspective of sequestration of carbon dioxide: role of TiO2 as photocatalyst. Sol. Energy **159**, 423–433 (2018)

A Blockchain Framework for Digital Asset Ownership and Transfer in Succession

Irenee Dondjio[1](✉) and Andreas Kazamias[2]

[1] University of Nicosia, Nicosia, Cyprus
dondjio.i@unic.ac.cy
[2] Platanus Services Ltd., Nicosia, Cyprus

Abstract. The accumulation of wealth and assets through inheritance forms the basis for future generations' prosperity, encompassing financial assets, physical possessions, and intangible riches like knowledge and skills. However, inheriting such multidimensional wealth presents complex technological, social, and legal challenges. To address these issues, the authors present a comprehensive exploration of technology-driven inheritance methods and the role of blockchain in asset management. Current research in this area is fragmented, lacking a unified conceptual framework, making it challenging to grasp the broader implications of blockchain in inheritance. This study aims to fill this gap by taking an exploratory approach to investigate the integration of blockchain and digital asset ownership and transfer in the context of Succession and Inheritance. The authors provide a conceptual framework to guide all stakeholders involved in the inheritance process. The study seeks to bridge existing knowledge gaps and offer a cohesive perspective on how blockchain technology can revolutionize inheritance practices through a combination of case studies, theoretical research, and practical implementation insights. The article is structured into four sections: an introductory section providing background context, an extensive review of relevant literature, including key principles and practical examples, an innovative conceptual framework, and a thorough analysis and synthesis of the research findings.

Keywords: Cryptocurrency · Digital assets · Blockchain · Inheritance · Smart contact · Wealth Management · NFTs · Trusts · e- wills

1 Introduction and Background

The industry crypto asset and decentralized finance (DeFi) continue to expand rapidly, marked by a combination of innovative concepts, risks, and regulatory challenges. The concept of inheritance has played a crucial role in the development of civilizations, as it allows for the accumulation of wealth, knowledge, and property over time, providing each subsequent generation with a foundation to build upon the successes of their predecessors [15]. In light of the prevalent trend of digital transformation across various spheres of everyday life, the use of blockchain technology has emerged as an effective means of secure and efficient transfer of assets between individuals [18]. One of the

M. Papadaki et al. (Eds.): EMCIS 2023, LNBIP 501, pp. 88–106, 2024.
https://doi.org/10.1007/978-3-031-56478-9_7

key challenges with digital assets is the issue of security. Digital assets, unlike physical ones, face susceptibility to theft or loss through hacking, fraud, or cybersecurity threats [7]. Hence, it's vital to establish strong security protocols, encompassing secure storage and access controls. Furthermore, digital assets pose an accessibility challenge, as their digital nature can hinder retrieval in the absence of clear owner instructions. This accessibility hurdle can lead to succession and inheritance complications, causing delays and disputes among heirs.

In light of these challenges, the incorporation of cryptocurrencies and digital asset ownership and transfer within the context of succession and inheritance necessitates a comprehensive conceptual framework. Such a framework must meticulously address the intricate interplay between the technical and legal dimensions involved in effectively managing these assets. This framework should take into account the security and accessibility of digital assets, the legal status of cryptocurrencies and other digital assets, and the potential tax implications of transferring these assets.

2 Research Methodology

As stated already, this paper aims to investigate the integration of blockchain technology in succession and inheritance, focusing on digital asset ownership and transfer. The authors set out to address the following key research questions:

- What are the challenges associated with digital asset ownership and transfer in the context of succession and inheritance?
- How can blockchain technology address security and accessibility issues in the transfer of digital assets?
- hat is the legal status of cryptocurrencies and digital assets in inheritance, and what are the potential tax implications?

Adopting an exploratory approach due to the fragmented nature of current research, the methodology involves an extensive literature review and the analysis of real-world blockchain implementations in inheritance. The ultimate goal is to develop a comprehensive conceptual framework that intricately considers technical and legal aspects, including security, accessibility, legal status, and tax implications of digital assets in the context of succession and inheritance.

3 Literature Review

3.1 Blockchain Technology

Blockchains are a specific kind of database, initially developed for data storage, that record and store the transactional history of the data. The data is stored in blocks that are interconnected and cannot be altered once they have been added to the chain. Sometimes referred to as a type of digital ledger technology (DLT), blockchain technology has been known to store vast amounts of transactional information in a digital format [9]. The rise of cryptocurrency markets has contributed to the widespread use of blockchain technology, showcasing its potential to revolutionize the financial sector. The decentralization, consensus-based operation, digital nature, immutability, chronology, and timestamping of blockchain technology have made it highly appealing, with significant potential to enhance various aspects of the corporate environment, such as reducing risk, increasing visibility of supply chain operations, eliminating fraudulent transactions, and promoting transparency [8].

3.2 Key Blockchain Characteristics

The section delves into the fundamental attributes of blockchain technology, highlighting its key characteristics that underpin its significance and functionality.

- Traceability allows users to trace transactions through the system, and to access important transaction information by reviewing a particular data block. This traceability is possible due to the close interlinking of each system block, enabling close monitoring of data.
- Transparency is another key feature of blockchain technology. It allows system members to view and regulate transactions, and to publish their own transactions. Additionally, transparency detects and prevents dubious transactions. By allowing stakeholders to choose which data is sent over the network, the system improves security and openness, safeguarding against unauthorized modification of network data.
- Immutability is a vital characteristic of blockchain technology, which guarantees that transactions cannot be modified or removed once validated. This ensures system reliability and builds stakeholder confidence. Furthermore, the decentralized nature of blockchain networks means that they are not reliant on centralized points of control, which helps to prevent system failure and build trust among stakeholders [8].

The delivery of blockchain functionality through maintaining decentralization, security, and scalability in a distributed ledger system poses a significant challenge. The blockchain trilemma posits that improving one of these elements comes at the cost of sacrificing one of the others. Decentralization refers to the fact that a blockchain network does not rely on centralized points of control, while security is the ability of a blockchain network to withstand and repel attacks like Distributed Denial of Service (DDoS). Finally, scalability refers to its ability to manage vast quantities of transactions [17] (Fig. 1).

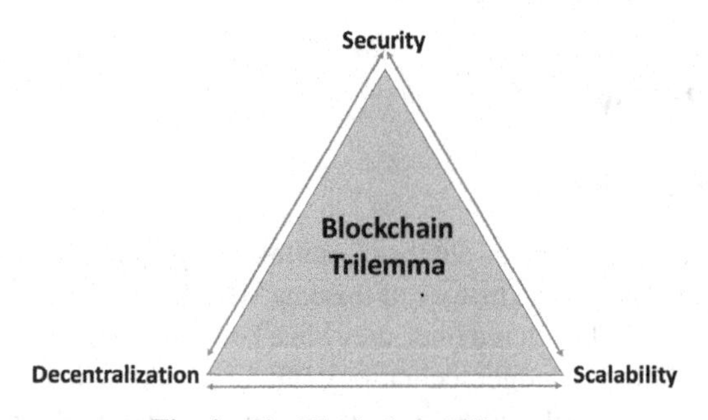

Fig. 1. The Blockchain Trilemma

"Blockchain technology facilitates secure and transparent transfer of assets and data. It operates on an open- source framework, making it accessible for all to participate. Through the use of a shared distributed ledger, all participants maintain an identical record, ensuring transparency and accountability. The immutability of transactions and blocks is guaranteed through encryption, preventing unauthorized alterations. By leveraging a peer-to-peer network, blockchain eliminates the need for intermediaries, fostering direct transactions between participants. Furthermore, blockchain empowers individuals to retain ownership and control over their own data, enabling data freedom."

3.3 Smart Contract

As shown in Fig. 2, smart contracts are automated and self-executing processes that facilitate contracts that embody transfer of ownership agreement terms in code, like those that apply when transferring an asset from a buyer to a seller. Blockchain hosts the code and contract agreements. Smart contracts allow anonymous individuals to trust one other without governance, legal system, central authority, or external enforcement. Ethereum is a popular platform for smart contract generation [1]. In fact, through its Turing programming language, Ethereum, as a flexible blockchain platform, allows the execution of complicated transactions as well as the construction of decentralized applications (DApps). It offers a solid framework for developers to create and implement their own smart contracts, allowing for more flexibility and creativity across sectors [8].

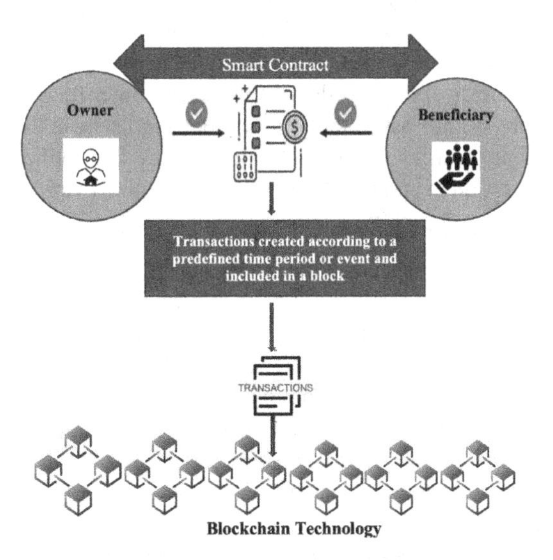

Fig. 2. The Smart contact within the blockchain

3.4 Digital Assets

The term "digital assets" describes a broad category of assets that includes anything created and exchanged on a blockchain. Typically, digital assets are grouped into five main categories [14].

- Crypto assets, also known as cryptocurrencies, are a type of digital asset that uses cryptographic techniques to secure financial transactions and control the creation of new units. They are decentralized and operate independently of a central bank or government. Cryptocurrencies such as bitcoin, ether, and litecoin are examples of crypto assets. They are often used as a medium of exchange, a store of value, or for speculative investment purposes. [Bullmann et al., 2019]
- Stablecoins are digital value units that employ various methods to stabilize the price of specific currencies and reduce fluctuations. They aim to provide quick stabilization of major currencies in the volatile crypto-asset market where there is no liable issuer, as well as in the broader economy [Bullmann et al., 2019]. Stablecoin prices are linked to fiat currencies, commodities, or other crypto assets. it can be used for cross-border payments, include foreign exchange, and other funds transfers.
- Non-fungible tokens (NFTs) **represent** ownership of a unique digital asset, such as an artwork, government ID, or manufacturing unit. NFTs establish that the owner may sell, exchange, or redeem the digital asset. NFTs can be used for proving identity and allowing access (virtual or physical), tokenizing the supply chain to monitor inventory and ownership of games, arts, avatars, virtual land ownership.
- Central Bank Digital Currencies (CBDCs) or Central Bank Digital Currencies, are digital assets that represent a country's fiat currency and are supported by the country's central bank. It's important to note that not all countries have issued CBDCs. CBDCs can be used for, domestic, cross-border payments and other funds transfers.
- Security tokens are digital assets that satisfy the requirements of a security or a financial investment, such as bonds and stocks. They are applicable to tokenized versions of stocks (equity) and bonds, as well as tokenized variations of real-world assets like as real estate, property, plant, and equipment [14].

The process of acquiring and transferring digital or crypto assets represents a significant change from traditional notions of owning and trading virtual items in the digital realm. This shift is underscored in Fig. 3, which highlights the transformation in how these assets are managed. It marks a move away from conventional ownership models, ushering in an era of decentralized, transnational ownership, and transfer, underpinned by advanced blockchain technology. These assets, which encompass cryptocurrencies like bitcoin and ether, as well as various tokens, rely on the foundational structure of blockchain technology. Ownership is established through cryptographic keys securely stored in digital wallets, granting individuals control over their assets without the need for intermediaries. Transferring these assets involves using private keys to digitally sign transactions, which are then recorded on the immutable blockchain ledger. This system ensures both security and transparency, enabling instant and borderless transactions. This innovative approach has the potential to disrupt traditional financial institutions by enabling direct peer-to-peer transactions globally, reducing costs, and minimizing delays [2]. However, challenges arise, including the need for robust security protocols

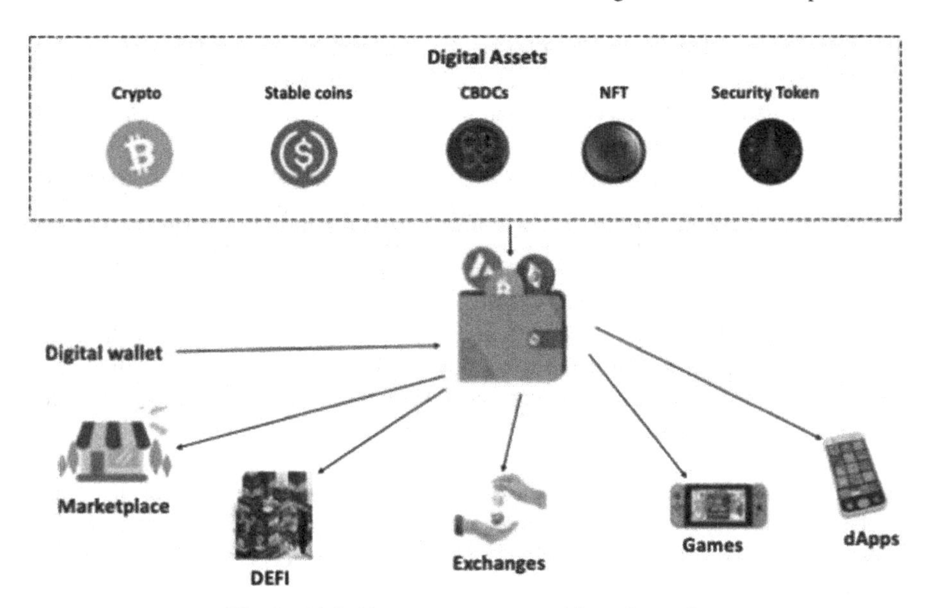

Fig. 3. Digital/crypto assets ownership and transfer

to prevent unauthorized access and the establishment of adaptive regulatory frameworks to navigate the dynamic digital asset landscape. The ongoing evolution of this field has the power to fundamentally reshape financial institutions, international remittances, and individuals' relationships with their financial holdings through the ownership and transfer of digital assets [3].

3.5 The Benefits and Challenges of Blockchain in the Inheritance Process

Benefits. The use of blockchain technology may facilitate the safe and transparent documentation of assets and transfers of ownership in the context of succession and inheritance. In their scholarly article, Sladić et al. (2021) outlined many benefits of using blockchain technology in the context of the inheritance process. They are presented below.

- Smart contracts. Smart contracts are contracts that automatically execute, with the parameters of the agreement between buyer and seller (owner or representative and recipient) inscribed directly into lines
- of code. In the context of succession and inheritance, smart contracts can be utilized to autonomously transfer ownership of assets, such as real estate or digital assets, to the decedent's successors. This eliminates the need for middlemen such as attorneys or institutions to manage the transfer of assets. In addition, it ensures transparency, confidence, and traceability throughout the process.

- Digital identity. Blockchain-based digital identity systems can provide a secure and tamper-proof method for authenticating the identities of successors and family members participating in the succession and inheritance process. This can help prevent fraud and property ownership disputes.
- Decentralized storage. Decentralized storage networks, such as IPFS or Filecoin, can be used to store and exchange essential inheritance-related documents, such as wills and trust agreements, in a secure manner. The decentralized nature of these networks ensures that the documents are not under the control of a single entity and are not susceptible to modification or censorship.
- Tokenization. Tokenization is the process of representing physical assets, such as real estate or artwork, as digital tokens on a blockchain. Tokenization can facilitate the transfer of assets to successors in the context of succession and inheritance, as tokens can be traded and transmitted without the need for intermediaries.

Challenges. As discussed in the preceding section, the adoption of blockchain and digital is crucial in succession and inheritance processes due to the unique characteristics of these assets. However, the decentralized nature of digital assets, coupled with complex laws for ownership transfer and tax plus regulations, can make it difficult for successors to access or transfer ownership. Individuals must therefore include their digital assets in their estate plans and take measures to ensure that their successors have access to and control over them. As suggested by Kharitonova [13], this may involve creating an inventory of all crypto assets, developing a clear plan for transferring ownership and control, and seeking professional guidance from experienced attorneys and financial advisors. The following section provides two case studies addressing these obstacles.

- Use Case 1: QuadrigaCX. The case of Gerald Cotten, the CEO of QuadrigaCX, a Canadian cryptocurrency exchange, serves as a stark example of the challenges surrounding digital asset management in the context of succession and inheritance. Cotten's sudden demise in December 2018 left behind approximately $190 million CAD in cryptocurrencies and digital assets. However, he failed to provide explicit instructions on accessing these assets, holding the sole knowledge of the encrypted wallet credentials. Consequently, QuadrigaCX could not access the funds, ultimately leading to the company's bankruptcy. Cotten's widow, Jennifer Robertson, claimed to have discovered the credentials written on paper in his office, though experts have raised doubts about this assertion. This case underscores the importance of secure storage and access controls to prevent unauthorized access to digital assets. It also highlights the need for clear instructions on accessing these assets in the event of the owner's demise. Additionally, it sheds light on the legal complexities surrounding digital assets in succession and inheritance planning. In Canada, there is no specific legal framework governing the transfer of digital assets upon an owner's death, leading to uncertainty and disputes, as seen in the QuadrigaCX case. In conclusion, the Gerald Cotten and QuadrigaCX case underscores the critical need for clear procedures and legal clarity when it comes to handling digital assets in succession and inheritance planning.
- Use Case 2: Mt. Gox. Mt. Gox was once the world's largest Bitcoin exchange, faced a catastrophic downfall in 2014 when it declared bankruptcy, having lost 850,000

bitcoins valued at around $450 million. This staggering loss resulted from a combination of inadequate security measures and internal wrongdoing. Mt. Gox stored its Bitcoins in a vulnerable "hot wallet" connected to the internet, making it susceptible to cyberattacks. Moreover, crucial security protocols like two-factor authentication were neglected, exposing accounts to potential hijacking. The exchange's CEO, Mark Karpeles, faced allegations of embezzlement and internal fraud. While initially disputing the charges, he was eventually convicted of data falsification and received a two-and-a-half-year prison sentence. The Mt. Gox case serves as a stark warning about the risks associated with digital asset ownership and exchange, emphasizing the critical need for robust security and control measures to protect these assets. It also underscores the importance of regulatory oversight in the digital asset industry to prevent fraudulent activities and ensure investor protection.

Lessons Learnt. The lessons learned from the Mt. Gox and QuadrigaCX case studies are crucial for individuals and organizations involved in the digital asset industry. Among the most important lessons to be gained from these cases are:

- Appropriate security measures are necessary: Due to insufficient security measures, both Mt. Gox and QuadrigaCX experienced significant losses. In the case of Mt. Gox, the Bitcoins were stored in a live wallet, which was susceptible to hijacking. The assets of QuadrigaCX were entrusted to a single
- individual, who failed to adequately secure them, resulting in their loss. Cold storage, two-factor authentication, and encryption, all operating within proper control mechanisms, are necessary to prevent such losses.
- Regulation oversight is required: Mt. Gox and QuadrigaCX demonstrate the need for regulatory supervision in the digital asset industry. In both instances, investors lost substantial sums of money because of fraudulent activity or inadequate security measures. Regulatory supervision can aid in preventing such losses by requiring companies to adhere to fundamental security protocols, implement appropriate risk management practices, and submit to regular audits.

In both instances, investors were kept in the dark about the true financial and operational state of the companies. Lack of transparency can result in suspicion and mistrust, which can ultimately undermine the industry's credibility as a whole. It is essential to maintain a high level of transparency with investors and stakeholders in order to develop trust and confidence.

Diversification is essential. The collapse of Mt. Gox and QuadrigaCX resulted in significant losses for many investors. To mitigate these risks, it is essential to diversify one's digital asset holdings across a variety of platforms and exchanges. This can help prevent losses caused by fraud or a single point of failure. Overall, the Mt. Gox and QuadrigaCX case studies highlight the significance of adequate security measures, regulatory supervision, transparency, and diversification in the digital asset industry. Individuals and organizations can contribute to the long-term viability and prosperity of the digital asset industry by learning from these cases and taking action to resolve the issues they highlight.

4 Conceptual Framework for Digital Assets Transfers via e-will

As society matures, inheritance conflicts are rising [10]. Electronic wills may be more secure and convenient than traditional wills. Creating an e-will involves several legal and technological issues. Legal guidance is required to ensure that the e-will is valid and enforceable. An experienced lawyer can help you understand and draft an e-will. They may also provide guidance on local legislation and guarantee that the e-will is legal. Lawyers may also help with e-will technicalities. They may advise on how to store and protect digital assets, as well as how to use secure digital signature technology to validate the e-will. Finally, local laws on e- will vary as well as vary between federal and state levels. As a result, a local lawyer is required. They can ensure that the e-will is valid and enforceable in your jurisdiction and meets all legal requirements. Chen et al. [5] presented a blockchain-based smart contract will mechanism. This system streamlines will asset management and distribution while assuring will security and non-tampering. Further, will assets be automatically dispersed to the beneficiaries without necessitating the intervention of middlemen. This reduces will delivery time and expense. The technology arbitrates will execution disputes and automates property distribution. This assures fair and fast dispute resolution without protracted legal battles. As presented in Fig. 4 below, the suggested system also offers data integrity, public verifiability, unforgeability, nonrepudiation, irreversibility, and counterfeiting resistance. These measures safeguard the will and its contents against tampering Chen et al. [5].

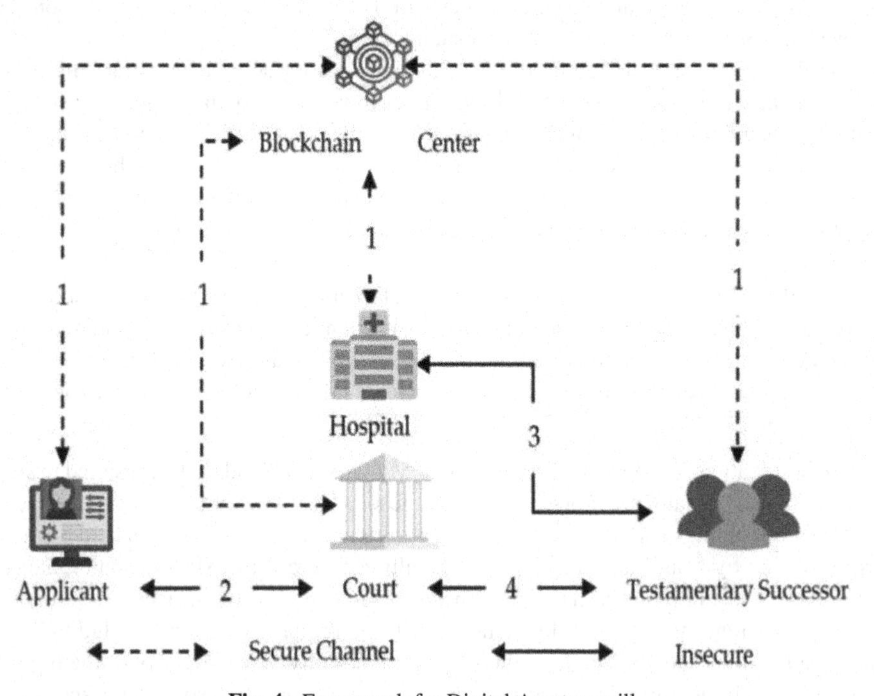

Fig. 4. Framework for Digital Assets e-will

The study presents a system architecture for an e-will system, involving five key actors: the will applicant, the overseeing court, the hospital providing death certificates, the testamentary successor representing beneficiaries, and the blockchain center responsible for executing smart contracts for transparent will execution. However, several limitations are noted due to the legal complexities of e-wills, varying state e-will laws, technical literacy requirements, potential conflict oversights in automated distribution, and challenges in maintaining privacy and security of personal data and digital assets on the blockchain while adhering to data protection laws. Legal and technical expertise is crucial for e-will creation, and its applicability may differ across jurisdictions. Technical proficiency is vital for participants, and conflicts beyond automated processes might require legal interpretation. Privacy and security concerns arise from using blockchain for sensitive information and assets, underscoring the need for careful consideration of legal, technical, and privacy factors. In addition to the study, Shah et al. [16] developed a blockchain-based Will execution mechanism aimed at providing a secure platform for digital asset transfers. The research project utilizes ERC-20 and ERC-721 tokens to map digital assets using blockchain technology, enabling Will owners to create and execute smart contracts with specific instructions and conditions. A well-defined consensus mechanism facilitates the contractual agreement and systematic asset transfer to selected beneficiaries. This solution leverages advanced technology and smart estate planning to expedite asset distribution while maintaining security and authenticity. However, it also faces limitations, such as potential weaknesses in the Proof of Vote for Death Verification technique, smart contract errors, privacy concerns, and challenges in accurately mapping various digital assets to tokens.

5 Proposed Framework for Digital Assets Ownership and Transfer

The authors of this article aim to address these limitations by providing a comprehensive framework. Their methodology involves two phases: first, establishing digital asset ownership and transfer through a trust, and second, focusing on transferring assets to beneficiaries after the settlor's death or other pre-defined event. This process emphasizes the framework's ability to streamline succession procedures while considering both legal and technical aspects. Additionally, the introduction of a strategic roadmap for electronic wills development underscores the authors' commitment to a holistic approach. This roadmap integrates legal and technological expertise, offering comprehensive guidance for effectively managing digital assets within succession frameworks. Overall, the study and its associated research contribute valuable insights and approaches to the evolving field of e-wills and digital asset inheritance.

5.1 Digital Assets Ownership and Transfer via a Trust

As illustrated in the Fig. 5 below, the use of a smart contract in such agreements adds a surprising amount of efficiency. By including the trust agreement's provisions and conditions directly into the smart contract, the total process becomes substantially quicker and more secure, with full traceability of activities. Furthermore, smart contracts remove the need for financial or operational middlemen, resulting in cost savings and simplified processes. Smart contracts' automation and self-executing nature guarantee that the agreed-upon conditions are automatically enforced whenever the pre-set criteria are satisfied. This removes the need for manual contract verification, supplementary documentation, and the risk for delays associated with conventional contract execution. Overall, the agreement becomes very efficient since activities are triggered and carried out in a smooth and timely way. Furthermore, the trust agreement is strengthened by the transparency and immutability inherent in blockchain technology. Every transaction and change made to the smart contract is recorded on the blockchain, resulting in an immutable and auditable trail of events. This traceability provides responsibility while also lowering the danger of fraud or manipulation.

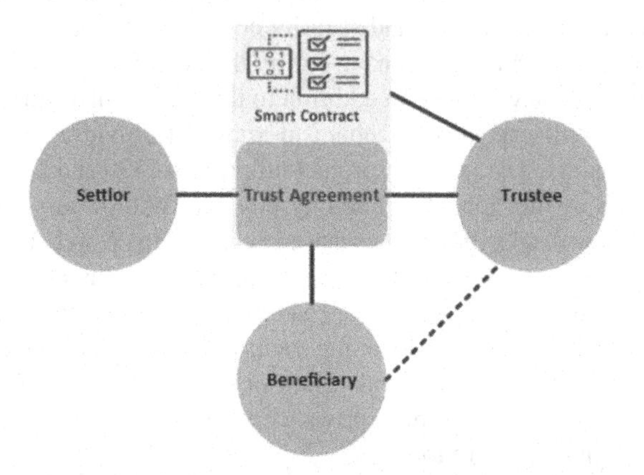

Fig. 5. Assets Ownership and Transfer via a Trust

Asset Transfer to Beneficiary upon the Settlor's Death. The following is a description of the seven crucial processes that comprise the complex process of transferring assets to a beneficiary after the death of the Settlor or other predefined event. See Fig. 6.

- Phase-1 Establishment of the Trust: The Settlor will consult with one or more legal specialists, as required, to have a trust agreement drafted. The objectives, conditions, and arrangements for the transfer of assets to the beneficiary following the death of the Settlor are outlined in the agreement. It contains information such as the assets that have been identified, the details of the recipient, and any particular requirements or limitations that are associated with the asset transfer.
- Phase-2 Identifying Assets: The trust agreement is used to establish which assets the settlor (the person who created the trust) intends to provide to the beneficiary (the person who will benefit from the trust). These assets may include financial holdings, real estate, intellectual property rights, or any other valuable items that may be lawfully transferred. Additionally, these assets may also include other types of identifiable things.
- Phase-3 Integration of Smart Contract: The terms and conditions of the trust agreement are converted into a smart contract in partnership with blockchain developers or experts on smart contracts. The terms, conditions and applicable regulations that regulate the transfer of assets are encoded into the smart contract code. This ensures that

the code can independently carry out the transfer in accordance with the circumstances that have been pre-set and consistently with the applicable laws.

- Phase-4 Verification and Deployment: The produced smart contract is put through extensive verification and testing before it is deployed. This is done to guarantee that it is accurate, secure, and complies with the necessary legal standards. In order to evaluate the functioning of the contract's code and ensure thatit complies with the applicable regulatory rules, it is examined by technical professionals, auditors, and legal specialists. After going through the verification process, the smart contract is then uploaded into a blockchain network that is appropriate for it, such as Ethereum or a private blockchain, in order to guarantee immutability and transparency.
- Phase-5 Transfer of Asset Ownership: The trust agreement outlines who is responsible for initiating the transfer of ownership of selected assets to the trust. This responsibility falls on the Settlor to ensure that requirements are clearly understood by the Trustee who is responsible for the execution and management of the trust. This phase often entails the creation of legal papers, such as deeds, certificates, or any other pertinent paperwork, in order to lawfully transfer ownership of the assets to the trust. This assures that the assets are retained formally by the trust and may be transferred to the beneficiary when the Settlor has passed away.
- Phase-6 Monitoring and Condition Fulfilment: Once the Settlor has passed away, the smart contract will automatically begin monitoring the pre-set conditions mentioned in the trust agreement. If any of these requirements are not met, the contract will terminate. The filing and certification of the Settlor's death certificate may be one of these requirements. Confirmation from authorized people or organizations may also be required, along with the fulfilment of any further particular criteria outlined in the trust agreement. In order to establish whether or not the transfer procedure may go forward, the smart contract will continually assess whether or not these requirements have been met.
- Phase-7 Transfer of Assets Automatically: Once the smart contract confirms that the predetermined requirements have been satisfied, it will immediately begin the process of transferring the assets from the trust to the beneficiary who has been specified. The transfer is carried out via the smart contract in accordance with the guidelines and requirements that were outlined in the trust agreement. The transfer of the asset is recorded on the blockchain, which provides an unchangeable and transparent record of the process of transferring the asset.

5.2 Proposed Framework for Structuring an e-will

The ownership and transfer of digital assets involves a complex interplay of legal, technical, and regulatory factors. The framework proposed in this section defines concrete actions, meaningful considerations, and best practices. It serves as a compass for those seeking to establish a trust-based basis for digital asset management. As previously stated, blockchain integration is critical for the succession of digital assets that are tailored to their specific characteristics. Nonetheless, the decentralized structure of crypto assets, along with complex tax legislation, causes difficulties in successor access and

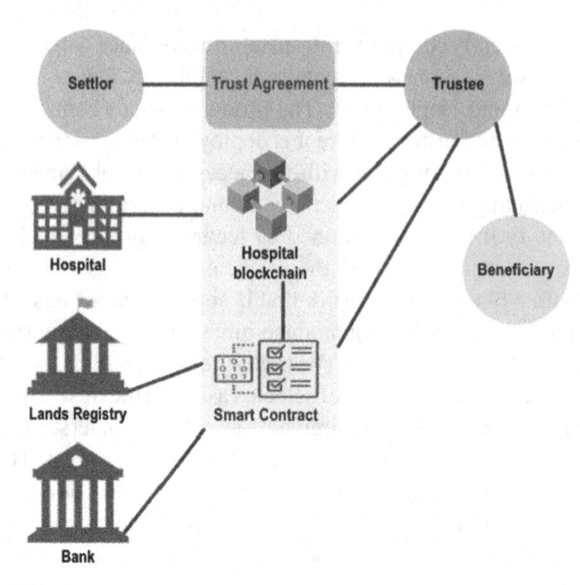

Fig. 6. Assets transfer to Beneficiary upon the Settlor's death

transfer. It is critical to include digital assets into estate planning and ensure successor control. Kharitonova [13] emphasizes the need of creating an inventory, developing clear transition strategy, and retaining competent legal and financial advice. The authors proposed a guideline in 6 major steps to assist individuals seeking to organize their e-will. See Fig. 7 below.

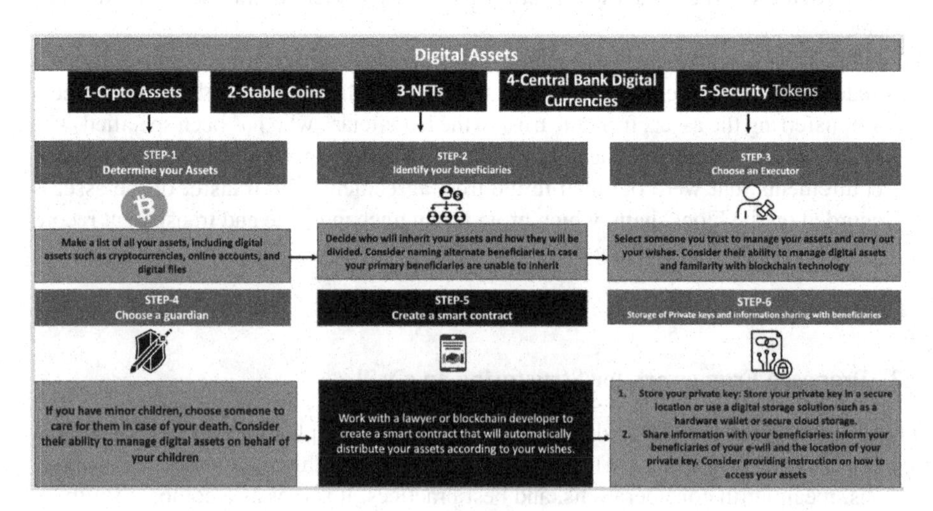

Fig. 7. A proposed framework to complete and secure an e-will

- Step-1 Determine your assets. Create a list of all assets, including digital assets, to get started. This includes digital assets, online investment accounts, digital files, and any other online accounts. For instance, a Coinbase account containing bitcoin and a Robinhood account containing stocks.
- Step-2 Identify your beneficiaries. Determine who should inherit the assets and how they should be divided. Consider designating alternative beneficiaries if primary beneficiaries cannot inherit. For instance, bequeath digital music collection to a closest friend or divide cryptocurrency holdings between two offspring.
- Step-3 Choose an executor. Choose a trusted individual to manage the assets and carry out the succession and inheritance (wealth management) instructions. Consider the individual's knowledge of blockchain technology and capacity to administer digital assets. For instance, choose a tech-savvy family member or friend, or a professional executor with experience in managing digital assets.
- Step-4 Choose a guardian. If minors may inherit, appoint a guardian on standby, if needed. Consider the guardian's capacity to administer digital assets on behalf of the minors until they become adults. For instance, choose a sibling or close acquaintance who has the required skills.
- Step-5 Create a smart contract. Collaborate with an attorney or blockchain developer to construct a smart contract that will autonomously distribute the assets in accordance with the stated requirements. A smart contract is a contract that is inscribed onto a blockchain network and can be used to transfer assets automatically based on predefined criteria. For instance, construct a smart contract that transfers bitcoin holdings to the beneficiaries at the occurrence of a predefined event.
- Step-6 Store your private key and share information with your beneficiaries. Your private key is a password necessary to gain access to specific digital assets. It is essential to use a secure location and/or a digital storage solution, such as a hardware wallet or secure cloud storage, to store the private key. The beneficiaries should be informed about the existence of the electronic will and where the private keys are stored. For instance, private keys may be stored in a safe deposit box at a bank with instructions to the bank to grant access to specific individuals upon one's death.

5.3 Legal and Technical Considerations for Digital Asset Transfer

Legal Consideration. In the realm of digital asset ownership and transfer, legal considerations play a critical role in ensuring compliance, establishing ownership rights, and protecting individuals and businesses from illicit activities. This section explores key legal considerations that stakeholders must address when dealing with digital assets.

KYC/KYB and AML Procedures. Know Your Customer (KYC), Know Your Business (KYB), and Anti- Money Laundering (AML) procedures are crucial components in the management and transfer of digital assets. The specific controls implemented will depend on various factors, including:

- Relevant Legislation. Different jurisdictions have different laws and regulations pertaining to digital assets and their transfer. Compliance with these regulations is essential, and the specific KYC/KYB and AML procedures will be influenced by the legal requirements of the jurisdiction in which the transactions occur.

- Type of Digital Asset. The nature of the digital asset being traded or transferred also plays a role in determining the applicable laws and procedures. Different digital assets, such as cryptocurrencies, security tokens, or utility tokens, may be subject to specific regulations and compliance obligations.
- Value of the Transactions. The value of each asset and the total can impact the level of scrutiny and controls required. Higher-value assets transfers typically attract more rigorous KYC/KYB and AML procedures to mitigate the risks associated with money laundering, terrorist financing, and other illicit activities.

In general, KYC procedures involve verifying the identity of customers or businesses involved in digital asset transactions, including transfers. This may require collecting identification documents, conducting identity verification checks, and maintaining up-to-date customer records. KYB procedures focus on understanding the businesses involved in digital asset transactions, including their structure, ownership, and legitimate business activities. AML procedures aim to prevent and detect money laundering activities by implementing robust monitoring and reporting mechanisms. This may involve transaction monitoring, screening against sanctions lists, conducting risk assessments, and reporting suspicious activities to relevant authorities. It's important for businesses and individuals involved in digital asset ownership and transfer to stay informed about the evolving legal and regulatory landscape in the relevant jurisdictions and to ensure compliance with the applicable KYC/KYB and AML procedures. Consulting with legal and compliance professionals is essential to establish and maintain effective control frameworks that align with the specific requirements of the legislation and the nature of the digital assets being traded. Furthermore, the following points should not be overlooked:

- Ownership and Property Rights. Determining ownership and property rights over digital assets can be complex, as traditional legal frameworks may not fully address these digital forms. Various legal theories and concepts, such as intellectual property rights, contract law, and property law, may be applied to establish ownership and enforce rights.
- Regulation and Compliance. Digital assets are subject to regulatory frameworks that vary from country to country. Some jurisdictions have specific regulations for cryptocurrencies and other digital assets, addressing issues like AML and KYC requirements. Compliance with these regulations is essential for businesses and individuals dealing with digital assets.
- Smart Contracts and Legal Validity. Smart contracts, which are self-executing contracts with terms written in code, present legal considerations. Parties must ensure that smart contracts comply with applicable laws, and legal systems need to recognize the validity and enforceability of such contracts.
- Data Protection and Privacy. Digital assets often involve the processing and storage of personal data. Compliance with data protection laws, such as the General Data Protection Regulation (GDPR), is crucial when handling personal information within digital assets.
- Jurisdictional Challenges. The borderless nature of digital assets can pose challenges when it comes to determining jurisdiction and resolving disputes. Legal frameworks

and cross-border cooperation need to be established to address these challenges effectively.

Technical Consideration. The successful adoption and implementation of blockchain technology for digital asset ownership and transfer require careful consideration of various technical aspects. This section explores key technical considerations that impact the efficiency, security, and interoperability of blockchain-based systems.

- Slow Adoption. Blockchain technology has been instrumental in enhancing trust, security, and transparency in digital asset ownership and transfer. However, widespread adoption of blockchain-based systems for digital/crypto assets has been slow. This can be attributed to various factors, such as regulatory uncertainties, lack of standardization, and limited awareness among the general population.
- Smart Contracts Security. In a blockchain-based systems, smart contracts are utilized to automate and enforce the terms of asset ownership and transfer. While smart contracts provide security against unauthorized access and tampering, they are also subject to vulnerabilities. Flaws in the contract code or exploitation of loopholes can lead to financial losses or theft of assets. Ensuring robust security measures and conducting thorough code audits are essential to mitigate these risks.
- Storage Issues. Digital/crypto assets generate a significant amount of data, including transaction records, ownership history, and asset metadata. Storing this data on the blockchain can present challenges due to the limited storage capacity of individual blockchain nodes. As the size of the blockchain grows with increasing adoption, scalability becomes a concern. Balancing the need for data retention with blockchain performance is crucial to maintain efficiency in asset ownership and transfer systems.
- Scalability. Stakeholders involved in digital/crypto asset ownership and transfer require high- performance systems that can handle a large volume of transactions. Blockchain scalability refers to the ability of the network to process transactions quickly and at a reasonable cost. As more users participate in the blockchain network, transaction throughput, execution time, and fees can become bottlenecks. Implementing solutions like layer-2 protocols or off-chain scaling mechanisms can help address scalability challenges and improve the overall user experience.
- Interoperability. The digital asset landscape is diverse, with various blockchain platforms and protocols supporting different types of assets. Interoperability between these platforms is crucial for seamless asset transfer and liquidity across different networks. However, achieving interoperability remains a significant challenge due to differences in protocols, consensus mechanisms, and technical standards. Developing cross-chain solutions and standardized protocols can facilitate interoperability and promote a more interconnected ecosystem for digital/crypto asset ownership and transfer.

6 Further Research

This framework provides a useful starting point for addressing the challenges associated with managing and transferring cryptocurrencies and digital assets during the succession and inheritance process. Individuals can ensure the proper management and transmission of their digital assets by emphasizing the significance of secure storage solutions and transparent access and transfer protocols. However, it is essential to recognize that digital assets continue to face significant legal, regulatory plus third-party technical integration challenges, and that additional measures may be required to effectively resolve them.

To address the research gaps mentioned above further research on managing and transferring cryptocurrencies and digital assets during the succession and inheritance process could focus on the following areas:

Legal and Regulatory Challenges. Explore the legal and regulatory complexities associated with the transmission of digital assets in different jurisdictions. Investigate the existing laws and regulations governing digital asset inheritance and succession and identify any gaps or inconsistencies. Research potential solutions for creating more uniform regulations that ensure compliance and facilitate the smooth transfer of digital assets across borders.

Fraud and Cybersecurity. Study the risks of fraud and cybercrime in relation to digital assets and identify effective measures to mitigate these risks. Examine real-world cases of digital asset theft or loss and identify common vulnerabilities. Research emerging technologies and practices that enhance security and resilience against cyber threats, such as advanced encryption methods, multi-factor authentication, and secure storage solutions.

User Education and Awareness. Investigate strategies to enhance user education and awareness regarding the management and transfer of digital assets during succession and inheritance. Develop educational resources, guidelines, and best practices to help individuals understand the risks, challenges, and proper procedures associated with digital asset ownership and transfer. Explore ways to promote responsible digital asset management and ensure that individuals have the knowledge and tools to protect their assets.

Ethical and Social Implications. Examine the ethical and social implications of digital asset inheritance and succession. Investigate issues related to digital asset distribution among beneficiaries, considerations of fairness and equality, and the potential impact on traditional inheritance practices. Research the societal and cultural aspects of digital asset management and inheritance and explore potential solutions to address emerging challenges.

7 Conclusion

The increasing popularity digital assets has created new challenges for managing and transferring these assets during the process of succession and inheritance. This framework provides a distinct set of steps for addressing these obstacles and ensuring that digital assets are managed and transferred appropriately. Its emphasis on developing secure storage solutions for digital assets is one of its primary strengths. Individuals can rest assured that their digital assets are maintained securely and are less susceptible to theft and fraud when advanced encryption and access controls are utilized. The framework's consideration on developing transparent access and transmission protocols is an additional strength. Individuals can ensure that their digital assets are transmitted according to their desires by identifying trusted family members or professionals who have access to the digital assets and providing them with explicit instructions. Even though this framework provides a defined set of steps for administering cryptocurrency and digital asset ownership and transfer in the succession and inheritance process, some may argue that it does not go far enough to address the unique challenges posed by these assets. Some may contend, for instance, that this framework does not adequately account for the legal and regulatory complexities associated with the transmission of digital assets. Given the lack of uniform regulations across jurisdictions, it can be challenging to ensure that the transmission of digital assets is compliant with all applicable laws and regulations. Moreover, some may contend that this framework does not adequately address the risk of fraud or cybercrime in relation to digital assets. While sophisticated encryption and access controls can help mitigate these risks, the loss or theft of digital assets is always a possibility.

References

1. Antonopoulos, A.M., Wood, G.: Mastering Ethereum: Building Smart Contracts and DApps. O'reilly Media, Sebastopol (2018)
2. Azar, P.D., et al.: The financial stability implications of digital assets. FRB of New York Staff Report (1034) (2022)
3. Bamakan, S.M.H., Nezhadsistani, N., Bodaghi, O., Qu, Q.: Patents and intellectual property assets as non-fungible tokens; key technologies and challenges. Sci. Rep. **12**(1), 2178 (2022)
4. Bullmann, D., Klemm, J., Pinna, A.: In search for stability in crypto-assets: are stablecoins the solution? Soc. Sci. Res. Netw. (2019).https://doi.org/10.2139/ssrn.3444847
5. Chen, C., Lin, C., Chiang, M., Deng, Y., Chen, P., Chiu, Y.: A traceable online will system based on blockchain and smart contract technology. Symmetry **13**(3), 466 (2021). https://doi.org/10.3390/sym13030466
6. Chohan, U.W.: Are cryptocurrencies truly trustless? In: Goutte, S., Guesmi, K., Saadi, S. (eds.) Cryptofinance and Mechanisms of Exchange. Contributions to Management Science, LNCS, pp. 77–89. Springer, Cham (2019). https://doi.org/10.1007/978-3-030-30738-7_5
7. Christodoulou, K., Katelaris, L., Themistocleous, M., Christodoulou, P., Iosif, E.: NFTs and the metaverse revolution: research perspectives and open challenges. In: Lacity, M.C., Treiblmaier, H. (eds.) Blockchains and the Token Economy. Technology, Work and Globalization, pp. 139–178. Palgrave Macmillan, Cham. (2022). https://doi.org/10.1007/978-3-030-95108-5_6
8. Dondjio, I.: The importance of blockchain for ecomobility in smart cities: a systematic literature review. In: Papadaki, M., Rupino da Cunha, P., Themistocleous, M., Christodoulou, K. (eds.) Information Systems. EMCIS 2022. LNBIP, vol. 464, pp. 165–184. Springer, Cham (2023). https://doi.org/10.1007/978-3-031-30694-5_13
9. Dondjio, I., Themistocleous, M.: Blockchain technology and waste management: a systematic literature review. In: Themistocleous, M., Papadaki, M. (eds.) Information Systems. EMCIS 2021. LNBIP, vol. 437, pp. 194–212. Springer, Cham (2022). https://doi.org/10.1007/978-3-030-95947-0_14
10. Handler, W.C.: Succession in family firms: a mutual role adjustment between entrepreneur and next-generation family members. Entrep. Theory Pract. **15**(1), 37–52 (1990). https://doi.org/10.1177/104225879001500105
11. Johnson, D.H., Menezes, A., Vanstone, S.A.: The elliptic curve digital signature algorithm (ECDSA). Int. J. Inf. Secur. **1**(1), 36–63 (2001). https://doi.org/10.1007/s102070100002

12. Kang, B., Shao, D., Wang, J.: A fair electronic payment system for digital content using elliptic curve cryptography. J. Algorithms Comput. Technol. **12**(1), 13–19 (2018). https://doi.org/10.1177/1748301817727123
13. Kharitonova, J.: Digital assets and digital inheritance. Law Digit. Technol. **1**(1), 19 (2021). https://doi.org/10.18254/s123456780015732-6
14. PricewaterhouseCoopers. (n.d.-b). Demystifying cryptocurrency and digital assets. PwC. https://www.pwc.com/us/en/tech-effect/emerging-tech/understanding-cryptocurrency-digital-assets.html
15. Prost, F.: Inheritance and Blockchain: Thoughts and Open Questions, 28 November 2022. arXiv.org. http://arxiv.org/abs/2212.01194v1
16. Shah, J.C., Bhagwat, M., Patel, D., Conti, M.: Crypto-Wills: transferring digital assets by maintaining wills on the blockchain. In: Bansal, J., Gupta, M., Sharma, H., Agarwal, B. (eds.) Communication and Intelligent Systems. ICCIS 2019. LNNS, vol. 120, pp. 407–416. Springer, Singapore (2020). https://doi.org/10.1007/978-981-15-3325-9_31
17. Teoh, B.P.C.: Navigating the blockchain trilemma: a supply chain dilemma. In: Ismail, A., Dahalan, W.M., Öchsner, A. (eds.) Advanced Maritime Technologies and Applications, LNCS. ASM, vol. 166, pp. 291–300. Springer, Cham (2022). https://doi.org/10.1007/978-3-030-89992-9_25
18. Volos, A.A.: Digitalization of society and objects of hereditary succession. Leg. Issues Digit. Age **3**(3), 68–85 (2022). https://doi.org/10.17323/2713-2749.2022.3.68.85
19. Yan, D., Liu, D.: Toward vulnerability detection for ethereum smart contracts using graph-matching network. Futur. Internet **14**(11), 326 (2022). https://doi.org/10.3390/fi14110326

Perspectives of Merchants Regarding Bitcoin's Role as a Currency and Its Utility as a Payment System

Alex Gutsche and Soulla Louca(✉)

University of Nicosia, 1700 Nicosia, Cyprus
Gutsche.a@live.unic.ac.cy, louca.s@unic.ac.cy

Abstract. In the last decade, Bitcoin has transformed from an experimental asset into a global store of value, emerging as a viable alternative to traditional finance. This study explores Bitcoin adoption among merchants in Panajachel (Guatemala), El Zonte (El Salvador), and Uvita (Costa Rica). Quantitative research gathered insights from 64 respondents, revealing that 64% of merchants believed they attracted more customers due to Bitcoin acceptance. Usage of Bitcoin earnings varied, with 44% holding it as an asset and 27% converting it to fiat. Additionally, 67% found Bitcoin easier than credit cards. The analysis was guided by the Technology Acceptance Model and the Diffusion of Innovations theory, focusing on perceived utility, ease of use, market advantage, and compatibility. This research highlights the value of localized studies in understanding Bitcoin's adoption, offering valuable insights for policymakers and businesses. It sheds light on Bitcoin's dual role as both a currency and a payment system, enabling users to transact outside of traditional finance.

Keywords: Bitcoin · Payment System

1 Introduction

Cryptocurrencies, led by the first cryptocurrency Bitcoin, have emerged as potential disruptors to the traditional banking system and offer a new paradigm for peer-to-peer value transfer and wealth storage [1]. While the developed world largely views Bitcoin through the lens of investment and speculation, a different technological use case exists in the developing world, particularly in Central America [1]. Most of the global south is unbanked, with little access to the traditional financial system. By providing an easy to use transactional and value storage medium, blockchain technology has the potential to improve financial inclusion in the developing world [2].

Historically, Bitcoin has acted a global value transfer system, typically facilitating the transfer of large values. However, the high fees associated with main chain transactions have rendered it unsuitable for small value transfers. With the emergence of layer 2 solutions such as the Lightning Network, small peer to peer Bitcoin transactions have become viable [3]. For any financial exchange to occur, it's essential for both parties, the payer and the merchant, to mutually accept the chosen currency.

M. Papadaki et al. (Eds.): EMCIS 2023, LNBIP 501, pp. 107–119, 2024.
https://doi.org/10.1007/978-3-031-56478-9_8

Virtual currency ecosystems, like Bitcoin, are made up of distinct parts, which include but are not limited to; (1) users (2) merchants (3) virtual currency exchanges (4) governance authorities (e.g. Bitcoin Core developers) (5) trade platforms (6) processing services providers (7) wallet providers/custodians (8) inventors (9) technical service providers (10) information providers and (11) miners [4]. These aspects make up a political economy and illustrate a sociological picture of the technology [4]. Understanding why businesses have decided to accept Bitcoin as a payment method is vital to facilitating widespread adoption.

Countries in Central America have experienced social instability, economic challenges, and in some cases, unstable fiat currencies. Because of these factors citizens in these countries may find Bitcoin's use case compelling [5].

The following study investigates the perceptions and experiences of Bitcoin-accepting merchants in developing world regions, focusing on three distinct communities: Panajachel, El Zonte, and Uvita. These towns are recognized as "self-proclaimed Bitcoin circular economies," indicating a significant Bitcoin user and merchant presence. The study's primary objective is to gain insights into local merchant adoption of Bitcoin and their perspectives on Bitcoin as a currency and payment system.

A survey comprising 13 questions was administered to merchants in each of these regions to collect data and insights regarding their Bitcoin experiences. The research delves into the experiences of Bitcoin-accepting merchants, uncovering both positive aspects and challenges they have faced in their Bitcoin adoption journey. In an era of rapid digital transformation, understanding the narrative of Bitcoin adoption among merchants holds promise in demonstrating how blockchain technology can address financial inequity in the developing world.

The paper's structure encompasses a review of related literature in Sect. 2, a detailed methodology in Sect. 3, presentation of results with acknowledgment of study limitations in Sect. 4, a discussion based on the findings in Sect. 5, and concludes by outlining directions for future research.

2 Literature Review

Bitcoin merchant adoption represents a relatively new field of study within the realm of cryptocurrency payment acceptance. Research in this field is relatively new, and as such, there is a lack of comprehensive research. Existing studies predominantly target e-commerce merchants who accept Bitcoin, with of a lack of information on merchants who accept Bitcoin in a physical setting.

The adoption of Bitcoin by merchants is influenced by a myriad of factors ranging from technical and sociocultural to legal. While there are clear benefits, barriers like technical complexity, regulatory ambiguity, and a lack of awareness/trust hinder its widespread acceptance. As the landscape evolves, understanding these factors becomes crucial for businesses, policymakers, and consumers.

2.1 Technical Factors

Technical aspects impacting Bitcoin adoption by consumers and businesses are significant. The Technology Adoption Model developed by Davis in 1989, emphasizes perceived usefulness and perceived ease of use, both relevant technological adoptions metrics [6].

Jonker highlights technical complexity as a deterrent for retailers, as cryptocurrency acceptance requires an initial technology investment and robust security [7]. Deloitte's survey points to market fragmentation and customer security as hindrances [8]. Harwick stresses the technical gap between digital currencies and the financial system [9]. However, Bitcoin's benefits, such as low fees, swift settlement, and user control, make it attractive as a payment method [10, 11]. Bitcoin's user-friendly interface simplifies transactions, though acquiring initial bitcoins can be challenging [11]. Its secure, irreversible transactions lower fraud risk [11]. Decentralization reduces reliance on central authorities [12]. Aspects of the technology that are perceived to be useful include productivity gains and cost-effective global transfers [13].

Bitcoin's innovation lies in introducing a new currency and payment method [9]. Decentralization was also shown to benefit merchants dealing with foreign currencies [7]. The ease of using Bitcoin's payment system is a key attraction [32]. Additionally, the blockchain's public, permissionless nature may attract merchants [15]. Digital currencies enable easy technical innovation [9]. And network effects impact blockchain payment system adoption [15].

2.2 Regulatory Factors

The evolving Bitcoin regulatory landscape raises uncertainty for merchants, with legal ambiguity and customer protection concerns influencing their decisions on cryptocurrency acceptance [16]. Regulatory challenges also affect cryptocurrency protocols, as networks cannot easily be shut down, but individual users can be targeted. Additionally, the increasing role of financial intermediaries allows regulatory systems to indirectly impact the network by targeting these supporting entities, potentially leading to market instability [7].

The legal status of digital currencies also affects consumer confidence, as countries like the USA and Germany tax them as assets as opposed to viewing them as legal tender [32]. Many tax authorities consider digital currency trading as taxable transactions, impacting their overall usefulness as a medium of exchange. The online and decentralized nature of digital currencies, a quality that makes them distinct from traditional finance, also influences how they are perceived and trusted [14, 32].

In the past, Bitcoin's pseudonymous transactions have made it a preferred choice for illicit activities, exemplified by its exclusive use on the Silk Road from 2011 to 2013 [17]. Virtual currencies like Bitcoin challenge territorial regulatory frameworks as they exist in a borderless virtual realm [4]. Bilotti emphasizes the importance of a regulatory approach ensuring fair competition, governing digital currency innovation, maximizing potential, and mitigating risks [18, 33].

Mazambani [19] emphasizes the need for regulators to foster consumer protection, social impact, and adoption through a supportive legal framework. South Africa's unique case underscores the importance of consumer onboarding. A favorable regulatory climate is essential for cryptocurrency market growth, requiring stable purchasing power to alleviate adoption concerns and promoting intermediation [7]. While governments cannot eliminate cryptocurrencies, they can hinder the development of financial institutions that facilitate adoption, crucial for cryptocurrency market advancement and challenging centralized financial systems [7].

2.3 Economic Factors

Bitcoin can significantly reduce costs for merchants by eliminating credit card processing fees and the risk of chargebacks [11]. Mendoza-Tello et al. [13] highlights the advantages of Bitcoin, including permissionless value transfers, global reach, and low transactional costs. Bitcoin also holds the potential to provide banking services to the unbanked, addressing the issue of insufficient access to banking services, especially for those without bank accounts [10]. Additionally, larger merchants may find Bitcoin transactions more cost-effective and efficient, potentially bypassing the high fees charged by traditional payment providers [20]. However, it is important to note that there are still transactional costs involved when converting Bitcoin to national currencies [11].

2.4 Sociocultural Factors

Educating both businesses and individuals about Bitcoin is essential for adoption. Baur et al. [21] note Bitcoin's potential in less-developed countries, while Marszk, Lechman, and Kato [22] discuss how technologies like Bitcoin have revolutionized society. The significance of educating the masses about Bitcoin is stressed by Presthus and O'Malley [23, 34–36], who argue that a clearer value proposition could facilitate wider acceptance. As an open-source, continuously evolving technology, Bitcoin can potentially connect global payment systems and transform the traditional financial system. How users perceive risk and trust is also discussed in reference to the use of cryptocurrencies. Users tend to reduce their risk when adopting new technology, and perceived risk is a source of uncertainty for consumers [13]. This is something merchants might need to consider when deciding whether to adopt Bitcoin as a medium of exchange.

2.5 Studies on Digital Currency Adoption

Jonker [7] surveyed 768 online Dutch retailers and discovered a low 2% cryptocurrency acceptance rate. The reasons included unfamiliarity (58%) and lack of trust (16%). The study focused on consumer adoption, transactional benefits, and retailer adoption effort.

Guych et al. [24] used the Technology Acceptance Model to predict Taiwanese hotel owners' intent to accept cryptocurrency payments. Their decision was influenced by perceived usefulness and ease of use.

Polasik et al. [25] collected data from online businesses accepting Bitcoin through a questionnaire. Company characteristics, use of other payment methods, customer knowledge about Bitcoin, and the economic environment significantly affected Bitcoin acceptance.

Shahzad et al. examined cryptocurrency adoption in mainland China [26]. Their survey targeted individuals involved in online gaming with Bitcoin experience. They found that awareness and perceived trustworthiness played a crucial role in Bitcoin adoption. Perceived ease of use and usefulness were also positively related to Bitcoin adoption.

2.6 Technology Adoption Models

This section describes two of the most well-known and widely used technology adoption models. Other lesser known technology adoption models were excluded from this analysis.

The Technology Acceptance Model (TAM) consists of two main aspects [6]:

- Perceived Usefulness: This evaluates whether merchants see value in accepting Bitcoin for their businesses. It considers whether Bitcoin can attract more customers, reduce transaction costs, accelerate transactions, or offer a competitive edge. Merchants are more inclined to adopt Bitcoin if they perceive these benefits.
- Perceived Ease of Use: This assesses whether merchants find Bitcoin user-friendly and easy to implement. It also considers the ease of converting Bitcoin to fiat currency if they prefer not to retain their value in Bitcoin. Merchants are more likely to accept Bitcoin if they find it straightforward to use.

The Diffusion of Innovations Theory (DOI) applies to Bitcoin merchant adoption in various ways [27]:

- Innovation-Decision Process: Merchants follow a similar process as individuals or entities when deciding to accept Bitcoin. They learn about it, assess its pros and cons, test it, and then decide.
- Relative Advantage: Merchants perceive Bitcoin as advantageous due to lower transaction fees, faster transaction times, and access to a global customer base.
- Compatibility: Bitcoin's compatibility with merchant values and needs influences its adoption.
- Complexity: The perceived complexity of Bitcoin can hinder adoption, as technical aspects may be daunting.
- Trialability: Merchants need to test Bitcoin as a payment system before full-scale implementation.
- Observability: The success of early adopters can influence other merchants.
- Categories of Adopters: Merchants can be classified into categories like Innovators, Early Adopters, Early Majority, Late Majority, and Laggards. The transition to the early majority is crucial for Bitcoin's broader adoption.
- Social Systems and Opinion Leaders: Opinion leaders in the Bitcoin ecosystem may influence merchants' decisions.
- Communication Channels: Information about Bitcoin spreads through various channels like seminars, workshops, and online forums, educating merchants about its potential as a form of payment.

3 Methodology

The study used quantitative research methods. It utilized systematic empirical research that dealt mainly with the collection and interpretation of numerical data. This methodology seeks to identify patterns, relationships, and associations among variables, allowing for statistical analysis to draw conclusions and make predictions.

A cross-sectional survey design was employed to gather information from merchants about their experiences, beliefs, and actions related to accepting Bitcoin. The survey was initially developed in English, and then translated into Spanish. Data collection occurred in late summer/early fall 2023 in Panajachel, Guatemala; El Zonte, El Salvador; and Uvita, Costa Rica. The survey included three types of questions: multiple choice, dichotomous, and scale/rating. These question types were used to collect comprehensive data about merchants' experiences with Bitcoin acceptance during a brief interview. The questions covered topics such as the duration of Bitcoin acceptance, the perceived ease of using the technology, the typical Bitcoin-using customer profile, the impact of accepting Bitcoin on their business, the influence of other merchants on their decision to use Bitcoin, ease of use compared to credit cards, and the size and sector of their business. The survey questions were designed with both the Technology Acceptance Model and the Diffusion of Innovations Theory in mind, aiming to use both technology adoption frameworks to explain the Bitcoin merchant adoption process.

The locations of the respondents were selected based on an online review of regions with self-proclaimed Bitcoin circular economies. Three regions were identified, each with a significant online presence related to Bitcoin acceptance. Merchants in these communities were identified based on listings on mobile applications including as Blink, BTC Map, and Bitcoin Jungle, or through visible signs indicating Bitcoin acceptance at their establishments. While these platforms were not exhaustive, they increased the likelihood of finding Bitcoin-accepting merchants. The surveys were conducted in person.

The questions were created to assess merchant perceptions, adoption, and usage of Bitcoin as both a payment method and a currency. After pretesting, the survey was further shortened to 13 questions to ensure brevity and simplicity while providing valuable insights into Bitcoin acceptance. The final survey was hosted on Jotform, a survey design and administration platform.

The study involved face-to-face surveys in three distinct regions: Panajachel in Guatemala, El Zonte in El Salvador, and Uvita in Costa Rica. To maximize participation, merchants were approached during their less busy hours. Data collection took place from September 3rd to September 21st, 2023, resulting in a total of 64 completed surveys. These surveys were distributed geographically as 27 in Panajachel, 16 in El Zonte, and 21 in Uvita. Interactive Google My Maps were created to visually pinpoint each Bitcoin-accepting merchant in the surveyed areas. Survey response rates were calculated by comparing completed surveys to the total number of listed merchants on the app, excluding those that were not found, closed, or did not accept Bitcoin. Across the three areas, 64 surveys were completed, resulting in a survey response rate of 78%.

This methodology provides valuable insights into Bitcoin's acceptance in Central American commerce. Through cross-sectional surveys, the study aims to reveal the practical implications of cryptocurrency for merchants. As digital transactions gain prominence, the findings of this research can inform the global discourse on Bitcoin. The perspective of merchants, often overlooked, plays a crucial role in shaping future policies and innovations.

3.1 Problems Encountered During Data Collection

The apps (Blink, BTC Map, and BTC Jungle) were unreliable indicators of Bitcoin acceptance, with various errors, such as pointing to non-Bitcoin-accepting businesses, businesses in unlocatable places, or incorrect business locations. Sample locations were based on these apps and visible Bitcoin acceptance signs, potentially missing Bitcoin-accepting businesses, leading to potentially inaccurate community acceptance data.

Moreover, some businesses declined to participate in the survey due to time constraints or lack of interest while some Bitcoin-accepting businesses were not surveyed due to technical issues or untrained staff, and were not categorized as Bitcoin-accepting businesses. In addition, s few businesses were closed during the survey period, making their Bitcoin acceptance status unclear. In some cases, businesses accepted Bitcoin but had no customers wanting to pay with Bitcoin, making many survey questions inapplicable as they required prior Bitcoin acceptance. Additionally, one survey question asked about Bitcoin's ease of use compared to credit cards, but some businesses had never used credit cards, therefore certain respondents could not make a direct comparison. However, most had explored credit card payment options and subsequently found them cost-inefficient.

4 Results and Discussion

The following chapter contains a comprehensive data analysis of the survey conducted across the three distinct regions. Focusing on the Technology Acceptance Model (TAM) and The Diffusion of Innovations Theory (DOI), this analysis aims to shed light on how different businesses and business sectors perceive and integrate emerging payment technologies, such as Bitcoin.

4.1 Data Analysis

Over half (55%) of surveyed Bitcoin accepting businesses belonged to the food and beverage sector, and 20% belonged to retail. Regarding Bitcoin acceptance timelines, 50% had been accepting Bitcoin for over 12 months and 24% for 6 to 11 months. Businesses can be categorized as follows in the DOI adoption paradigm:

- Innovators & Early Adopters (50%): The pioneers who integrated Bitcoin early.
- Early Majority (24%): Adopted Bitcoin after seeing its benefits.
- Late Majority (around 26.5%): Started accepting Bitcoin later.
- Laggards are not represented in this survey, but this category could include merchants still considering Bitcoin adoption or those unaware of its potential benefits.

Most survey respondents were small businesses, with 63% having 1 to 5 employees, 19% having 5 to 9 employees, and only 17% having 10+ employees. The prevalence of small businesses is noteworthy because they are often considered flexible and innovative, capable of easily adapting their practices compared to larger corporations. The business size could influence the compatibility aspect of the DOI. Larger enterprises may have different compatibility perceptions compared to smaller businesses and individual entrepreneurs. Smaller businesses tend to be more agile in embracing new technologies and may be more open to exploring innovative payment methods to distinguish themselves in the market. Consequently, the data suggests that Bitcoin payments may align better with smaller establishments.

54% of merchants initially found Bitcoin confusing, suggesting a learning curve. On the other hand, 46% did not encounter confusion, indicating either prior familiarity or ease of comprehension. This split in responses highlights varying technical expertise and underscores the need for education and user-friendly tools to promote Bitcoin adoption, addressing both the ease of use (TAM) and complexity (DOI) aspects.

91% of merchants found it easy to integrate Bitcoin as a payment method, indicating the technology is generally perceived as straightforward. This relates to both the perceived ease of use of TAM and the complexity aspect of the DOI aspects, as merchants believed the technology was both user-friendly and not overly complex.

Approximately 64% of merchants stated that accepting Bitcoin attracted more customers, showing they found the technology beneficial to their business. This group of merchants also believed accepting Bitcoin gave them a relative advantage over their peers, aligning with the DOI. In contrast, 33% of merchants might not see the technology as useful since they have not noticed an increase in customers. They do not believe Bitcoin provides a competitive edge over their peers and may question its future utility.

67% of respondents found Bitcoin transactions simpler than credit card transactions, providing a direct comparison with widespread, incumbent payment technology. For these merchants using Bitcoin provided a relative advantage over credit cards. It should also be noted that businesses were not questioned as to whether they had used credit cards, only if they thought using Bitcoin was easier than using credit cards, potentially influencing the results.

78% of merchant's believed foreigners and tourists primarily used Bitcoin for payments. This data indicates that Bitcoin is currently used as a niche payment technology for foreigners and has yet to become widely used by locals in the three communities.

44% of merchants choose to keep their Bitcoin earnings, while 27% exchanged their Bitcoin for fiat currency. This indicates that almost half of merchants thought that Bitcoin functioned as a store of value. Those who convert Bitcoin to fiat may not view Bitcoin as a reliable store of value but still find it useful as a payment method. Merchants holding Bitcoin believe it provides a relative advantage over fiat.

Approximately 50% of merchants tested Bitcoin before full integration, showing that the trialability aspect of the DOI was important for half of respondents. This suggests that the other half did not find testing necessary. These merchants may have had prior experience with using Bitcoin before deciding to accept it in their businesses.

A noteworthy 75% of merchants were inspired to accept Bitcoin after seeing others do the same. However, 22% made the decision independently, possibly due to internal motivations or self-education. This highlights the observability (DOI), innovation decision process (DOI), and perceived usefulness (TAM) aspects. Positive experiences shared by merchants can encourage further adoption among observers who see its utility.

52% of merchants reported receiving Bitcoin payments "sometimes," suggesting occasional customer use. Another 38% noted that Bitcoin payments were 'rare', indicating that there were a limited number of customers that paid in Bitcoin. Only 5% experienced 'frequent' Bitcoin payments. These terms ('sometimes', 'rare' and 'frequent') were not codified, and further research should attempt to assess how often merchants received Bitcoin payments, either through more robust surveys or through the analysis of regional transaction volumes. Additionally, comparing Bitcoin payment frequency with conventional methods like credit cards and cash would help gauge its prevalence.

4.2 Limitations of the Study

General Limitations. Determining Bitcoin accepting businesses was challenging as the apps were unreliable indicators of Bitcoin acceptance. Businesses not accessible through these methods were excluded, and the accuracy depended on these apps. The survey used a dichotomous format due to language and time constraints and as such lacked a Likert scale for deeper analysis. Examining merchant perspectives on accepting Bitcoin versus cash or other digital payment methods would have been valuable. Additionally, a larger sample size with hundreds of merchant surveys would have yielded more survey data, improving the veracity of data analysis. In addition, the survey focused solely on merchant behavior and did not consider customer motivations for paying in Bitcoin. Future research should explore consumer perspectives on using Bitcoin as a payment method.

Research Limitations. The survey only had 64 respondents, which is not a sufficient number for broad adoption generalizations. A more robust approach would involve a full census of all merchants in each location. This could be used to determine the total number of businesses and the percentage of businesses in each community that accepted Bitcoin as payment. Originally, the research aimed for a comparative analysis among the three regions, but the lack of responses precluded this type of analysis. Differences between the communities could be analyzed directionally, but the small sample size would not provide meaningful insights. Future research could benefit from a longitudinal study to track evolving merchant opinions and views on acceptance over a longer period.

Self-reporting Limitations of the Study. Some respondents may not have had full knowledge of Bitcoin acceptance in their business, potentially affecting response accuracy. Future research should aim to survey business owners or decision makers that were directly involved in the decision to accept Bitcoin.

Language Limitations. The survey was conducted in Spanish, and language barriers may have affected understanding during the conversations between the surveyor and respondents. Translated questions could also have had inaccuracies. The cultural differences among respondents from Guatemala, El Salvador, and Costa Rica may have influenced responses.

Time Limitations. The study was a snapshot in time and did not account for merchants evolving views regarding Bitcoin acceptance. The survey was also done without considering Bitcoin's price fluctuations, which could impact adoption. Future research could compare merchant's attitudes with the price of Bitcoin, illuminating how Bitcoin's price impacts user sentiment. Time constraints also limited merchant sample size as some businesses were closed during the visitation period, and revisits were not always possible.

5 Key Findings and Recommendations

Most businesses surveyed had been accepting Bitcoin for over a year, signifying they had considerable experience with utilizing Bitcoin as both a payment system and as a currency. By accepting Bitcoin early on these adopters may have gained a competitive edge in the market, as 64% of merchants reported an increase in customer numbers after Bitcoin adoption. 78% of businesses believed that tourists predominantly paid with Bitcoin, indicating that Bitcoin is currently a niche currency utilized by foreigners. Small businesses with 1 to 5 employees represented more than half of Bitcoin accepting business, suggesting their adaptability to new technologies and willingness to innovative may provide them with a competitive advantage over their peers. 67% of respondents noted that using Bitcoin was easier than using credit cards, showing that for some Bitcoin provides a relative advantage over existing payment methods.

When it comes to Bitcoins utility as a store of value 44% of merchants believed the asset was worth holding on to, relative to converting to fiat, believing in the upward volatility of the asset. On the other hand, 27% converted their Bitcoin to fiat, enjoying either the stability or the liquidity that fiat currency offers. Finally, 75% of merchants were influenced by others when they decided to begin accepting Bitcoin as payment, indicating that a significant network effect exists across the three Bitcoin accepting communities.

Recommendations for increasing Bitcoin acceptance among merchants requires:

- Sector-Specific Focus: Concentrate efforts on sectors with the highest Bitcoin adoption, namely food and beverage and retail, as they see distinct advantages.
- Educate New Merchants: Offer workshops and resources to educate newer merchants about the benefits of Bitcoin, especially focusing on those not currently accepting it.
- Support Small Businesses: Small businesses (1–5 employees) find it easy to adopt Bitcoin. Establish an educational ecosystem with easy-to-use Point of Sale (POS) systems and streamlined technological training.
- Showcase Success Stories: Create informational campaigns highlighting businesses' success with Bitcoin adoption. Use case studies, testimonials, and short videos as persuasive tools to encourage other merchants.
- Tourist-Focused Campaigns: Develop campaigns targeting tourists in these tourist-centric areas, promoting Bitcoin use and encouraging merchants to accept it. Consider using pamphlets and digital campaigns.
- Safeguards Against Volatility: Address concerns about Bitcoin's volatility by offering solutions or educational initiatives for hedging. Introduce platforms for instant conversion to fiat or stablecoins and promote low-fee Bitcoin ATMs.
- Engage with Late Adopters: Understand and address the concerns of late adopters through tailored outreach programs. Explore the reasons behind their hesitance to accept Bitcoin.
- Feedback Loop: Maintain regular contact with Bitcoin-accepting merchants to address their evolving needs, concerns, and challenges. Use surveys, focus group discussions, or informal meet-ups to refine adoption strategies and build a sense of community.

- Incentives for Adoption: Partner with Bitcoin wallets or platforms to offer promotional incentives, such as discounted transaction fees, cashback, or marketing promotions to encourage merchants.
- Infrastructure Enhancement: Collaborate with local governments or business associations to improve infrastructure for Bitcoin transactions, including better internet connectivity, wider Bitcoin ATM availability, and local Bitcoin information centers.
- Comparative Studies: Investigate other self-proclaimed Bitcoin circular economies in different parts of the world to compare adoption across cultures and understand cultural differences affecting adoption.
- Community Initiatives: Understand the organizations and individuals behind community-led initiatives for onboarding merchants into Bitcoin acceptance.
- Education Programs: Expand education programs focused on teaching interested individuals about Bitcoin's multifaceted aspects, alleviating fears and misinformation, and facilitating digital currency adoption.

6 Conclusion

In summary, Central American merchants exhibit cautious optimism towards Bitcoin adoption. While many see its potential benefits and have integrated it into their operations, there ss a clear indication of an adoption learning curve and the importance of education. Merchants generally believe that accepting Bitcoin has enhanced their business by attracting more customers and providing a cost-free digital payment system. Most respondents also consider using Bitcoin easier than using credit cards. The data also underscores the significance of social influence, with many merchants looking to their peers when considering Bitcoin adoption. Understanding these perceptions and behaviors is crucial for Bitcoin's broader acceptance in the commercial landscape.

Furthermore, the findings reveal that Bitcoin adoption among merchants is on the rise in the three surveyed locations. Businesses are embracing the benefits of this digital currency, including its global reach and potential for reduced transaction fees. They use applications like Blink, BTC Map, and Bitcoin Jungle to signal their acceptance of Bitcoin, and some display signs indicating their willingness to accept it as payment. Across these regions, a common thread is the curiosity and willingness of merchants to adopt innovative payment methods.

A notable takeaway from this study is the community aspect of Bitcoin adoption. It is not just a business decision; it influences other merchants and customers, fostering conversations and collective curiosity. Community organizations in these areas play a vital role in educating citizens about Bitcoin use, which can impact merchant payment acceptance based on customer demand.

The growth of Bitcoin in these regions mirrors its global trajectory, providing valuable insights that bridge the gap between global trends and local realities. In conclusion, localized studies like this are crucial for assessing Bitcoin adoption. They offer a real-world perspective on Bitcoin's challenges and successes, providing lessons that can inform broader strategies and decisions. The experiences and insights from these three regions may play a role in shaping the narrative around Bitcoin and its place in the global economy.

References

1. Gladstein, A.: Financial freedom and privacy in the post-cash world digital currencies: risk or promise? Cato J. **41**(2), 271–294 (2021)
2. Schuetz, S., Venkatesh, V.: Blockchain, adoption, and financial inclusion in India: research opportunities. Int. J. Inf. Manag. **52**, 101936 (2020). https://doi.org/10.1016/j.ijinfomgt.2019.04.009
3. Divakaruni, A., Zimmerman, P.: Ride the Lightning: Turning Bitcoin into Money, Rochester, NY (2020). https://doi.org/10.2139/ssrn.3514125
4. Lastra, R.M., Allen, J.G.: Virtual currencies in the eurosystem: challenges ahead. Int. Lawyer **52**(2), 177–232 (2019)
5. Alzahrani, S., Daim, T.U.: Analysis of the cryptocurrency adoption decision: literature review. In: 2019 Portland International Conference on Management of Engineering and Technology (PICMET), pp. 1–11 (2019). https://doi.org/10.23919/PICMET.2019.8893819
6. Davis, F.D.: Perceived usefulness, perceived ease of use, and user acceptance of information technology. MIS Q. **13**(3), 319–340 (1989)
7. Jonker, N.: What Drives Bitcoin Adoption by Retailers, Rochester, NY (2018). https://doi.org/10.2139/ssrn.3134404
8. Deloitte: Merchants Getting Ready for Crypto (2022)
9. Harwick, C.: Cryptocurrency and the problem of intermediation. Indep. Rev. **20**(4), 569–588 (2016)
10. Saiedi, E., Broström, A., Ruiz, F.: Global drivers of cryptocurrency infrastructure adoption. Small Bus. Econ. **57**(1), 353–406 (2021). https://doi.org/10.1007/s11187-019-00309-8
11. Folkinshteyn, D., Lennon, M.: Braving bitcoin: a technology acceptance model (TAM) analysis. J. Inf. Technol. Case Appl. Res. **18**(4), 220–249 (2016). https://doi.org/10.1080/15228053.2016.1275242
12. Baron, J., O'Mahony, A., Manheim, D., Dion-Schwarz, C.: National Security Implications of Virtual Currency, Rand Corporation (2015)
13. Mendoza-Tello, J.C., et al.: Disruptive innovation of cryptocurrencies in consumer acceptance and trust. Inf. Syst. e-Bus. Manag. **17**, 195–222 (2019). SpringerLink
14. Themistocleous, M., Cunha, P., Tabakis, E., Papadaki, M.: Towards cross-border CBDC interoperability: insights from a multivocal literature review. J. Enterp. Inf. Manag. **36**(5), 1296–1318. Emerald Publishing Limited (2023). 1741–0398. https://doi.org/10.1108/JEIM-11-2022-0411
15. Toufaily, E., Zalan, T., Ben Dhaou, S.: A framework of blockchain technology adoption: an investigation of challenges and expected value. Inf. Manag. **58**(3), 103444 (2021)
16. Kumpajaya, A., Dhewanto, W.: The acceptance of bitcoin in Indonesia: extending TAM with IDT. J. Bus. Manag. (2015)
17. Luther, W.J.: Bitcoin and the future of digital payments. Indep. Rev. **20**(3), 397–404 (2016)
18. Cunha, P.R., Soja, P., Themistocleous, M.: Blockchain for development: a guiding framework. Inf. Technol. Dev. **27**(3), 417–438 (2021). https://www.tandfonline.com/doi/full/10.1080/02681102.2021.1935453
19. Mazambani, L., Mutambara, E.: Predicting FinTech innovation adoption in South Africa: the case of cryptocurrency. Afr. J. Econ. Manag. Stud. **11**(1), 30–50 (2019). https://doi.org/10.1108/AJEMS-04-2019-0152
20. Bolt, W., Van Oordt, M.R.C.: On the value of virtual currencies. J. Money Credit Bank. **52**(4), 835–862 (2020). https://doi.org/10.1111/jmcb.12619
21. Baur, A.W., et al.: Cryptocurrencies as a disruption? Empirical findings on user adoption and future potential of bitcoin and co. In: Janssen, M., et al. (eds.) Open and Big Data Management and Innovation. I3E 2015. LNCS, vol. 9373, pp. 63–80. Springer, Cham (2015). https://doi.org/10.1007/978-3-319-25013-7_6

22. Marszk, A., Lechman, E., Kato, Y.: Information and communication technologies for financial innovations. In: Marszk, A., Lechman, E., Kato, Y. (eds.) The Emergence of ETFs in Asia-Pacific, LNCS, pp. 53–81. Springer, Cham (2019). https://doi.org/10.1007/978-3-030-12752-7_3
23. Themistocleous, M., Christodoulou, K., Iosif, E., Louca, S., Tseas, D.: Blockchain in academia: where do we stand and where do we go? In: Proceedings of the Fifty-third Annual Hawaii International Conference on System Sciences, (HICSS 53), 7–10 January 2020. Maui, Hawaii, USA, IEEE Computer Society, Los Alamitos, California, USA (2020). https://scholarspace.manoa.hawaii.edu/handle/10125/64398
24. Guych, N., et al.: Factors influencing the intention to use cryptocurrency payments: an examination of blockchain economy (2018). https://mpra.ub.uni-muenchen.de/99159/
25. Polasik, M., et al.: Price fluctuations and the use of bitcoin: an empirical inquiry. Int. J. Electron. Commer. **20**(1), 9–49 (2015). https://doi.org/10.1080/10864415.2016.1061413
26. Shahzad, F., et al.: An empirical investigation on the adoption of cryptocurrencies among the people of mainland China. Technol. Soc. **55**, 33–40 (2018). https://doi.org/10.1016/j.techsoc.2018.05.006
27. Rogers, E.M.: Diffusion of Innovations, 4th edn. Simon and Schuster, New York (2010)
28. Alfonsi, S.: Bitcoin Beach: How a town in El Salvador became a testing ground for bitcoin (2022). https://www.cbsnews.com/news/bitcoin-beach-el-salvador-60-minutes-2022-04-10/
29. Buckler, N.: Inside Bitcoin Jungle: The Latest Hotspot For Crypto Travellers, The Chainsaw (2023). https://thechainsaw.com/defi/bitcoin-jungle-the-new-holiday-destination-for-crypto-lovers/
30. Creswell, J.W.: Research Design: Qualitative, Quantitative, and Mixed Methods Approaches. Sage Publications, Thousand Oaks (2014)
31. Gándara, N.: Bitcoin Lake: Guatemala's Crypto Enclave Launches, Bloomberg Línea (2022). https://www.bloomberglinea.com/english/bitcoin-lake-guatemalas-crypto-enclave-launches/
32. Zarifis, A., Efthymiou, L., Cheng, X., Demetriou, S.: Consumer trust in digital currency enabled transactions. In: Abramowicz, W., Kokkinaki, A. (eds.) Business Information Systems Workshops. BIS 2014. LNBIP, vol. 183, pp. 241–254. Springer, Cham (2014). https://doi.org/10.1007/978-3-319-11460-6_21
33. Bilotta, N., Botti, F.: Libra and the Others: The Future of Digital Money, Istituto Affari Internazionali (IAI) (2018). https://www.jstor.org/stable/resrep19691
34. Themistocleous, M.: Teaching blockchain: the case of the MSc in blockchain and digital currency of the university of Nicosia. In: Paliwoda-Pękosz, G., Soja, P. (eds.) Supporting Higher Education 4.0 with Blockchain, pp.150–165, Routledge, New York (2024). ISBN: 9781003318736
35. Themistocleous, M., Christodoulou, K., Katelaris, L.: An educational metaverse experiment: the first on-chain and in-metaverse academic course. In: Papadaki, M., Rupino da Cunha, P., Themistocleous, M., Christodoulou, K. (eds.) Information Systems. EMCIS 2022. LNBIP, vol. 464, pp. 678–690. Springer, Cham (2023). https://doi.org/10.1007/978-3-031-30694-5_47
36. Presthus, W., O'Malley, N.O.: Motivations and barriers for end-user adoption of bitcoin as digital currency. Procedia Comput. Sci. **121**, 89–97 (2017). https://doi.org/10.1016/j.procs.2017.11.013

Digital Governance

A Chatbot Generator for Improved Digital Governance

Christos Bouras[1](✉), Damianos Diasakos[1], Chrysostomos Katsigiannis[1], Vasileios Kokkinos[1], Apostolos Gkamas[2], Nikos Karacapilidis[3], Yannis Charalabidis[4], Zoi Lachana[4], Charalampos Alexopoulos[4], Theodoros Papadopoulos[4], Georgios Karamanolis[5], and Michail Psalidas[5]

[1] Computer Engineering and Informatics Department, University of Patras, 26504 Patras, Greece
{bouras,kokkinos}@upatras.gr, {up1084632,up1072490}@upnet.gr
[2] University Ecclesiastical Academy of Athens, 14561 Athens, Greece
gkamas@aeavellas.gr
[3] Department of Mechanical Engineering, University of Patras, 26504 Patras, Greece
karacap@upatras.gr
[4] Department of Information and Communication Systems Engineering, University of Aegean, 83200 Samos, Greece
{yannisx,zoi,alexop,t.papadopoulos}@aegean.gr
[5] Crowd Policy, 18345 Athens, Greece
{george,michael}@crowdpolicy.com

Abstract. Chatbots, the pioneering conversational artificial intelligence (AI) agents, have experienced remarkable growth and integration in various domains. In modern societies, chatbots have emerged as transformative digital entities, revolutionizing the way humans interact with technology. These conversational AI agents have transcended their initial applications to become integral parts of various industries and daily life. One of the most prominent roles of chatbots is in customer service, where they offer round-the-clock assistance, swift issue resolution, and personalized interactions. By handling routine queries and tasks, chatbots free up human agents to focus on complex and specialized issues, thus optimizing overall efficiency and customer satisfaction. To this end, this paper aims to present and describe the architecture of a novel chatbot generator with improved functionality in terms of quality of communication with end users and level of provided services, with a specialized infrastructure understanding the Greek language. The chatbot generator was developed in the framework of a research project and will be pilot tested by two end-users, the National Bank of Greece (NBG) and the General Secretariat for Information Systems & Digital Governance (GSIS-DG).

Keywords: Artificial Intelligence · Natural Language Processing · Machine Learning · Virtual Assistants

M. Papadaki et al. (Eds.): EMCIS 2023, LNBIP 501, pp. 123–134, 2024.
https://doi.org/10.1007/978-3-031-56478-9_9

1 Introduction

Chatbots are sophisticated artificial intelligence-driven software applications designed to simulate human-like conversations with users. These conversational agents utilize cutting-edge technologies such as Natural Language Processing (NLP) and Machine Learning (ML) to understand and interpret user input, enabling seamless interactions through text or voice commands. At the heart of their functionality lies NLP, which allows chatbots to comprehend natural language, identify user intent, and generate appropriate responses [1]. Through ML, chatbots continuously learn from user interactions, improving their accuracy and efficiency over time.

Generally speaking, chatbots offer a wide range of benefits, including efficient and instantaneous responses to queries, round-the-clock availability, and the ability to handle large volumes of interactions concurrently. By automating routine tasks and providing instant assistance, they optimize processes, save time, and enhance user experiences. While they excel in various applications, the development of chatbots is not without challenges. Issues like context comprehension, sentiment analysis, and avoiding biases in responses remain areas of active research to ensure more meaningful and contextually relevant interactions.

Except from NLP and ML, chatbots leverage an additional array of cutting-edge technologies and functions to offer intelligent and dynamic conversational experiences, including:

- Intent Recognition: By employing sophisticated ML models, chatbots can accurately discern user intent from their queries. This allows chatbots to understand user requests, directing them to the most appropriate responses or actions.
- Sentiment Analysis: To enhance user experience, chatbots can analyze user sentiment and emotions in the input text. This analysis helps chatbots respond with empathy or appropriately address negative feedback, leading to more meaningful interactions.
- Dialog Management: Chatbots employ dialog management systems to maintain coherent and context-aware conversations with users. They can retain information from previous interactions, ensuring smoother and more natural exchanges.
- Personalization: Through ML-driven algorithms, chatbots can tailor responses and recommendations based on user preferences, history, and behavior. Personalization enhances user engagement and satisfaction, creating a more seamless experience.
- Knowledge Base Integration: Chatbots can integrate with knowledge bases and databases, providing access to vast amounts of information in real-time. This integration allows chatbots to deliver accurate and up-to-date responses to user queries.
- Context Understanding: Improved context comprehension empowers chatbots to maintain more coherent and natural conversations. They can remember past interactions, follow-up on ongoing tasks, and adapt responses based on the conversation's context.
- Task Automation: Chatbots excel at automating routine tasks, such as appointment scheduling, order tracking, or information retrieval. By automating these processes, chatbots save time for users and optimize operational efficiency.

As technology continues to evolve, chatbots will likely incorporate more advanced functionalities, enabling them to further enhance user experiences and expand their applications across various industries. With responsible development and ethical considerations, chatbots have the potential to reshape how we interact with technology and provide valuable assistance in our daily lives.

Based on the above, the goal of this paper is to present a chatbot architecture that is targeted at efficiency-oriented digital governance. The proposed architecture takes into account the particular needs and difficulties that organizations encounter when implementing chatbot technology. In the heart of the proposed architecture lies the chatbot generator that enables system administrators to automate the process of creating chatbots without requiring extensive coding knowledge or technical expertise. The paper is part of the PYTHIA project which is co-financed by the European Union and Greek national funds through the Operational Program "Competitiveness, Entrepreneurship and Innovation", under the call "RESEARCH - CREATE - INNOVATE (2nd Cycle)".

The PYTHIA project invests in the technology of Chatbots, AI agents that support natural language conversations and exploit next-generation internet technologies. The primary objective of the project is to significantly improve the functionality of chatbots, in terms of quality and service in communication for end users, by developing a specialized infrastructure for understanding the Greek language. The development of this infrastructure is based on the use of NLP technologies, combined with ML techniques and the integration of Argumentation, Logic, and Structured Dialogue models [2]. The subsystems resulting from this research implement a complete system/platform that supports a new service model called Bots-as-a-Service.

The remainder of the paper is structured as follows: In Sect. 2, we introduce the project requirements that guided the implementation of the proposed architecture. Section 3 presents in detail the proposed chatbot generator architecture. Finally, Sect. 4 outlines concluding remarks and sketches future work directions.

2 Project Requirements

For the needs of this paper, we formulated specific questionnaires, in order to derive the necessary information about the functionalities that the cooperating institutions needed, i.e., the National Bank of Greece (NBG) and the General Secretariat for Information Systems & Digital Governance (GSIS-DG). Based on the partners' response, Table 1 presents the main requirements for each partner, which were complemented by specific needs, such as the requirements for security and privacy.

Table 1. Main requirements of each partner.

Operator	Requirements
NBG	Utilization of Chatbot in the form of Q&A for the circular system "Athena" -NBG Intranet Portal, for use by the employees of the organization
GSIS-DG	Use of the Chatbot to serve citizens on the different platforms/services of the institution

2.1 Functional Requirements

Chatbots have automation options and features that can significantly speed up users' processes and direct them appropriately to the information they need. However, not all chatbot software solutions are equally effective. There are specific requirements that have a substantial impact on chatbot performance while meeting customer needs [3, 4]. A list of the most important requirements of chatbot software is presented below:

- Complex Dialogues: To effectively engage in conversations, chatbot software should possess NLP capabilities that allow it to analyze conversational context. Essential functions of this chatbot include discerning question intent, delivering precise answers, and offering suggestions to confirm or resolve the matter at hand. Proactively seeking information and asking clarifying questions, the best chatbots possess advanced conversational capabilities, ensuring a non-linear conversation.
- Flexible Data Interfaces: Flexible data interfaces in chatbots refer to the capability of chatbot systems to interact with and process data from various sources and formats in a versatile manner. These interfaces allow chatbots to integrate with diverse data repositories and external systems, making them more powerful and adaptable in handling user queries and providing relevant information.
- Multi-channel Capability: Multi-channel capabilities in chatbots refer to their ability to interact and communicate with users across various communication channels and platforms. Instead of being limited to a single interface, chatbots with multi-channel capabilities can engage with users through multiple touchpoints, providing a seamless and consistent user experience across different channels. Multi-channel capabilities could be considered as a non-functional requirement but it is worth mentioning.
- Fast Onboarding: Fast onboarding refers to the process of rapidly integrating users into a system, service, or application with minimal effort and time required on the user's part. In the context of chatbots or digital applications, fast onboarding is crucial for ensuring a smooth and efficient user experience from the very beginning.
- Easy Handling: Chatbots' or digital applications' easy handling is crucial for providing a positive user experience and encouraging user engagement.
- Ongoing Optimization: It involves regularly monitoring the chatbot's interactions, analyzing data, and making iterative adjustments to ensure that the chatbot remains effective, relevant, and aligned with user needs and preferences.
- Analytics & Reporting: In chatbots analytics and reporting play a crucial role in understanding user behavior, evaluating performance, and optimizing the user experience.

- Performance and Protection of Personal Data: Performance and protection of personal data are two essential aspects when it comes to the operation of chatbots and digital applications, particularly concerning user privacy and the quality of user experience.

Regarding the issue of performance in the context of chatbots, it refers to the effectiveness, efficiency, and reliability of the chatbot in delivering accurate and timely responses to user queries and tasks. A high-performing chatbot should be able to understand user intent, provide relevant information, and complete tasks efficiently, resulting in a smooth and satisfactory user experience. Continuous monitoring, optimization, and regular updates are necessary to maintain and enhance the performance of chatbots over time.

Another important requirement is the protection of personal data. As chatbots interact with users and gather information, there is a need to ensure the protection and privacy of users' personal data. Personal data includes any information that can identify an individual, such as names, contact details, or financial information. Chatbot developers must implement robust security measures and data encryption to safeguard this sensitive information from unauthorized access, breaches, or misuse. Compliance with relevant data protection regulations, such as the General Data Protection Regulation (GDPR) in the European Union, is crucial to ensure that personal data is handled ethically and responsibly.

2.2 Technical Requirements

Successful web chat integrations supported by chatbots depend on meeting important technical requirements. The main goal is to establish smooth communication with the chatbot's backend service and generate a unique user ID for personalized interaction while enabling seamless messaging between the user and the chatbot. Such technical requirements include:

- Communication with the Chatbot Backend Service: The main technical requirement to run a web chat (text window) is to be able to communicate with the chatbot's backend service. This communication is required to send messages from users to the chatbot (or vice versa) and to create a unique user ID for each user communicating with the chatbot.
- Communication for loading the required HTML: To integrate the chatbot into a website, it is essential to include a specific line of code within the <head> section of the site's HTML. By including the necessary code, the chatbot widget is created and seamlessly integrated into the website, enabling users to access and interact with the chatbot functionality.
- System Requirements: The system requirements for the application include an Ubuntu 20.04 (LTS) Server or a more recent version. The application is designed to be compatible with this operating system to ensure optimal performance and functionality.

3 Platform Architecture

This section describes the proposed chatbot generator architecture that follows the partner-specific, the functional and the technical requirements presented in Sect. 2. The main components of the architecture are also described. It is worth mentioning, that the conceptual design of the architecture was based on similar ideas put forth in the literature (e.g., [5–7]), with the necessary addition of elements that automate the chatbot development process.

3.1 The PYTHIA Platform Architecture

Figure 1 illustrates in detail the platform architecture proposed in the context of the PYTHIA project.

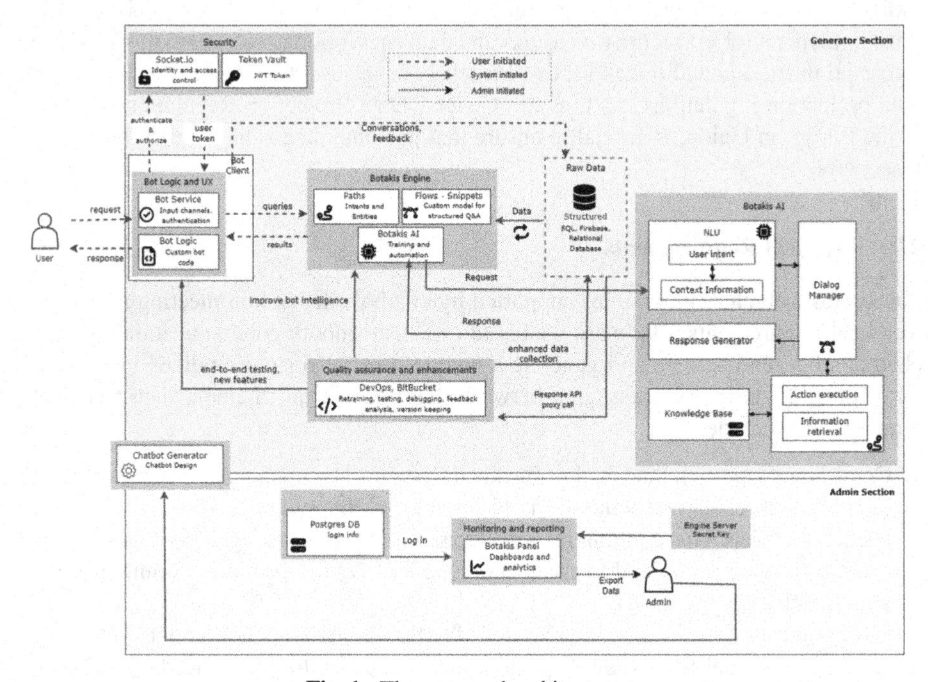

Fig. 1. The proposed architecture.

At the end-user side, native webchat can be added to the website with the use of a script in the HTML head section. The native script includes:

- The project's HTML (Hyper Text Language).
- The JavaScript that creates the build for the Webchat functions.
- The Vendor JavaScript file which includes the NPM Modules that have been used to build the project.
- The generated CSS (Cascading Styles Sheets) that contains all the styling of the Webchat.

- Additional CSS file which includes any styling changes that may occur.

The procedure to interact with the native webchat includes:

- User login to the service's website or the digital assistant's page.
- Through the script on the page, a call (GET) is made to the page to return the native webchat and it is displayed on the site.
- The script contains all the HTML of the project while adding all the generated files build to it.
- Upon the user choosing to initiate a chat via native webchat, a cookie is created and stored on the page which includes a JavaScript Object Notation (JSON) Web Token which is used to achieve user uniqueness. If previously created and not deleted, it is recognized by native webchat and returns the history of the user's conversation with the digital assistant.
- After the page call, native webchat contacts the digital assistant's engine (mentioned as Botakis engine in Fig. 1) to return the appropriate content.

User Message Flow

A series of steps are performed after the user interacts with the chatbot. These steps include:

- Authentication: Users start with authenticating their identity by making a post call to the designed host URL of the chatbot to verify that the user exists in the database and is eligible to access his/her chat history with the chatbot. In case the user is visiting the system for the first time, a new JSON web token will be created which will be stored in the user's browser cookies as well to identify the user the next time he/she enters the chatbot environment again.
- Digital assistant message: In case the user is contacting the digital assistant chatbot for the first time, a welcome message is sent by the engine. This message is not repeated the next time the same user enters the chatbot.
- User message: After authentication, the user will be able to send messages to the chatbot. The chatbot reads the message and internally forwards it to the recognition mechanism. This step captures intents (what the user wants to do) and entities (what the user is interested in). Chatbots understand user intent and then generate a response based on that outcome. For best results while using the chatbot and generated queries, the content is trained by content managers to identify the user's intent more accurately in subsequent messages.
- Response: At this point, the chatbot determines the best answer and sends it to the user. If the chatbot returns a fallback or suggestion message, it means it was unable to respond appropriately.
- Recording: When a user message is received or a response is sent, all chat actions should be logged anonymously in a log database along with performance metrics and common errors.
- Feedback: Gathering customer feedback and satisfaction scores is another important technique. The user is prompted to give the process a rating after the chatbot's final response. Feedback aids in the resolution of issues with natural language comprehension and enhances the precision of chatbot responses.

In Sects. 3.2–3.7 we will introduce and elaborate the main components of the proposed Chatbot architecture. These include the Chatbot Generator, Natural-Language Understanding, Dialogue Management, Information Retrieval System, Backend and Response Generation.

3.2 Chatbot Generator

The main novelty of the proposed architecture is the "chatbot generator". The chatbot generator is a tool that automates the process of creating chatbots without requiring extensive coding knowledge or technical expertise. It allows users to build custom chatbots for various purposes, such as customer service, lead generation, or informational support, using a user-friendly interface and predefined templates.

The chatbot generator offers a range of features and functionalities that users can customize to tailor the chatbot to their specific needs. These features include:

- Pre-built Templates: Chatbot generator provides pre-built templates for different industries or use cases, making it easier for users to get started quickly.
- Drag-and-Drop Interface: Users can design the conversation flow and structure of the chatbot by using a simple drag-and-drop interface, without writing complex code.
- Natural Language Processing (NLP): The chatbot generator includes NLP capabilities, enabling chatbots to understand and respond to user inputs more intelligently. A combination of statistical and rule-based NLP techniques is used to improve the conversational capabilities of chatbots. The chatbot builder understands the syntactic structure of the user's input using a rule-based grammar, allowing it to recognize patterns and extract meaningful elements from the user.
- Multi-Channel Integration: Chatbot generators allow the chatbot to be integrated with various communication channels, such as websites, messaging apps, or social media platforms.
- Analytics and Reporting: The chatbot generator offers built-in analytics and reporting features to track chatbot performance and user interactions.
- Data Collection and User Management: The Chatbot generator can collect user data and manage user profiles to provide personalized experiences.

3.3 Natural-Language Understanding

Natural Language Understanding (NLU) is a crucial component of a chatbot due to enabling it to comprehend and interpret user input effectively. By leveraging NLU technology, the system can accurately extract appropriate information from user queries and create a comprehensive representation of their underlying meaning [8]. NLU accomplishes this through three fundamental tasks: dialogue act classification, intent classification, and information extraction [9]. These processes collectively empower the system to better engage with users and provide more relevant and contextually appropriate responses. In detail:

- Classification of dialog operations involve identifying the purpose of user input, or more precisely, connecting user input with different types of dialog operations. This input can be categorized either as a question, remark, suggestion, or other types of interactive acts. Knowing the interactivity being performed is important to better understand user requests and determine appropriate responses [10].

- Intent classification identifies the user's primary goals. Intent varies largely by industry. For example, industry inquiries such as food orders, hotel reservations, and weather forecasts. An agent's intent in the hotel booking industry may be to book, cancel, or modify a reservation, and an agent's intent in the grocery ordering industry may be similar, to submit, query, or modify an order.
- Extracting information by survey is the final step in NLU. The chatbot further extracts the necessary details and combines them with interaction actions and intents to fully understand the user's request.

To fully understand user input in the pilot version, a combination of statistical and neural methods was used. Statistical methods such as probabilistic modeling have been used for tasks such as conversational behavior classification, where the system determines the intent of a user's input by associating it with various types of conversational activities. Different dialogues such as requests for information, comments or suggestions. On the other hand, to classify user intent, neural methods, such as deep learning algorithms, have been used.

3.4 Dialogue Management

The Dialogue Management (DM) subsystem processes information from other subsystems, controls and updates the context of the conversation, and regulates chatbot behavior. Designing a robust DM strategy is challenging because it is difficult to predict which system behaviors will lead to high user satisfaction. Here, we discuss two common DM design problems: interaction strategy and confirmation strategy selection [11].

An interaction strategy determines who controls the conversation. Conversations can be user-driven, system-driven, or mixed-driven. In a user-driven interaction, the user takes the initiative, and the system merely executes the user's requests and instructions. In a system-driven dialogue, the system assumes control and merely complies with its demands. Both the user and the system can take over if there is mixed initiative in the interaction. [11].

As far as the confirmation strategies are concerned, the are two types of strategies: explicit confirmation and implicit confirmation. When using the explicit confirmation strategy, the system confirms understanding by asking the user another question. In the implicit acknowledgment strategy, the system includes some of the received information in the response [11].

For the pilot testing purposes, the mixed-driven interaction strategy and the explicit confirmation strategy have been used. However, we plan to extend the functionality by developing the remaining strategies and allowing admins to select their preferred strategy.

3.5 Information Retrieval System

The Information Retrieval System (IRS) has two main purposes. The first one is to achieve quality and effective results. The second one is about how quickly the requested information is retrieved; it is about efficiency. In order to send a request to the system, the user uses an interface, for example a browser, connected to the IRS via the HTTP communication protocol. Specifically, user requests through the interface are processed using NLU mechanisms and converted into a format suitable for IRS.

However, the following problems are often observed: Some of the answers returned by the IRS are not very relevant to the query. For this reason, relevance feedback techniques were developed and used to improve the quality of results. Using this method gives the user the opportunity to select some of the IRS's suggested answers that are more relevant than others. Therefore, the system can redefine the answer based on user selection.

3.6 Backend

The chatbot gets the information it needs from the backend to perform the required task and forwards the message to the dialog management and response generation subsystems [12]. A rule-based chatbot requires a knowledge base (KB) to store manually created rules [13]. The chatbot retrieves previous conversations using a relational database (RDB). By considering previous information, the chatbot can communicate more consistently, accurately, and reliably [14]. KB development is done by humans and can be time consuming and difficult. To overcome this difficulty, we have automatically created a new KB from the partners' existing KB [15].

3.7 Response Generation

Once the appropriate information is obtained, the next step for the dialog system is to determine what the response will be and how to best present it. The Response Generation subsystem is responsible for generating responses in a format that the user can understand. The specific subsystem includes five processing steps: signal analysis, data interpretation, document design, micro-design, and implementation.

4 Conclusions

This paper presented in detail the chatbot generator architecture developed in the framework of the PYTHIA project. The overall architecture was based on the requirements and specifications defined by two of the project partners, i.e., the National Bank of Greece and the General Secretariat for Information Systems & Digital Governance. The main components of the architecture were presented, and their functionality was analyzed. Overall, the chatbot generator simplifies the process of creating and deploying chatbots, making it accessible to a broader audience and enabling businesses and individuals to implement chatbot solutions efficiently without significant development resources.

The requirements and specifications described in this paper will guide the full implementation of the overall system. Then, the chatbot generator will undergo a thorough pilot testing process with multiple end-users across diverse applications, and its overall performance and efficiency will be evaluated. As stressed in [16], the application of the foreseen platform must be evaluated through a set of Key Performance Indicators

(KPIs), aiming to measure its usefulness and ease of use. The focus of the evaluation would be to assess, for various types of contexts and associated questions (of varying complexity, ambiguity and uncertainty), to what extent it can provide useful and relevant information as well as support for conducting relevant transactions.

Acknowledgements. This research has been co-financed by the European Union and Greek national funds through the Operational Program "Competitiveness, Entrepreneurship and Innovation", under the call "RESEARCH - CREATE - INNOVATE (2nd Cycle)" (project id: T2EDK-01921).

References

1. Doshi, J.: Chatbot user interface for customer relationship management using NLP models. In: International Conference on Artificial Intelligence and Machine Vision (AIMV), pp. 1–4, Gandhinagar, India (2021)
2. Karacapilidis, N., Papadias, D.: A Computational approach for argumentative discourse in multi-agent decision making environments. AI Commun. J. **11**(1), 21–33 (1998)
3. Online Website. https://onlim.com/en/top-requirements-for-chatbot-software/
4. Lau, L., Yang-Turner, F., Karacapilidis, N.: Requirements for big data analytics supporting decision making: a sensemaking perspective. In: Karacapilidis, N. (eds.) Mastering Data-Intensive Collaboration and Decision Making. Studies in Big Data, LNCS, vol. 5, pp. 49–70. Springer, Cham (2014). https://doi.org/10.1007/978-3-319-02612-1_3
5. Avishka, W., Kuhaneswaran, B., Gunasinghe, H.: A novel conceptual chatbot architecture for the Sinhala language – a case study on food ordering scenario. In: Proceedings of the 2022 2nd International Conference on Advanced Research in Computing (ICARC), pp. 254–259. Belihuloya, Sri Lanka (2022)
6. Maher, S.K., Bhable, S.G., Lahase, A.R., Nimbhore, S.S.: AI and deep learning-driven chatbots: a comprehensive analysis and application trends. In: Proceedings of 2022 6th International Conference on Intelligent Computing and Control Systems (ICICCS), pp. 994–998, Madurai, India (2022)
7. Van Thin, D., Quang, T.M., Toan, P.M., Thien, V.M., Hung, L.M., Quan, T.T.: A human-like interactive chatbot framework for Vietnamese banking domain. In: Proceedings of 2022 9th NAFOSTED Conference on Information and Computer Science (NICS), pp. 41–46, Vietnam (2022)
8. Reiter, E.: An architecture for data-to-text systems. In: Proceedings of the ENLG-2007 11th European Workshop on Natural Language Generation, pp. 97–104. Saarbrücken, Germany (2007)
9. Firdaus, M., Golchha, H., Ekbal, A., Bhattacharyya, P.: A deep multi-task model for dialogue act classification, intent detection and slot filling. Cogn. Comput. **13**, 626–645 (2020)
10. Tur, G., Deng, L.: Intent determination and spoken utterance classification. In: Tur, G., de Mori R. (eds.) Spoken Language Understanding: Systems for Extracting Semantic Information from Speech, pp. 93–118. Wiley, Chichester (2011)
11. McTear, M., Callejas, Z., Griol, D.: The Conversational Interface Talking to Smart Devices. Springer, Cham (2018). https://doi.org/10.1007/978-3-319-32967-3
12. Adamopoulou, E., Moussiades, L.: Chatbots: history, technology, and applications. Mach. Learn. Appl. **2**, 100006 (2020)
13. Khanna A., Pandey B., Vashishta K., Kalia K., Bhale P., Das T.: A study of today's A.I. through Chatbot and rediscovery of machine intelligence. Int. J. u- e-Serv. Sci. Technol. **8**, 277–284 (2015)

14. Abdul-Kader, S.A., Woods, J.: Survey on Chatbot design techniques in speech conversation systems. Int. J. Adv. Comput. Sci. Appl. **6**, 72–80 (2015)
15. Arsovski, S., Osipyan, H., Oladele, M.I., Cheok, A.D.: Automatic knowledge extraction of any Chatbot from conversation. Expert Syst. Appl. **137**, 343–348 (2019)
16. Androutsopoulou, A., Karacapilidis, N., Loukis, E., Charalabidis, Y.: Transforming the communication between citizens and government through AI-guided chatbots. Gov. Inf. Q. **36**(2), 358–367 (2019)

A Structured Analysis of Domain-Specific Linked Open Vocabularies (LOV): Indicators for Interoperability and Reusability

Maria Ioanna Maratsi(✉), Charalampos Alexopoulos, and Yannis Charalabidis

University of the Aegean, University Hill, 81100 Mytilene, Greece
{ioanna.m,alexop,yannisx}@aegean.gr

Abstract. The concept of linking data, in its core, is one of the cornerstones of the Semantic Web, and the design principles of Linked Data aim to establish standardization for knowledge representation. The rationale behind this research is to analyze Linked Open Vocabularies (LOV) to elicit useful insights regarding the status and potential of interoperability of existing vocabularies, organized in thematic areas (domains). The goal behind the latter is to identify how semantic interoperability of vocabularies (and datasets which are described by the relevant knowledge schema) could be improved, while the expected outcome is an enhanced overview of the existing knowledge representation in the LOV ecosystem, with emphasis put on domain-specific vocabularies and their reuse. In order to gain a general perspective of the domain-specific vocabularies that currently exist, a mapping between the 13 European Data Portal (EDP) thematic categories and the categorization of LOV was initially performed. Afterwards, an analysis framework to capture the information from this desk-based research was included. The analysis framework included: i) the connections of the identified (from the initial mapping) domain-specific vocabularies to core vocabularies in the LOV ecosystem, ii) the connections of domain-specific vocabularies to vocabularies within the same domain, and iii) the connections of domain-specific vocabularies to the other domains. The previous three dimensions were then combined to create an initial assessment on how reusable and thus, interoperable (and beyond) the vocabularies under analysis were, acting as an indicator for domain interoperability readiness.

Keywords: Open data · Linked Data · Vocabularies · Linked Open Vocabularies · LOV · Open Government data · knowledge graphs · semantics · semantic interoperability · metadata · reusability

1 Introduction

The concept of Linked Data, in its core, is one of the cornerstones of the Semantic Web, creating links between datasets with the purpose of enhancing knowledge integration among disparate sources, following a number of design principles. The design principles of Linked Data aim to establish standardization for data representation and propagate the usage of hyperlinks among the different data sources [1]. The envisioned transition from

M. Papadaki et al. (Eds.): EMCIS 2023, LNBIP 501, pp. 135–152, 2024.
https://doi.org/10.1007/978-3-031-56478-9_10

the "Web of linked documents" to the "Web of linked data" involves the availability of data (and metadata) in machine-readable and machine-processable format, its connection using Uniform Resource Identifiers (URIs), and the ability for users to search for this data using HTTP requests [2]. In parallel, measuring the potential reuse of vocabularies across different but related knowledge domains is a growing field of research and pivotal to ensuring semantic interoperability of data. The importance of standardization and reuse has been emphasized by the World Wide Web Consortium (W3C) [3] in the Best Practices regarding data vocabularies, specifically, Best Practice 15 mentioning the contribution of vocabulary reuse on interoperability and the emergence of Linked Open Vocabularies (LOV) as a "testbed" for research in this regard. The Linked Open Vocabularies (LOV) objective is to provide an ecosystem of vocabularies, by making explicit the ways they are linked to each other and by providing metrics on how they are used in the linked data cloud, but also to improve their usability, visibility, and quality [4]. LOV is an increasingly useful tool for both users who might need to find the appropriate vocabularies to describe their data or receive an overview of the existing vocabularies, and vocabulary managers who are kept informed on best practices regarding the data they publish but also feedback on their published material, its quality, and usability. The LOV registry provides a user interface with search options, a SPARQL endpoint, and an API REST, allowing for various ways to access the data available, having as raison d'être the fact that one of the main challenges to the deployment of linked data is the difficulty to determine which vocabularies to use in order to describe semantics [5]. In other words, to identify the vocabulary that can provide the "semantic glue" which can add value to the data and its links to other data.

Following the same paradigm, the rationale behind this research is to analyze Linked Open Vocabularies, by looking into the existing, most basic, domain-specific vocabularies and ontologies per thematic category, and elicit useful insights regarding the usability and reusability of the existing vocabularies per domain, and to make an assessment of potential inter-domain usability (or interoperability) of the vocabularies under analysis. The goal behind the latter is to identify how semantic interoperability of vocabularies (and datasets which are described by the relevant knowledge schema) could be enhanced and the expected outcome of this study is an overview of the existing knowledge representation in the LOV ecosystem, with emphasis put on domain-specific vocabularies and potential of expanding from intra-domain to inter-domain reusability for knowledge representation. The study is structured as follows: In the Background Section, the basic concepts and relations relevant to this research are introduced, the Methodology Section presents the proposed methodological approach and steps conducted, the Results Section provides an overview of the analysis outcome, while the Discussion Section provides the contextualization and explanation of the latter. Finally, the last Section of this study includes the conclusion, possible future directions, and the limitations of the presented research.

2 Background

The structure of data and information is achieved through conceptual constructs such as vocabularies, taxonomies, ontologies, and knowledge graphs. While in the context of Linked Open Vocabularies, the notions of ontology and vocabulary are mostly used interchangeably, for the sake of completeness it is noteworthy to set the background basis for the central concept of this piece of research, the Linked Open Data and vocabularies, and additively make a brief but essential reference to the related concepts, including their similarities and differentiations.

2.1 Linked (Open) Data

Commonly, Linked Data is available as Linked Open Data and its content is structured in a way which allows for direct machine-readability and discovery on the Web [6]. At the same time, publishing Open Data has attracted the interest of the research community, however, the data being published is generally facing various problems, among others, low data quality, something which, to a great extent, discourages contribution and reuse [7]. Generally, the term "Open Data" refers to data and information beyond just governmental institutions and includes those from other relevant stakeholder groups such as business/industry, citizens, NPOs and NGOs, science or education, and must comply with a set of principles, such as completeness, timeliness, accessibility, machine-processability, be non-proprietary and license-free, and more [8]. Moreover, interoperability and context are key aspects of the potential benefits of Open (Government) Data, and here the role of Linked Open Data (LOD) becomes critical. Bauer [8] pinpointed the significance of the LOD transition for both data publishing and consuming and demonstrated the power of the envisioned concept and its applications.

Similarly, various approaches have been proposed, utilizing Linked Open Data for numerous purposes and in a variety of contexts. Escobar [7] designed a framework, including a Linked Open Data assessment step, with the purpose of facilitating the publication of statistical data from public repositories using RDF and SPARQL, and allowing for reduced complexity to the end user when interacting with the data, based on the RDF Data Cube Vocabulary. Fibriani [9] used Linked Open Spatial Data to assess land suitability for agricultural activities based on parameters available from spatial data, such as soil type, texture, landform, rock condition, water absorption, and more. The research conducted by Fibriani [9] also resulted in the contribution of the combined data, which the authors used, as a reusable source for land suitability, in the form of LOD on the Semantic Web (connecting the sources using URIs to the vocabulary data used for this domain). In the Cultural Heritage sector, the Europeana service [10] aggregates cultural data in Europe and makes it available as Linked Open Data by providing semantic enrichment and abiding by the requirements for Linked Open Vocabularies. The Linked Open Data importance is also emphasized by the Getty Conservation Institute [11], showcasing the Linked Art paradigm, a homogenized model to provide the "semantic glue" and enable interoperability and reusability for data publishing among cultural heritage institutions. In a slightly different context, Oh [12] deployed Linked Open Vocabularies to enhance library data and make access to various open library datasets easier, but also search for further potential utilization of LOV to improve Linked Library Data quality and reusability. Apart from domain-specific applications, LOV has also been used to facilitate data modeling. Schaible [13] proposed a support method for ontology engineers, namely LOVER, which gives recommendations to the Linked Data modeler, based on existing, active vocabularies and best practices. The challenge of finding suitable ontologies for a given domain of application is also mentioned by Vila-Suero [14], who made use of the Linked Open Data cloud to compare ontology similarity by applying topic modeling techniques and extracting relevant (to the desired domain) topics. The Linked Open Vocabularies (LOV) community-based ecosystem attracts attention due to its semantically rich potential, as well as the perspective of knowledge reuse. Vandenbussche [15] provided an overview of the LOV initiative and presented the multifaceted role the adoption and integration of the LOV catalogue can play in various aspects of data description, quality, publishing, and reuse.

2.2 Domain Knowledge Representation

In general, the relationships between data, as well as the data itself, are represented in the form of a defined structure, such as a knowledge graph, ontology, or vocabulary. On many occasions these concepts are used interchangeably, and although very closely connected, they are subject to several differences. Jia [16] performed conceptual analysis to distinguish these concepts and analyze their composition on a deeper level but also to show how they evolved and relate to each other. According to Jia [16], vocabularies mainly focus on the initial formation of knowledge, putting extra weight on conceptual, cognitive completeness, while linked data relates knowledge directly to the real world by using formal, semantic representations for enhanced logical expressibility [16].

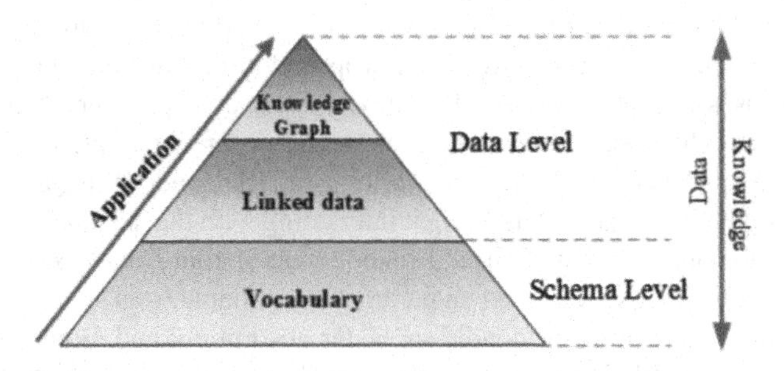

Fig. 1. The evolution of vocabularies, linked data, and knowledge graphs (Jia, 2020) [16]

Finally, knowledge graphs extend linked data and use vocabularies for their data schema construction and semantic infrastructure, thus transforming data into knowledge (Fig. 1).

According to Pauwels [17] an ontology is "a formal, explicit specification of a shared conceptualization", linked data refers to best practices for publishing data on the Web in a structured way, and a knowledge graph puts data in context via links and semantic metadata, thus providing a framework for integration, unification, analytics, and sharing [17]. Pan [18] mentions that a knowledge graph is a set of typed entities and relationships which, in their turn, are defined in schemata (ontologies), and that those defined types are the vocabularies. The Resource Description Framework (RDF) standard is a means to represent knowledge graphs [18]. Tiwari [19] conducted a characterization of the existing knowledge graphs through a systematic review of the state-of-the-art, and, among others, their most significant applications (semantic searching, domain-specific applications, knowledge sharing, recommender systems etc.), which are shown in Fig. 2.

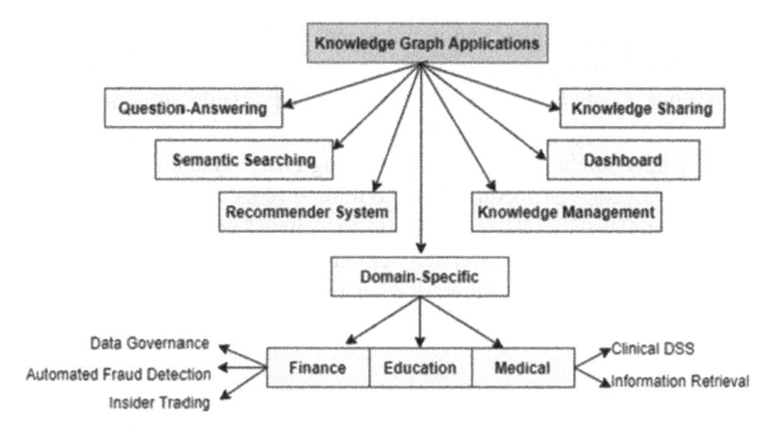

Fig. 2. Applications of knowledge graphs by Tiwari et al., (2021) [19]

As described earlier, knowledge graphs and other knowledge representation structures, might have applications on general knowledge structuring or domain-specific contexts, such as libraries and digital humanities [20], geographical and geoscience data [21] and more, however, each knowledge domain bears its own challenging specificities (e.g., the heterogeneity of research outputs in the humanities domain, or the organized work around a specific individual or group [20]), thus making it impossible to imagine a universal or global knowledge graph to fit all. Instead, it becomes clear that the connections and possible "bridges" among different domains and data, are essential to enable re-use of information and knowledge. Linked Data principles pose an approach towards the growth of publicly accessible interlinked graph-structured knowledge bases [22], and for this reason, they are of pivotal interest for the presented study.

3 Methodology

In this Section, the method followed for the analysis is described. The main method used is desk-based research, and an analysis framework which addresses the most important aspects of this study.

The Linked Open Vocabularies (LOV) registry provides over 800 vocabularies (in the context of LOV a vocabulary is synonymous to an ontology), based on a number of quality requirements regarding publication, formats, metadata, versioning policy etc., and whose terms are connected (linked) inside a vocabulary but also across vocabularies [23]. The LOV User Interface allows for navigation through the available vocabularies using a search bar, an alphabetical list of the vocabularies, a menu with vocabularies organized in themes (named "Tags"), and a Language menu, which organizes the available vocabularies per language.

The vocabularies are visually presented using a coloured schema, different circle sizes and a number of incoming and outgoing links, the former indicating information about the purpose, quality or status of the vocabulary, the second indicating the vocabulary's popularity (although it cannot be considered a measure due to the fact that the size is also in relevance to other vocabularies; small circles could still indicate a popular vocabulary in comparison to other non-popular ones [23]), while the latter directly refers to the relationships among the vocabularies. Incoming links show the dependencies or references of other vocabularies to a specific vocabulary, while outgoing links demonstrate the dependencies and references of this vocabulary towards the other vocabularies it is linked to. Understandably, since LOV is using an ecosystemic approach of interdependencies, the number of incoming and outgoing links of a vocabulary is a strong indicator of its popularity, of its proximity to the related vocabularies, but also of its potential usage in the context of its related domains. An instance showing the described visualization of links and colours of relations, is presented in Fig. 1, taking the example of one of the most basic vocabularies, FOAF (Friend of a Friend) (Fig. 3).

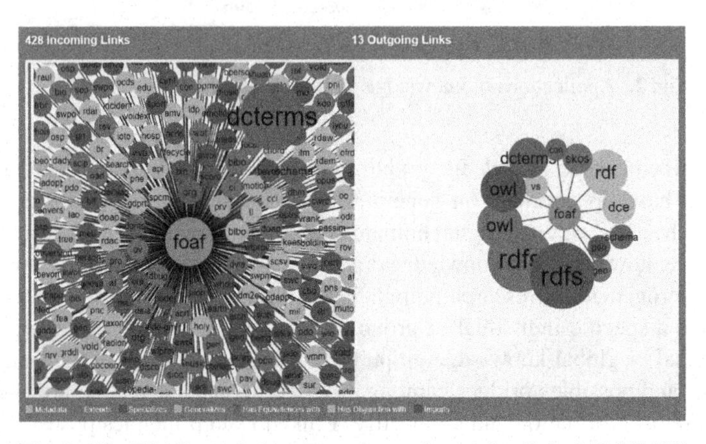

Fig. 3. The FOAF vocabulary incoming and outgoing links in LOV [23]

FOAF has 428 incoming links, and 13 outgoing links, the former demonstrating the dependencies of other vocabularies on FOAF (which are understandably a lot, considering the general purpose of this vocabulary), and the latter showing the dependencies of FOAF on other basic vocabularies, such as SKOS (Simple Knowledge Organization System), RDFS (The RDF Schema Vocabulary), GEO (WGS84 Geo Positioning), and DCE (Dublin Core Metadata Element Set). Regarding the outgoing links, FOAF is specialized by RDFS, OWL, and GEO, extended by RDF, SKOS, OWL, and GEO, and its metadata is described using RDF, VS, and DCE.

In the context of this study, the main goal is to make an initial assessment on how re-usable the information of a vocabulary is, and which vocabularies someone should choose when deciding upon the selection of an appropriate domain-specific vocabulary to make it more valuable in terms of reusability and interoperability. In this light, the following dimensions of the LOV ecosystem are taken into account; basic vocabularies in LOV, the most basic, domain-specific vocabularies organized in thematic areas, their connection to the basic (core) vocabularies, the links among vocabularies which belong to the same theme/domain, and the links among vocabularies of different themes/domains.

Initially, in order to gain a better understanding of the domain-specific vocabularies that exist, a mapping between the 13 European Data Portal thematic categories and the categorization of LOV was made. The concept of mapping in order to retrieve the most basic vocabularies per domain is shown in Table 1. For each EDP theme, the equivalent LOV Tag and keywords were used.

Table 1. The mapping steps between the EDP thematic categories and LOV to identify basic domain-specific vocabularies.

European Data Portal Theme	Linked Open Vocabularies Tags/Keywords
Economy and Finance	Industry, eBusiness, Services tag + keywords "economy", "finance"
Education, Culture and Sport	Academy tag + keywords "culture", "sport"
Energy	Energy tag
Environment	Environment tag
Government and Public Sector	Government tag
Health	Health tag + keywords "medical", "medicine"

The second step included in this study is to study the links among the previously mentioned vocabularies. First of all, the pink, outgoing links for each vocabulary, which denote the links to vocabularies which the vocabulary under analysis is using to describe its metadata, is included. This corresponds to the column "Core Vocabularies Direct Links" of the analysis framework, as these links show how many domain vocabularies are using the core ones. The column "Inter-Domain Linked Vocabularies" shows how many vocabularies from other domains/themes are using the vocabulary under investigation, while "Intra-domain Linked Vocabularies" are the number of connected vocabularies for each examined vocabulary within the same domain/theme. The links to core vocabularies are excluded from the last two columns as they are mentioned separately and do not provide further information for this point.

So, the connections included in the analysis framework include i) the connections of domain-specific vocabularies to core vocabularies, ii) the connections of domain-specific vocabularies to vocabularies of the same domain, and iii) the connections of domain-specific vocabularies to other domains (themes). The previous three columns are then advised to produce the last column of the analysis framework "Domain Interoperability Readiness", which is an initial assessment (based on the aforementioned) on how reusable and thus, domain interoperable (and beyond) the examined vocabularies are. More specifically, the "Domain Interoperability Readiness" consists of two indicators. The first is the sum of the (unique) domain vocabularies which are linked internally (intra-) or externally (inter-) to other LOV divided by the total amount of vocabularies per theme/domain (the number indicated in the first column). The second is the sum of links to other vocabularies and the core vocabularies (inter-, intra-, and core).

The methodological approach of this study is illustrated in Fig. 4, emphasizing that the "Domain Interoperability Readiness" column is the result of indications and calculations based on three elements of the analysis framework, making it the biggest focus of this study due to its strong relationship with interoperability aspects. The results obtained following the aforementioned methodology are presented in the next Section.

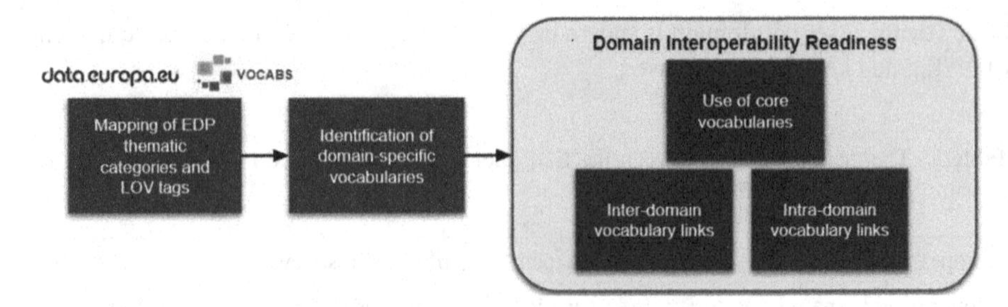

Fig. 4. The methodological approach for the Linked Open Vocabularies (LOV) analysis

4 Results

4.1 Domain-Specific Vocabularies in LOV per Theme

As described earlier in the methodology Section, in order to acquire the most basic, existing domain-specific vocabularies in LOV, an initial mapping to the EDP's thematic categories was added as a prior step. The obtained vocabularies are listed and organized in the corresponding thematic category. It is worth mentioning again that among the vocabularies acquired per thematic category, many popular vocabularies (e.g., RDF, RDFS, XSD etc.) are exempt from inclusion in the second column of the analysis framework due to their general-purpose usage and connection to the majority of vocabularies in LOV, which itself alone is not offering a lot of new information. The list of included vocabularies is not exhaustive; however, the most representative and well-connected (compared to the rest of the links) vocabularies are considered. When no further links are mentioned under the column "Links", it is implied that the vocabulary in word is only linked to the basic vocabularies in LOV, or none.

In this Section, the results which were directly relevant to the focus of this research are presented. A synopsis of the results of the analysis framework previously described in the Methodology Section is shown in Table 2.

The vocabularies obtained by applying the described method provide an initial overview of the existing, domain-specific vocabularies (although, understandably, a few of them might be reused in other domains and the list is not exhaustive but limited to the results which this specific method retrieved) and they are shown under the column "LOV Vocabularies" of Table 2. For Economy and Finance, 45 vocabularies/ontologies were identified, including vocabularies such as the AMLO project ontology that extends the Financial Industry Business Ontology (fibo) with some concepts to describe the Anti Money Laundering (aml) knowledge and facts (amlo-core), the Ontology for Consumer Electronics Products and Services (ceo), the European Union Company Mobility Ontology (cbcm), the Corporate Financial Reports and Loans Ontology (cfrl), the Ontology for the description of customizable products (cold), The euBusinessGraph ontology (ebg), the vocabulary for describing tickets for events, transportation, or points of interest for e-commerce (tio), the ontology to describe people and requests for timebank (tb), and the Vehicle Sales Ontology for Semantic Web-based E-Commerce (vso). For Education, Culture, and Sport 15 vocabularies were identified, including the Academic Institution Internal Structure Ontology (aiiso), the Scientific People Ontology (scip), the Scholarly

Table 2. The results of the conducted analysis framework in LOV

Theme/ Domain	LOV Vocabularies	Core Vocabularies Direct Links	Inter-Domain Linked Vocabularies	Intra-domain Linked Vocabularies	Domain Interoperability Readiness
Economy and Finance **(45)**	amlo-core, dk, dot, gleif-elf, ofrd, rami, saref, sto, traffic, veo, vvo, bcom, bto, ccp, cdc, cocoon, datex, dio, docso, dpn, esco, ioto, uiot, ceo, coo, core, dfc-b, acco, cbcm, cfrl, cold, ebg, tb, tio, uco, vso, bot, arco, lifecycle, otl, sbeo, seasto, ssso, td, trao	dcterms/dce (44) vann (30) void (1) rdf/rdfs (45) skos (6) cc (14) foaf (6) gr (4) vs (8) voaf (3) schema (3) gleif-base (1) prov (2) **# core vocabs (13)**	dk (2) gleif-elf (1) saref (1) veo (1) vvo (1) bto (1) ccp (1) cdc (1) esco (1) uiot (2) cbcm (1) vso (1) sbeo (5) seasto (1) trao (1) **# links (21)**	sto (1) veo (1) vvo (3) coo (1) uco (1) vso (3) sbeo (1) ssso (1) **# links (12)**	# vocabs with links (÷) by the total # domain vocabs: **(0.4)** Σ of the core, inter- and intra-domain links: **(46)**
Education, Culture and Sport **(15)**	aiiso, crsw, edu, edupro, frapo, istex, m4i, meb, scip, scoro, seo, swc, swrc, teach, vivo	dcterms/dce (12) vann (5) prv (1) rdf/rdfs (15) skos (3) foaf (3) vs (2) bibo (2) obo (1) dbpedia-owl (1) soap (1) sioc (1) **# core vocabs (12)**	swc (1) swrc (2) **# links (3)**	swc (1) swrc (2) **# links (3)**	# vocabs with links (÷) by the total # domain vocabs: **(0.1333)** Σ of the core, inter- and intra-domain links: **(18)**
Energy **(7)**	reegle, eepsa, s4ener, opm, seas, topo, omg	dcterms/dce (6) vann (5) rdf/rdfs (5) bibo (1) cc (3) **# core vocabs (5)**	reegle (1) eepsa (2) opm (1) seas (1) omg (1) **# links (6)**	opm (1) seas (1) omg (1) **# links (3)**	# vocabs with links (÷) by the total # domain vocabs: **(0.714)** Σ of the core, inter- and intra-domain links: **(14)**
Environment **(21)**	aws, bimerr-op, BRK, cff, cfp, dogont, earth, eepsa, geosp, hw, moac, opm, reegle, seasb, seasbo, seasfo, seasto, shw, sosa, w3c-ssn, veo	dcterms/dce (17) vann (10) rdf/rdfs (18) cc (4) skos (2) foaf (1) vs (2) bibo (1) ucum (1) schema (1) soap (1) qu (1) prv (1) **# core vocabs (13)**	aws (1) eepsa (2) opm (2) reegle (1) seasbo (1) seasto (1) sosa (3) **# links (11)**	**# links (0)**	# vocabs with links (÷) by the total # domain vocabs: **(0.3333)** Σ of the core, inter- and intra-domain links: **(24)**

(continued)

Table 2. (*continued*)

Government and Public Sector **(11)**	gc, cgov, gd, ctorg, drm, ei2a, fea, pay, odapp, wai, oslo	dcterms/dce (11) vann (6) rdf/rdfs (10) skos (3) foaf (5) vs (1) void (1) adms (2) qb (1) xhv (1) voaf (1) spin (1) wdrs (1) **# core vocabs (13)**	ctorg (1) wai (2) oslo (2) **# links (5)**	ctorg (1) drm (1) fea (1) **# links (3)**	# vocabs with links (÷) by the total # domain vocabs: **(0.454)** Σ of the core, inter- and intra-domain links: **(21)**
Health **(10)**	tmo, obo, demlab, eeo, dicom, cochrane, hosp, biotop, medred, aos	dcterms/dce (10) vann (5) rdf/rdfs (10) skos (1) foaf (2) vs (1) ru (1) obo (2) spin (1) schema (1) swrc (1) **# core vocabs (11)**	obo (4) aos (2) **# links (6)**	tmo (1) obo (2) **# links (3)**	# vocabs with links (÷) by the total # domain vocabs: **(0.3)** Σ of the core, inter- and intra-domain links: **(20)**
Agriculture, Fisheries, Forestry and Foods **(7)**	s4agri, aws, qu, food, spfood, fowl, dfc	dcterms/dce (5) vann (2) rdf/rdfs (6) adms (1) foaf (1) **# core vocabs (5)**	aws (1) qu (1)	**# links (0)**	# vocabs with links (÷) by the total # domain vocabs: **(0.285)** Σ of the core, inter- and intra-domain links: **(7)**
Transport **(10)**	atd, atm, eqp, lc, mv, nas, rto, uiote, trao, uiot	dcterms/dce (5) vann (6) rdf/rdfs (10) skos (2) foaf (1) vs (2) cc (2) void (1) bibo (1) **# core vocabs (9)**	uiote (2) uiot (1) **# links (3)**	atd (2) atm (3) eqp (1) nas (3) uiote (1) trao (1) uiot (1) **# links (12)**	# vocabs with links (÷) by the total # domain vocabs: **(0.7)** Σ of the core, inter- and intra-domain links: **(24)**
Science and Technology **(8)**	modsci, lsc, opus, npg, ecrm, nlon, opmo, cdc	dcterms/dce (7) vann (4) rdf/rdfs (8) skos (3) foaf (1) vs (2) cc (2) void (1) schema (1) prv (1) **# core vocabs (10)**	opus (1) cdc (1) **# links (2)**	**# links (0)**	# vocabs with links (÷) by the total # domain vocabs: **(0.25)** Σ of the core, inter- and intra-domain links: **(12)**

(*continued*)

Table 2. (*continued*)

Regions and Cities **(8)**	ccr, ngeo, faldo, gci, s4ity, km4c, turismo, open311	dcterms/dce (5) vann (2) rdf/rdfs (7) foaf (2) vs (1) spin (1) **# core vocabs (6)**	ccr (1) **# links (1)**	**# links (0)**	# vocabs with links (÷) by the total # domain vocabs: **(0.125)** Σ of the core, inter- and intra-domain links: **(7)**
Population and Society **(29)**	arco, akt, cdesc, ddesc, esco, gdprov, gov, oo, oslo, seasbo, sioc, swc, agrelon, aos, whois, gleif-base, bio, bperson, cco, con, emotion, foaf, gen, ic, opo, plink, rel, sor, wai	dcterms/dce (27) vann (15) rdf/rdfs (29) foaf (5) vs (6) cc (5) skos (5) swrc (2) sioc (1) soap (1) ov (2) spin (1) adms (1) wdrs (1) xhv (1) ctlog (1) voaf (2) ont (1) **# core vocabs (18)**	arco (1) akt (1) cdesc (1) esco (1) gov (1) sioc (1) gleif-base (3) aos (3) swc (2) oslo (2) seasbo (1) foaf (11) opo (1) wai (2) **# links (31)**	arco (1) bio (2) gen (1) opo (1) plink (1) rel (2) **# links (8)**	# vocabs with links (÷) by the total # domain vocabs: **(0.62)** Σ of the core, inter- and intra-domain links: **(57)**
Justice, Legal System and Public Safety **(13)**	eli, cbcm, gov, rov, gleif-L2, gleif-L1, loc, gleif-ra, gleif-repex, gleif-geo, gleif-base, gleif-elf, ctorg	dcterms/dce (12) vann (10) rdf/rdfs (11) voaf (1) wdrs (1) foaf (1) skos (7) vs (1) gleif-base (5) prov (1) cc (2) **# core vocabs (11)**	cbcm (1) gov (1) rov (1) gleif-base (2) gleif-elf (1) ctorg (1) **# links (7)**	gleif-L1 (3) loc (1) gleif-repex (1) gleif-base (8) **# links (13)**	# vocabs with links (÷) by the total # domain vocabs: **(0.692)** Σ of the core, inter- and intra-domain links: **(31)**
International Issues **(5)**	ocds, ids, ipo, rico, ic	dcterms/dce (5) vann (4) rdf/rdfs (5) foaf (2) cc (2) skos (2) adms (1) void (1) **# core vocabs (8)**	ic (1) **# links (1)**	**# links (0)**	# vocabs with links (÷) by the total # domain vocabs: **(0.2)** Σ of the core, inter- and intra-domain links: **(9)**

Contributions and Roles Ontology (scoro), the Education Ontology (edu), the Semantic Web Conference Ontology for describing academic conferences (swc), the Funding, Research Administration and Projects Ontology (frapo), and the Teaching Core Vocabulary Specification (teach). The Energy domain included 7 vocabularies, among others, the

Renewable Energy and Energy Efficiency vocabulary (reegle), the Energy Efficiency Prediction Semantic Assistant ontology (eepsa), and the Ontology for Managing Geometry and descriptions of objects (omg). The Environment domain included 21 vocabularies, among them the Ontology for Meteorological sensors (aws), the Climate and Forecast features Ontology (cff), the SEAS Battery ontology, which defines batteries and their state of charge ratio property (seasb), the GeoSpecies Ontology (geosp), the Linked Earth Ontology (earth), and the Vehicle Emissions Ontology (veo). The Government and Public Sector included 11 vocabularies, among them the oeGOV Government Core Ontology (gc), the Central Government Ontology (cgov), the Federal Enterprise Architecture Vocabulary (fea), and the Roles and Profiles Ontology (wai). The Health domain included 10 vocabularies, among which the Translational Medicine Ontology (tmo), the Hospital Vocabulary (hosp), the Upper-level ontology for Biology and Medicine (biotop), the Experimental Evaluation Ontology (eeo), and the Appearances Ontology Specification (aos). The Agriculture, Fisheries, Forestry and Foods domain included 7 vocabularies, among which the Food Ontology (food), and the Data Food Consortium Ontology (dfc). Transport included 10 vocabularies, among them being the Air Traffic Data Ontology (atd), the US National Airspace System (nas), the Rail Topology Ontology (rto), the Aircraft Equipment Vocabulary (eqp), and the Transport Administration Ontology (trao). Science and Technology included 8 vocabularies, among which the Linked Science Core Vocabulary (lsc), the Ontology of Computer Science Publications (opus), and the Nature Core Ontology (npg). Regions and Cities included 8 vocabularies among them being the NeoGeo Geometry Ontology (ngeo), and the Global City Indicator Foundation Ontology (gci). Population and Society with 29 vocabularies, among which the vocabulary for biographical information (bio), the Cognitive Characteristics Ontology (cco), the Emotion Ontology for Context Awareness (emotion), the Vocabulary for Linked Genealogical Data (gen), the International Contact Ontology (ic), and the vocabulary for describing relationships between people (rel). Justice, Legal System and Public Safety with 13 vocabularies included the European Legislation Identifier (eli), the Global Legal Entity Identifier Foundation Registration Authority Ontology (gleif-ra), the Ontology for modelling historic administrative information (gov), and the Registered Organization Vocabulary (rov). Finally, the International Issues domain included 5 vocabularies, among which the Issue Procedure Ontology (ipo), and the International Council on Archives Records in Contexts Ontology (rico).

As mentioned previously, the column "Core Vocabularies Direct Links" includes the links of the domain vocabularies to the LOV core ontologies, so this column denotes the number of domain-specific vocabularies which are linked to the core ontologies/vocabularies. The identified, basic (core) vocabularies from this analysis are described in further detail in the following Sect. 4.2.

The column "Inter-Domain Linked Vocabularies" shows how many vocabularies from other domains/themes are using the vocabulary under investigation. This column aims to give an impression on the number of links from each vocabulary of this theme to vocabularies of other themes and it could be an indication of how "reusable" vocabularies of one theme are across other themes/domains (externally). Under the column "Intra-domain Linked Vocabularies" is the number of connected vocabularies for each examined vocabulary within the same domain/theme. This number could be an indication of how well connected the vocabularies of the same domain are among them (internally). The columns "Inter-Domain Linked Vocabularies" and "Intra-domain Linked Vocabularies" together can provide an initial overview of connections/links of the identified vocabularies within their domain but also across domains/themes. The last column of Table 2, "Domain Interoperability Readiness", consists of 2 indicators and shows:

- The # of (unique) domain vocabularies which have links (inter- and intra-) to the other domain-specific vocabularies, divided by the total number of domain vocabularies (column 1). This is an indication of the percentage of domain vocabularies that get reused by other vocabularies in the same or different domains.
- The sum of the # of existing links (inter-, intra-, and core links). This indicator aims to show comparatively how well-connected each domain is in the LOV ecosystem.

4.2 Core (Basic) Vocabularies in LOV

In this Section, the core vocabularies deriving from the previous analysis are mentioned in further detail for context and reference to the results of the analysis.; it is noteworthy to refer to them as they are linked to a big number of other vocabularies, either as core vocabularies/ontologies, or to extend, specialize, or describe the other vocabularies' metadata.

- A Vocabulary for Annotating Vocabulary Descriptions (vann): VANN is the vocabulary/ontology that describes other vocabularies with examples and usage notes [4]. It has 399 incoming links and 3 outgoing ones (rdfs, rdf, dcterms), the former showing that the vast majority of related vocabularies use vann to describe their metadata (which is logical since vann is widely used for vocabulary descriptions), and the latter demonstrating how vann itself is specialized and expressed using the RDF schema and utilizing Dublin Core and RDF to describe its own metadata. VANN is used in 19 datasets in LOD.
- Simple Knowledge Organization System (skos): SKOS is one of the most widely used vocabularies in LOV and the model used for sharing and linking knowledge organization systems in the Semantic Web [4]. It has 312 incoming links, indicating vocabularies which are extended, specialized, generalized, and imported by SKOS, or using it for their metadata descriptions. It has 4 outgoing links, it is specialized by RDFS, extended by RDF and using DCTERMS and RDF to express its own metadata. SKOS is used in 152 datasets in LOD.
- DCMI Metadata Terms (dcterms)/ Dublin Core Metadata Element Set (dce): This is a specification of all metadata terms maintained by the Dublin Core Metadata Initiative, including properties, vocabulary encoding schemes, syntax encoding schemes, and classes [4]. It has 705 incoming links in a variety of relations to other vocabularies, however, with a dominant view of describing other vocabularies' metadata, and 8 outgoing links, showing its relation to other core vocabularies, such as skos, rdf, foaf, and rdf. Dcterms is used in 327 datasets in LOD. Similarly, dce, the DC metadata element set is used for resource description, it has 464 incoming links (the majority for metadata description), and 3 outgoing links to dcterms, rdf, and skos. Dce is used in 178 datasets in LOD.

- The RDF Concepts Vocabulary (rdf)/ The RDF Schema Vocabulary (rdfs): This is the RDF Schema for the RDF vocabulary terms in the RDF Namespace, as defined in the RDF 1.1 concepts [4]. RDF has 899 incoming links with other vocabularies which are specialized and extended by RDF, while most of the links refer to vocabularies which use RDF to describe their metadata. RDF has 4 outgoing links, showing it is generalized, specialized, and extended by RDFS, while its own metadata is described using DCE. RDF is used in 504 datasets in LOD. Similarly, RDFS is the RDF Schema, which provides a data-modeling vocabulary for RDF data and is an extension of the basic RDF vocabulary [4]. RDFS has 1673 incoming links, most of them referring to vocabularies which are specialized or extended by RDFS or using RDFS for their metadata description. RDFS has 5 outgoing links, showing it is extended, specialized, and generalized by RDF, while for its metadata description it uses RDF and DCE. RDFS is used in 452 datasets in LOD.
- Data Catalog Vocabulary (dcat): DCAT is the RDF vocabulary designed to facilitate interoperability between data catalogs published on the Web [4]. DCAT has 32 incoming links and 17 outgoing links. The incoming links refer to vocabularies which are specialized by DCAT, such as Asset Description Metadata Schema (adms), DDI-RDF Discovery Vocabulary (disco) for documenting research and survey data, VoCaLS: A Vocabulary and Catalog for Linked Streams (vocals), IDS Information Model (ids) for international data spaces, Physical Data Description (phdd), RiverBench metadata ontology (rb), and DataID (dataid) for LOD datasets, vocabularies which are extended by DCAT, such as CDC: Construction Dataset Context ontology (cdc), BotDCAT-AP - Data Catalogue vocabulary Application Profile for chatbots (bot), AMV:Algorithm Metadata Vocabulary (amv), Ontologie du projet Istex pour la diffusion de la littérature scientifique (istex), and The visualization vocabulary for LOD applications (dvia), vocabularies which have equivalences with DCAT, such as the Schema.org vocabulary (schema), the Data Provider Node ontology (dpn), and the Dataset Ontology (donto), and vocabularies which use DCAT to describe their metadata, such as the TempO - Temporal Ontology (tempo), the General Transit Feed Specification (gtfs), the I-ADOPT Framework ontology (iadopt), the OntoUML Metamodel Vocabulary (ontouml), and the TAXREF-LD Ontology (taxref-ld) for the representation of the TAXREF taxonomic register as Linked Data. The outgoing links show the vocabularies by which DCAT is specialized, such as RDFS, FOAF, SKOS, DCTYPE, and DCTERMS, the vocabularies which it is extended by, such as RDFS, VCARD, XSD, SKOS, FOAF, and DCTERMS, the vocabulary which it is imported by, which is DCTERMS, and the vcabularies DCAT uses for it own metadata, RDF, SKOS, DCTERMS, FOAF, and SCHEMA. DCAT is not used in any datasets in LOD.

Other basic vocabularies identified in Sect. 4.1 but used to a smaller extent, include the SemWeb Vocab Status Ontology (vs), which is an RDF vocabulary for relating Semantic Web vocabulary terms to their status [4], the Creative Commons Rights Expression Language (cc), which describes copyright licenses in RDF, the Friend of a Friend Vocabulary (foaf), which was described in the Methodology Section, and the Vocabulary of a Friend (voaf), which describes linked data vocabularies and their relations.

5 Discussion

The results of Sect. 4.1 are an overview of the current status which holds for each knowledge domain and its corresponding linked open vocabularies. Some knowledge domains such Economy and Finance, or Population and Society, represent a big number of relevant vocabularies and links to other vocabularies, without, however, necessarily having the highest percentage of vocabulary reuse. Domains/themes with fewer identified vocabularies were characterized by tighter connections among them, such as Justice, Legal System and Public Safety, Energy, and Transport the last two having the biggest percentage of vocabulary reuse in the domain (around 0.7 or 70%) internally and externally, and another domain/theme following closely after with 0.69 or 69%. It is noteworthy that some themes/domains have very few but tightly linked vocabularies (such as Transport and Energy), while closely related domains such as Environment, Regions and Cities, or Science and Technology have much lower Interoperability Readiness scores. This could potentially be attributed to an unsuccessful (non-exhaustive) mapping in the methodology part, but it could also be an indication of low vocabulary/ontology reuse across proximal knowledge domains. Consequently, ontology mapping/reuse among themes/domains with common or overlapping elements can result in higher interoperability. The themes/domains with the biggest number of links are Population and Society, and Economy and Finance, followed by Justice, Legal System and Public Safety, Environment, Transport, and Government and Public Sector. Interestingly, the themes/domains with the highest number of links are not necessarily the ones with the highest reuse of their own vocabularies. In addition, domains/themes pairs which share overlapping or very closely related concepts such as Environment, Energy and Science and Technology, or Population and Society and Regions and Cities, do not necessarily share similar interoperability readiness scores. This can be an indication of low cross-domain reuse of the existing vocabularies/ontologies and disparate sources of knowledge representation. On the other hand, standalone domains/themes such as Health, would be expected to have a higher internal connection among the existing ontologies/vocabularies, since concepts are more domain-specific and could easily be reused, however, this is not the case, a potential explanation being the retrieval of very few niche vocabularies using this methodology, but further investigation could reveal a need for further enhancement of intra-domain ontology reuse in the medical and healthcare sector. In addition, as can be seen in columns 4 and 5 of Table 2, most vocabularies have zero, one or very few links (in contradiction to their consistent high linking to core vocabularies) to other domain vocabularies, an indication of a need for further linking of the existing representation schemata to boost their reusability within and across domains. The themes of International Issues and Agriculture, Fisheries, Forestry and Foods are not analyzed further in this context due to the lack of proper definition of relevant results and the limited turnout.

As far as the connection to core vocabularies is concerned, most domains use quite some (relative to their size) core vocabularies to describe their elements, the most prevalent ones presented in Sect. 4.2. This implies that the domain-specific linked open vocabularies could be sufficiently reusing core vocabularies to describe their information, however, further work needs to be done in terms of link ability to proximal vocabularies of the same domain or relevant vocabularies of other domains. An interesting and slightly surprising result is the absence of DCAT from the identified core vocabularies, which is something with potential implications on data interoperability (e.g., on data portals such as the EDP or similar, which structure their information using this metadata standardization).

6 Conclusion and Future Directions

The presented study acts as a step towards assisting the enablement of an enhanced, linked open data ecosystem by providing an overview of domain-specific LOV and how loosely or tightly they are currently linked to other vocabularies of the same but also of other domains, as well as the basic, core vocabularies/ontologies. In its turn, this can facilitate decisions regarding the choices among vocabularies or datasets to be linked (according to the knowledge domain the data belongs to), in order to achieve improved interoperability of the data.

However, the conducted study is subject to some limitations. In essence, the mapping between the EDP thematic categories and LOV results in a non-absolute match, which consequently hinders an exhaustive fit and implies that other mappings could produce slightly different results. In this case, the mapping is intuitive and aims to give an overall impression of existing knowledge domains and their specificities, however, different methodologies could be experimented with in order to complement the results of this research. A similar situation applies to the examined links among the identified vocabularies of this study, which are not exhaustive but an attempt to capture significant information about their relations. It is also worth mentioning that the identified domain-specific vocabularies are the ones existent in LOV, so the ontologies/vocabularies which exist in other sources are not considered. Furthermore, the notion of "Domain Interoperability Readiness" is an approach for a preliminary assessment of the conducted analysis and it could further be validated using more official means, e.g., statistical analysis or other, combined methods.

Future directions for this study could involve the application of algorithmic techniques to process the information obtained from the LOV ecosystem in a complete manner and use machine learning to recommend appropriate courses of action dependent on the parameters set to be improved or attained, and propose links with significant impact on the interoperability, reusability, and findability of data. In addition, the presented research could shift focus towards one specific knowledge domain and examine in further detail its relationship and links to other domains in order to enhance intra- and inter-domain semantic interoperability.

Acknowledgments. This project has received funding from the European Union's Horizon 2020 research and innovation programme under the Marie Skłodowska-Curie grant agreement No 955569.

Disclosure of Interests. The authors have no competing interests to declare that are relevant to the content of this article.

References

1. LinkedData. LinkedData - W3C Wiki. (n.d.). https://www.w3.org/wiki/LinkedData#:~:text=The%20term%20Linked%20Data%20rfers,can%20look%20up%20those%20names
2. Presentation metadata - europa.eu. (n.d.). https://data.europa.eu/sites/default/files/d2.1.2_training_module_1.2_introduction_to_linked_data_en_edp.pdf
3. Data on the web best practices. W3C. (n.d.). https://www.w3.org/TR/dwbp/#dataVocabularies
4. About linked open vocabularies. Joinup. (n.d.). https://joinup.ec.europa.eu/collection/linked-open-vocabularies/about
5. Linked open vocabularies (LOV). (2018, January 3). https://datos.gob.es/en/blog/linked-open-vocabularies-lov
6. Admin, H.: Using Linked and Other Open Vocabularies. Hedden Information Management, 24 September 2018. https://www.hedden-information.com/using-linked-and-other-open-vocabularies/
7. Escobar, P., Candela, G., Trujillo, J., Marco-Such, M., Peral, J.: Adding value to linked open data using a multidimensional model approach based on the RDF data cube vocabulary. Comput. Stand. Interfaces **68**, 103378 (2019). https://doi.org/10.1016/j.csi.2019.103378
8. Bauer, F., Kaltenböck, M.: Linked open data: the essentials: a quick start guide for decision makers (2012)
9. Fibriani, C., Ashari, A., Riasetiawan, M.: Linked open spatial data for evaluation of land suitability. J. Theor. Appl. Inf. Technol. **101**(6) (2023)
10. Empowering digital change for the Cultural Heritage Sector. Europeana PRO, 25 July 2023. https://pro.europeana.eu/
11. Getty: Resources for Visual Art and Cultural Heritage. (n.d.-a). https://www.getty.edu/research/tools/vocabularies/newbury_sanderson_itwg_2017.pdf
12. Oh, S., Yi, M., Jang, W.: Deploying linked open vocabulary (LOV) to enhance library linked data. J. Inf. Sci. Theory Pract. **3**, 6–15 (2015). https://doi.org/10.1633/JISTaP.2015.3.2.1
13. Schaible, J., Gottron, T., Scheglmann, S., Scherp, A.: LOVER: support for modeling data using linked open vocabularies. In: International Conference on Extending Database Technology (2013)
14. Vila-Suero, D., Gracia, J., Gómez-Pérez, A.: Topic Modeling for Linked Open Vocabularies (2015)
15. Vandenbussche, P.-Y., Atemezing, G., Poveda-Villalón, M., Vatant, B.: Linked open vocabularies (LOV): a gateway to reusable semantic vocabularies on the Web. Semant. Web **8**, 437–452 (2017). https://doi.org/10.3233/SW-160213
16. Jia, J.: From data to knowledge: the relationships between vocabularies, linked data and knowledge graphs. J. Doc. ahead-of-print (2020). https://doi.org/10.1108/JD-03-2020-0036
17. Pauwels, P., Costin, A., Rasmussen, M.H.: Knowledge graphs and linked data for the built environment. In: Bolpagni, M., Gavina, R., Ribeiro, D. (eds.) Industry 4.0 for the Built Environment. Structural Integrity, LNCS, vol. 20, pp. 157–183. Springer, Cham (2022). https://doi.org/10.1007/978-3-030-82430-3_7
18. Pan, J.Z., Vetere, G., Gomez-Perez, J.M., Wu, H.: Exploiting Linked Data and Knowledge Graphs in Large Organisations (2017). https://doi.org/10.1007/978-3-319-45654-6
19. Tiwari, S., Al-Aswadi, F.N., Gaurav, D.: Recent trends in knowledge graphs: theory and practice. Soft Comput. **25**, 8337–8355 (2021). https://doi.org/10.1007/s00500-021-05756-8
20. Haslhofer, B., Isaac, A., Simon, R.: Knowledge Graphs in the Libraries and Digital Humanities Domain (2018)
21. Cheatham, M., et al.: The GeoLink knowledge graph. Big Earth Data **2**(2), 131–143 (2018). https://doi.org/10.1080/20964471.2018.1469291

22. Haller, A., Fernández, J., Kamdar, M., Polleres, A.: What are links in linked open data? A characterization and evaluation of links between knowledge graphs on the web. J. Data Inf. Qual. **12**, 1–34 (2020). https://doi.org/10.1145/3369875
23. Linked open vocabularies (LOV). Linked Open Vocabularies. (n.d.). https://lov.linkeddata.es/dataset/lov/

Predicting Digital Winners and Losers in Economic Crises Using Artificial Intelligence and Open Government Data

Euripidis Loukis(✉) and Mohsan Ali

University of Aegean, Samos, Greece
{eloukis,Mohsan}@aegean.gr

Abstract. In market-based economies often appear significant decreases of economic activity, which lead to recessionary economic crises. These economic crises have quite negative consequences for firms, as they lead to significant decrease of their sales revenues; firms respond by decreasing on one hand their production and in general operational activities and expenses, personnel employment and materials' procurement, and on the other hand their investments in production equipment, digital technologies, etc., which leads to technological obsolescence. This reduction of investments, and especially of the ones in digital technologies, due to their importance for firms' efficiency, effectiveness, and innovation, can have quite negative impact on their future competitiveness, and even put at risk their survival. However, these negative consequences of economic crises differ significantly among firms: some of them exhibit a lower vulnerability to the crisis, so they have less negative consequences, while some other firms exhibit a higher vulnerability, and have more negative consequences; so the competitive position of the former is significantly strengthened with respect to the latter, and finally the former are the 'winners' of the crisis, while the latter are the 'losers'. This paper proposes a methodology for predicting the winner and loser firms of future economic crises with respect to a highly important class of technologies: the digital technologies. In particular, the proposed methodology enables the prediction of the multi-dimensional 'pattern of digital vulnerability' of an individual firm to a future economic crisis, which consists of the degrees of reduction of the main types of 'digital investments' as well as 'digital operating expenses' in a future economic crisis. For this purpose, we are using Machine Learning algorithms, in combination with the Synthetic Minority Oversampling Technique (SMOTE), in order to increase their performance, which are trained using open government data from Statistical Authorities. Furthermore, a first application/validation of the proposed methodology is presented, using open data from the Greek Statistical Authority for 363 firms for the severe Greek economic crisis period 2009–2014, which gave satisfactory results concerning the prediction of nine different aspects of digital vulnerability to economic crisis (five of them concerned the main types of digital investment, and the other four concerned the main types of digital operating expenses).

Keywords: Economic Crises · Digital Technologies · Artificial Intelligence · Machine Learning

M. Papadaki et al. (Eds.): EMCIS 2023, LNBIP 501, pp. 153–166, 2024.
https://doi.org/10.1007/978-3-031-56478-9_11

1 Introduction

In market-based economies often appear significant decreases of economic activity, which lead to recessionary economic crises of different intensities, geographical scopes and durations [1–6]. In [3] are briefly described numerous economic crises that appeared during the previous century, as well as their origins and negative consequences for the society and the economy; in the beginning of this century initially we had the severe 2007 Global Financial Crisis [1, 7], shortly after the end of it we had the COVID-19 that gave rise to another economic crisis [8], while recently the Ukraine war resulted in big increases in the prices of oil, gas, wheat and other goods, which spark another economic crisis.

These economic crises have quite negative consequences for firms, as they lead to decrease of the demand for their products and services, and therefore of their sales; firms respond to this by decreasing on one hand their production and in general operational activities and expenses, as well as personnel employment and materials' procurement, and on the other hand their investment in production equipment, digital technologies, etc., which leads to technological obsolescence. This reduction of investment, and especially of the ones in digital technologies, due to their high importance for firms' efficiency, effectiveness, and innovation, can have quite negative impact on the future competitiveness of firms, or even put at risk the survival of some of them [1, 2, 9–12]. However, these negative consequences of economic crises differ significantly among firms: some of them exhibit a higher capacity to cope with the crisis, so they have less negative consequences, and therefore higher resilience to the crisis; on the contrary, some other firms exhibit a lower capacity to cope with the crisis, so they have more negative consequences, and therefore lower resilience to the crisis [1–3, 13]. This results in a strengthening of the competitive position of the former with respect to the latter, so finally the former are the 'winners' of the crisis, while the latter are the 'losers'. Governments, in order to mitigate the negative consequences of these economic crises for the firms and the citizens undertake huge interventions, such as large-scale economic stimulus programs, which include the provision to firms of tax rebates, financial assistance, subsidies, financial support for investments, low-interest (or even no-interest) loans, etc. [14–17].

This paper proposes a methodology for predicting the winner and loser firms of future economic crises with respect to a highly important class of technologies for firms: the digital technologies. In particular, the proposed methodology enables predicting the multi-dimensional 'patterns' of digital vulnerability to economic crisis of individual firms. This digital vulnerability pattern of a firm includes the degrees of reduction of the main types of 'digital investment', e.g., in ICT hardware, software, staff training, etc., due to economic crisis. Furthermore, since the benefits that a firm obtains from the use of digital technologies depends not only on the magnitude of the above digital investments, but also on its relevant 'digital operational expenses', e.g., for ICT personnel, for cloud computing, etc. So, the pattern of digital vulnerability to economic crisis of an individual firm has to include also the degrees of reduction of the main types of 'digital operating expenses'.

For this purpose, we employ Artificial Intelligence (AI) algorithms from the area of Machine Learning (ML) [18–20], which are used in order to construct a set of prediction models of the digital vulnerability of a firm concerning the abovementioned main types of digital investment as well as digital operating expenses; as independent variables are used the characteristics of each individual firm (e.g. concerning human resources' size and quality (education and skills), production equipment, use of ICT, innovation, exports, strategic directions, comparative financial performance vis-à-vis competitors, etc.). For the training of these prediction models are used open government data (OGD) [21–23] from Statistical Authorities. The great potential of using AI in order to make advanced analysis of OGD, and extract from them valuable knowledge and prediction models, which can be quite useful for addressing the serious challenges that modern societies face, is highlighted and analyzed in a recent report of UNESCO [24]; in this report are also suggested some guidelines for governments in order to promote and accelerate this advanced exploitation of OGD through AI. However, at the same this report mentions that the use of OGD for training AI models can pose important challenges, which concern data privacy, data ethics, legal limitations, data infrastructures, data governance, as well as unrepresentative datasets with limited data from marginalized groups.

Furthermore, in this paper a first application/validation of the proposed methodology is presented, using OGD from the Greek Statistical Authority for 363 firms for the severe Greek economic crisis period 2009-2014, which gave satisfactory results concerning the prediction of nine different aspects of digital vulnerability to economic crisis (five of them concerned the main types of digital investment, and the other four concerned the main types of digital operating expenses).

The proposed methodology can be useful for government agencies that design and implement various types of interventions, such as large-scale economic stimulus programs, for mitigating the negative consequences of economic crises, in order to focus their digital actions (e.g., financial support or subsidies for firms' digital investment, employment of ICT personnel or digital operating expenses in general) on the most digitally vulnerable firms, which most need government support. Also, our methodology can be useful for banks and investment firms, in order to grant loans to and invest in firms that will not show high digital vulnerability to economic crisis, and therefore are not at risk of digital obsolescence (which can have negative consequences for their future competitiveness or even for their survival), in a future economic crisis. Finally, our methodology can be useful for ICT firms as well as consulting firms offering ICT-related services during economic crises, enabling them to identify firms that will not show high digital vulnerability to the crisis, in order to focus on them their marketing and sales activities (thus increasing the likelihood of substantial sales revenue).

The remaining of the paper has been structured as follows: the second section explains the background of our research; the third section describes the proposed methodology, and then the fourth section presents the abovementioned application of it; lastly, the conclusions are summarized in the fifth section.

2 Background – AI in Government

Though AI remained for long time mainly in a research stage in university research labs, in the last decade there has been a sharp increase in its 'real-life' application and exploitation, initially by private sector firms, and recently by government agencies as well [25–30]. The huge amounts of data possessed by government agencies motivated them to proceed to a more advanced and sophisticated exploitation of them using AI techniques, mainly from the area of ML, in order to automate, support or enhance/augment more sophisticated mental tasks than the simpler routine ones that are currently automated, supported or enhanced/augmented by the traditional information systems of government agencies.

There have been some first cases of successful AI use in various domains of government activity, for instance in tax administration for the fight against tax evasion (e.g. for identifying citizens and firms with high tax evasion risk, in order to conduct more targeted tax audits); in healthcare for supporting diseases' diagnosis and prevention, as well as treatment planning and support of clinical decision making; in social policy for fraud detection, for the prediction of higher-risk youth with respect to future criminal activity, in order to target prevention interventions; in policing for the prediction of persons or geographical areas with high criminal activity risk (e.g. forecast spatial crime patterns based on socio-demographic factors for directed patrolling in order to increase effectiveness in crime prevention), or for face recognition from images or video data (e.g., from cameras); for improving the interaction with citizens through chatbots, etc. With respect to the main directions/objectives of the first AI uses in government a recent relevant study of the AI Watch of the European Union [28] concludes that most cases of AI use in the public sector of the member states aim at the improvement of public services (e.g., through the provision of personalized and citizen-centered information and services), followed by the improvement of internal administrative efficiency (e.g., through improved management of public resources as well as increase of the quality and decrease of the cost of internal processes). Furthermore, there is an intense debate concerning the role that AI can play in government; we can distinguish two main roles for AI [30], which correspond to two quite different approaches to/ways of exploiting AI in government:

a) Use of AI for automating some administrative tasks (in order to achieve cost savings and efficiencies, which is quite useful due to the fiscal constraints that most governments face but might lead to public sector workforce substitution and job losses.
b) AI use for augmenting some administrative tasks, mainly higher-level ones associated with decision-making and policy design (by providing to them some additional useful inputs, which can increase the quality and the effectiveness of the execution of these tasks).

Comprehensive reviews of the research that has been conducted concerning the use of AI in government are provided in [25, 27, 29] and [30].

However, it is widely recognized that the research and practice in the area of AI use in government has exploited only a small part of the large potential that the wide spectrum of AI techniques have for automating, supporting or enhancing/augmenting various government functions and activities; also, most cases of AI use in government concern low or medium importance and financial magnitude activities [25, 27] and [30]. Therefore, extensive further research is required in order to exploit more this large potential, developing new ways and methodologies of using AI in government, especially in its more important and costly activities for increasing their efficiency and effectiveness. Also, further research is required also concerning the use of AI for analyzing OGD in order to achieve higher levels of exploitation of them, by extracting from them valuable knowledge and prediction models, which can be useful for both public as well as private sector decision-making, following the recent recommendations of the UNESCO [24]. The present study makes a contribution in both these directions, by developing a methodology of using AI/ML for making an advanced exploitation of OGD, for the development of prediction models in order to support and enhance/augment one of the most important and costly kinds of interventions that governments have to make: the interventions and especially the large-scale economic stimulus programs they undertake in tough times of economic crises in order to reduce their negative consequences, focusing on the 'digital' interventions.

3 Proposed Methodology

The proposed methodology aims to predict the multi-dimensional 'pattern of digital vulnerability' to an economic crisis (DVEC) of an individual firm, which is defined as a vector that has as components the degrees of reduction of the main types of digital investments (e.g. in hardware, software, etc.) as well as digital operating expenses (e.g. for ICT staff payroll, external ICT services); they can be measured either in a continuous scale or in a 5-levels Likert scale: not at all, small, moderate, large, very large), during an economic crisis (Fig. 1):

$$\mathrm{DVEC} = \left[\mathrm{DVEC}_1,\ \mathrm{DVEC}_{2,\dots\dots,}\mathrm{DVEC}_N\right]$$

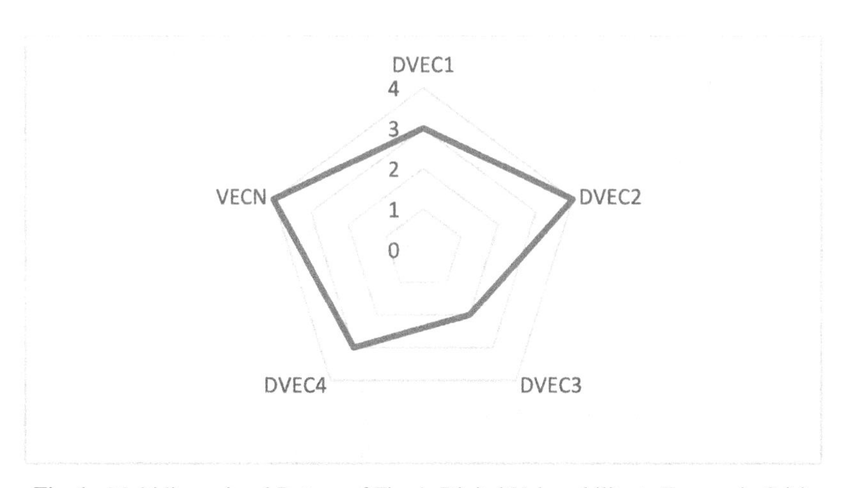

Fig. 1. Multidimensional Pattern of Firm's Digital Vulnerability to Economic Crisis

Firms predicted to have a high degree of reduction of the above main types of digital investments as well as digital operating expenses in a future economic crisis have a risk of becoming 'digitally obsolete'; they will be the 'digital losers' of the crisis, and this might have negative impact on their future competitiveness, or even put at risk their survival. On the contrary, firms predicted to have a low degree of reduction of the above main types of digital investments and digital operating expenses in a future economic crisis, will be the 'digital winners' of the crisis, and this will strengthen their competitive position with respect to the 'digital losers'.

For each of the abovementioned components/dimensions of firm's digital vulnerability to economic crisis $DVEC_1$, $DVEC_2$,......, $DVEC_N$ a prediction model of it is constructed, having it as dependent variable. In order to determine the appropriate independent variables of these prediction models we have been based on management science research, which has been developed several frameworks of the main elements of a firm that determine its performance; the 'Leavitt's Diamond' framework is the most widely recognized one, which includes five main elements: strategy, processes, people, technology, and structure [31, 32]. We can expect that these five main elements of the 'Leavitt's Diamond' framework will be the main determinants of the performance of a firm not only in normal economic periods but also in economic crisis ones as well.

So, the prediction models of firm's digital vulnerability concerning the main types of digital investments and digital operating expenses $DVEC_i$ will include five corresponding groups of independent variables concerning:

a) strategy (e.g., degree of adoption of the main competitive advantage strategies defined in relevant strategic management literature [33], such as cost leadership, differentiation, focus, innovation, etc.)
b) processes (e.g., main characteristics of firm's processes, such as complexity, flexibility, etc.)
c) people (e.g., shares of firm's human resources having different levels of education or specific skills, certifications, etc.)
d) technology (e.g., use of various production technologies, digital technologies, etc.)
e) structure (e.g., main characteristics of firm's structure, degree of adoption of 'organic' forms of work organization, such as teamwork, etc.)

and also, a sixth group of independent variables concerning general information about the firm, such as size, sector, comparative performance vis-à-vis competitors, etc.

For the construction of each of these prediction models we can use/try the main supervised ML algorithms described in relevant literature [18–20], such as Decision Tree (DT), Random Forest (RF), Logistic Regression (LR), Support Vector Machines (SVM), and Multilayer Perceptron (MLP), compare their prediction performances, and finally select the one with the highest prediction performance. For training them we can use relevant OGD for economic crisis periods provided by Statistical Authorities; the available dataset is divided into two parts: the 'training dataset', which is used for constructing the prediction model, and the 'test dataset', in which the prediction performance of this model is evaluated, by calculating its prediction accuracy, precision, recall, and F-Score are calculated. The formals for these terms are given below in the Eqs. 1, 2, 3, 4; they are calculated based on the numbers of TP (True positive), TN (True negative), FP (False positive), and FN (False negative).

$$Accuracy = \frac{TN + TP}{TN + FP + TP + FN} \tag{1}$$

$$Precision = \frac{TP}{TP + FP} \tag{2}$$

$$Recall = \frac{TP}{TP + FN} \quad (3)$$

$$F - score = \frac{TP}{TP + FN} \quad (4)$$

However, the abovementioned datasets we use for the construction of the models usually a) are too noisy because of the missing values, b) their size can be not large enough to train a healthy and accurate ML model, and c) are unbalanced with respect to the classes (they include small numbers of observations/samples for some of the classes, and much larger numbers of observations/samples for some other classes); these can result in prediction models with lower prediction performance and also biased. In order to address problems b) and c) our methodology includes a pre-processing of these datasets using the Synthetic Minority Oversampling Technique (SMOTE) [34]; this technique increases the number of samples of the dataset using the existing samples of the classes (oversampling), balances the dataset with respect to the number of samples of each class, fills missing values, which enable the estimation of better prediction models with higher prediction performance. Furthermore, our methodology includes an initial Exploratory Data Analysis (EDA) in order to get a first insight of the data through visualizations, and make some necessary transformations, and then a Principal Component Analysis (PCA) in order to analyze the importance of the features, and select the eliminate the important ones, and eliminate the ones that are not important, which helps to improve performance and reduce training time.

4 Application

A first application/validation of the proposed methodology was made using a dataset that was released by the Greek Statistical Authority on request by the authors, and after signing an agreement concerning its use, so it constitutes OGD freely available (under specific terms) for research purposes. The full dataset was comprised of 363 instances, with each instance being an independent firm. It included for each firm the following features/variables:

a) Nine variables concerning different dimensions of firm's digital vulnerability to the severe economic crisis that Greece experienced between 2009 and 2014 (dependent variables). Five of them concerned digital investment: degree of reduction of the total hard and soft ICT investment (in ICT hardware, ICT software, ICT training of staff and ICT consulting), ICT hardware, ICT hardware, ICT staff training and ICT consulting; and the remaining four concerned digital operating expenses: degree of reduction of total ICT operating expenses (for ICT personnel and ICT services), ICT staff expenses, ICT outsourcing expenses and cloud computing services' expenses. All these variables were measured in a 6-levels Likert scale (increase, negligible impact, small decrease, moderate decrease, large decrease, very large decrease), and were then converted to binary ones (with the first three values being converted to 'non-vulnerable' and the other three being converted to 'vulnerable'). The corresponding questions of the Greek Statistical Authority questionnaire are shown in the Appendix.

b) Forty variables concerning various firm's characteristics with respect to strategy, personnel, technology (focusing on the use of various digital technologies), structure (focusing on the use of organic' forms of work organization, such as teamwork) and also some general information about the firm (size measured through the number of employees, sector (services or manufacturing), comparative financial performance in the last three years in comparison with competitors) (independent variables).

The dataset was noisy because of missing values and was converted to noise free by filling the missing value with the most suitable value based on the type of each variable (for instance, if the variable was ordinal, then the missing values were filled with the highest relative frequency value of this variable). Then exploratory data analysis (EDA) was applied in order to gain a better insight into the data through visualizations and make some necessary transformations. As a next step, since the size of our data set (which included data for 363 firms as mentioned above) did not allow constructing/training a good, supervised ML prediction model (with good prediction performance), we used the abovementioned (in section 3) oversampling and class-balancing algorithm SMOTE. Finally, the dataset was divided into a training dataset including 66% of the samples and a testing dataset including 33% of the samples. The former was used for the training of ML models with the following algorithms: Decision Tree (DT), Random Forest (RF), Logistic Regression (LR), Support Vector Machines (SVM), and Multilayer Perceptron (MLP). The latter was used for assessing the prediction performance of these ML models; we can see the results (prediction accuracy, precision, recall and F-score in Table 1 (with bold are shown for each dimension of digital crisis vulnerability the results for the best performing algorithm).

Table 1. Prediction Performance of AI/ML Algorithms for each Digital Crisis Vulnerability Dimension

Dimension of Digital Crisis Vulnerability	AI /ML Algorithm	Accuracy	Precision	Recall	F1-score
Degree of Reduction of Total Hard and Soft ICT Investment	Decision Tree	0.76	0.76	0.76	0.76
	Random Forest	0.84	0.84	0.84	0.83
	Logistic Regression	0.67	0.65	0.67	0.65
	SVM	**0.86**	**0.86**	**0.86**	**0.86**
	MLP	0.71	0.70	0.71	0.69
Degree of Reduction of Investment in ICT Hardware	Decision Tree	0.78	0.79	0.78	0.78
	Random Forest	**0.88**	**0.89**	**0.88**	**0.87**
	Logistic Regression	0.76	0.75	0.76	0.74
	SVM	0.86	0.86	0.86	0.85
	MLP	0.72	0.70	0.72	0.68
Degree of Reduction of Investment in ICT Software	Decision Tree	0.77	0.77	0.77	0.77
	Random Forest	**0.87**	**0.88**	**0.87**	**0.86**
	Logistic Regression	0.67	0.64	0.67	0.65
	SVM	0.87	0.88	0.87	0.86
	MLP	0.71	0.77	0.71	0.62

(continued)

Table 1. (*continued*)

Dimension of Digital Crisis Vulnerability	AI /ML Algorithm	Accuracy	Precision	Recall	F1-score
Degree of Reduction of Investment ICT Training of Staff	Decision Tree	0.80	0.80	0.80	0.80
	Random Forest	**0.90**	**0.91**	**0.90**	**0.90**
	Logistic Regression	0.72	0.70	0.72	0.70
	SVM	0.90	0.90	0.90	0.90
	MLP	0.76	0.75	0.76	0.75
Degree of Reduction of Investment ICT Outsourcing	Decision Tree	0.78	0.78	0.78	0.78
	Random Forest	**0.85**	**0.85**	**0.85**	**0.85**
	Logistic Regression	0.74	0.73	0.74	0.73
	SVM	0.85	0.85	0.85	0.84
	MLP	0.74	0.73	0.74	0.73
Degree of Reduction of Total ICT Operating Expenses	Decision Tree	0.76	0.76	0.76	0.76
	Random Forest	**0.90**	**0.90**	**0.90**	**0.89**
	Logistic Regression	0.73	0.72	0.73	0.72
	SVM	0.85	0.85	0.85	0.84
	MLP	0.72	0.71	0.72	0.68
Degree of Reduction of ICT Staff Expenses	Decision Tree	0.79	0.79	0.79	0.79
	Random Forest	0.91	0.91	0.91	0.91
	Logistic Regression	0.74	0.73	0.74	0.73
	SVM	**0.91**	**0.91**	**0.91**	**0.91**
	MLP	0.77	0.76	0.77	0.76
Degree of Reduction of ICT Outsourcing Expenses	Decision Tree	0.76	0.77	0.76	0.76
	Random Forest	0.88	0.88	0.88	0.87
	Logistic Regression	0.75	0.74	0.75	0.74
	SVM	**0.88**	**0.88**	**0.88**	**0.88**
	MLP	0.72	0.70	0.72	0.70
Degree of Reduction of Cloud Computing Services Expenses	Decision Tree	0.81	0.81	0.81	0.81
	Random Forest	**0.92**	**0.92**	**0.92**	**0.92**
	Logistic Regression	0.70	0.69	0.70	0.69
	SVM	0.89	0.90	0.89	0.89
	MLP	0.78	0.77	0.78	0.77

We can see that for all dimensions of firm's digital vulnerability to economic crisis we have high prediction performances, with prediction accuracies (for the best perforing AI/ML algorithm for each dimension) ranging between 0,85 and 0,92. Overall, the results of this first application of the proposed methodology (concerning prediction performances) can be regarded as quite satisfactory, taking into account the small size of the dataset we have used (data from 363 firms), and provide a first validation of this methodology. We expect that using a larger dataset (as governments have such data for quite large numbers of firms) will allow training crisis-vulnerability prediction models with higher prediction performances.

5 Conclusion

In the previous sections we have described a methodology for predicting the whole pattern of digital vulnerability to economic crisis of individual firms with respect to the main types of digital investments as well as digital operating expenses. It is based on AI/ML techniques, in combination with SMOTE in order to increase their performance, which are trained using OGD from Statistical Authorities. This is in line with the recent recommendations of UNESCO for advanced exploitation of OGD using AI [24]. Our methodology enables a prediction of the degree of reduction of the main types of digital investment as well as digital operating expenses of an individual firm in a future economic crisis, based on its characteristics with respect to human resources, technologies, strategic orientations, processes and structure; this enables predicting the digital winner as well as the digital loser firms of future economic crises. The proposed methodology has as theoretical foundation the widely recognized 'Leavitt's Diamond' framework [31, 32] from management science.

Furthermore, a first application/validation of the methodology has been presented, using OGD from the Greek Statistical Authority about 363 firms for the severe Greek economic crisis period 2009-2014, which gave quite satisfactory results concerning the prediction of nine different aspects of digital vulnerability to economic crisis: five of them concerned the main types of digital investment, and the other four concerned the main types of digital operating expenses.

The research presented in this paper has some interesting implications for research and practice. With respect to research, it makes a significant contribution to two highly important research streams. First, it makes a contribution to the growing research stream concerning the use of AI in government, by developing a novel approach for a highly beneficial use of AI/ML for supporting and enhancing/augmenting a critical activity of government, which is characterized by quite high economic/social importance and financial magnitude: the interventions, and especially the large-scale economic stimulus programs, for mitigating the negative consequences of economic crises (focusing on the digital interventions/actions of these programs). Second, it makes a contribution to the OGD research stream, by providing an approach for increasing the economic/social value generation form the OGD through advanced processing of them using AI/ML techniques. With respect to practice, it provides a useful tool for government agencies responsible for the above interventions and programs that aim to mitigate the negative consequences of economic crises, as well as for banks and investment firms, for enhancing/augmenting their process of making important decisions; for the former the critical decisions they have to make concerning the selection of the firms that will receive financial support or subsidies for digital investment, employment of ICT personnel or digital operating expenses in general (enabling the focus of the scarce financial resources on the most digitally vulnerable firms, which are in greatest need of government support); for the latter the critical decisions they have to make concerning the selection of the firms they will grant loans to, or firms they will invest in (enabling them to grant loans to and invest in firms that will not show high digital vulnerability to economic crisis, and therefore are not at risk of digital obsolescence, which can have negative consequences for their future competitiveness or even for their survival). Furthermore, our methodology provides a useful tool for ICT firms, as well as consulting firms offering ICT-related services, for enhancing/augmenting their process of making important decisions concerning the selection of firms they will focus on during economic crises (enabling them to focus their marketing and sales activities on firms that will not show high digital vulnerability to economic crisis, thus increasing the likelihood of substantial sales revenue).

However, further research is required in the following directions: a) further application of the proposed methodology, using larger datasets, with more dependent variables (i.e. more types of digital investment as well as digital operating expenses) and more independent variables (i.e. more firms' characteristics), in various national and sectoral contexts (experiencing different types and intensities of economic crises), and even for different categories of investments and operating expenses (beyond the digital ones); b) investigation of the use of other pre-processing algorithms (e.g., oversampling and class-balancing algorithms) as well as AI/ML algorithms (e.g., deep learning ones) for achieving higher prediction performance; c) investigation of the combination of OGD with other sources of data about firms (e.g. from other government agencies, from private firms, such as private business information databases), in order to obtain more information about firms that might improve the performance of the prediction of their pattern of vulnerability to economic crises; d) investigation of the large-scale application of the proposed methodology, and the main challenges it may pose (possibly using as a starting point the challenges of using OGD for training AI models mentioned in [24]).

Acknowledgments. This project has received funding from the European Union's Horizon 2020 research and innovation programme under the Marie Skłodowska-Curie grant agreement No 955569.

Appendix - Dependent Variables' Questions

How important was the impact of the economic crisis during the period 2009-2014 on the following categories of ICT related investment and operating expenses in your company?

	Increase	Negligible Impact	Small Decrease	Moderate Decrease	Large Decrease	Very Large Decrease
Total hard and soft ICT investments (in ICT hardware, ICT software, ICT training and ICT consulting)						
Investments in ICT hardware						
Investments in ICT software						
Investments in ICT training of your staff						

(continued)

(*continued*)

	Increase	Negligible Impact	Small Decrease	Moderate Decrease	Large Decrease	Very Large Decrease
Investment in ICT consulting						
Total ICT operating expenses (ICT personnel and services)						
ICT staff expenses						
ICT outsourcing expenses						
Cloud Computing services expenses						

References

1. OECD, Responding to the Economic Crisis Fostering Industrial Restructuring and Renewal. OECD Publishing (2009a)
2. Keeley, B., Love, P.: From Crisis to Recovery - The Causes Course and Consequences of the Great Recession. OECD Publishing, Paris (2012)
3. Knoop, T.A.: Recessions and Depressions: Understanding Business Cycles - 2nd edn. Praeger Santa Barbara, California (2015)
4. Allen, R.E.: Financial Crises and Recession in the Global Economy, 4th edn. Edward Elgar Publications, Cheltenham, UK (2017)
5. Saifi, S., Horowitz, J.: Blackouts and soaring prices: Pakistan's economy is on the brink. Paris, https://edition.cnn.com/2023/02/02/economy/pakistan-economy-crisis/index.html
6. Kottika, E., et al.: We survived this! What managers could learn from SMEs who successfully navigated the Greek economic crisis. Ind. Market. Manage. **88**, 352–365 (2020). https://doi.org/10.1016/j.indmarman.2020.05.021
7. Loukis, E., Arvanitis, S., Myrtidis, D.: ICT-related behaviour of Greek banks in the economic crisis. Inf. Syst. Manage. **38**(1), 79–91 (2021)
8. Baldwin, R., Di Mauro, B.W.: Mitigating the COVID Economic Crisis: Act Fast and do Whatever it Takes. Center of Economic Policy Research Press, London (2020)
9. Oliveras, L., et al.: Energy poverty and health: trends in the European union before and during the economic crisis, 2007–2016. Health Place **67**, 102294 (2021). https://doi.org/10.1016/j.healthplace.2020.102294
10. Dagoumas, A., Kitsios, F.: Assessing the impact of the economic crisis on energy poverty in Greece. Sustain. Cities Soc. **13**, 267–278 (2014). https://doi.org/10.1016/j.scs.2014.02.004
11. Stylidis, D., Terzidou, M.: Tourism and the economic crisis in Kavala, Greece. Annals of Tourism Research **44**, 210–226 (2014). https://doi.org/10.1016/j.annals.2013.10.004
12. Das, M.: Economic crisis in Sri Lanka causing cancer drug shortage. Lancet Oncol. **23**(6), 710 (2022). https://doi.org/10.1016/S1470-2045(22)00254-6
13. Arvanitis, S., Loukis, E.: Factors explaining ICT investment behavior of firms during the 2008 economic crisis. Inf. Syst. Manage. (2023). (in-press)

14. Khatiwada, S.: Stimulus Packages to Counter Global Economic Crisis – A Review. Int. Institute Labour Stud. Geneva (2009)
15. Coenen, G., Straub, R., Trabandt, M.: Gauging the Effects of Fiscal Stimulus Packages in the Euro Area. Working Paper 1483, European Central Bank, Frankfurt am Main, Germany (2012)
16. Kalinowski, T.: Crisis management and the diversity of capitalism: fiscal stimulus packages and the East Asian (neo-) developmental state. Econ. Soc. **44**(2), 244–270 (2015)
17. Taylor, J.: Fiscal Stimulus Programs During the Great Recession. Economics Working Paper 18117, Hoover Institution, Stanford, CA (2018)
18. Witten, I.H., Frank, E., Hall, M.A., Pal, C.J.: Data mining - practical machine learning tools and techniques. Morgan Kaufmann, Amsterdam, London (2017)
19. Blum, A., Hopcroft, J., Kannan, R.: Foundations of Data Science. Cambridge University Press, Cambridge (2020)
20. Russell, S., Norvig, P.: Artificial Intelligence. A Modern Approach, 3rd edn. Pearson, Essex, UK (2020)
21. Charalabidis, Y., Zuiderwijk, A., Alexopoulos, C., Janssen, M., Lampoltshammer, T., Ferro, E.: The World of Open Data - Concepts, Methods, Tools and Experiences. Springer International Publishing AG, Switzerland (2018)
22. Ali, M., Alexopoulos, C., Charalabidis, Y.: A comprehensive review of open data platforms, prevalent technologies, and functionalities. In: Proceedings of the 15th International Conference on Theory and Practice of Electronic Governance, New York, NY, USA, pp. 203–214 (2022). https://doi.org/10.1145/3560107.3560142
23. Gao, Y., Janssen, M., Zhang, C.: Understanding the evolution of open government data research: towards open data sustainability and smartness. Int. Rev. Admin. Sci. **89**(1), 59–75 (2013)
24. UNESCO, Open data for AI, United Nations Educational, Scientific and Cultural Organization (2023)
25. DeSousa, W.G., DeMelo, E.R.P., De Souza Bermejo, P.H., Sous Farias, R.A., Gomes, A.O.: How and where is artificial intelligence in the public sector going? A literature review and research agenda. Gov. Inf. Quart. **36**(4), 101392 (2019)
26. Zuiderwijk, A., Chen, Y.C., Salem, F.: Implications of the use of artificial intelligence in public governance: a systematic literature review and a research agenda. Gov. Inf. Quart. **38**(3), 101577 (2021)
27. Medaglia, R., Gil-Garcia, R., Pardo, T.A.: Artificial intelligence in government: taking stock and moving forward. Soc. Sci. Comput. Rev. (2021). (in-press)
28. Manzoni, M., Medaglia, R., Tangi, L., Van Noordt, C., Vaccari, L., Gattwinkel, D.: AI Watch. Road to the Adoption of Artificial Intelligence by the Public Sector. Publications Office of the European Union, Luxembourg (2022)
29. Van Noordt, C., Misuraca, G.: Artificial intelligence for the public sector: results of landscaping the use of AI in government across the European Union. Government Information Quarterly (2022). (in-press)
30. Madan, R., Ashok, M.: AI adoption and diffusion in public administration: a systematic literature review and future research agenda. Gov. Inf. Quart. **40**(1), 101774 (2023)
31. Leavitt, H.J.: Applied organizational change in industry: Structural, technological and humanistic approaches. In: March, J.G. (ed.) Handbook of Organizations --, 3rd edn., pp. 1144–1170. Rand McNally & Company, Chicago, IL (1970)
32. Scott-Morton, M.S.: The Corporation of the 1990s. Oxford University Press, New York (1991)

33. Whittington, R., Regner, P., Angwin, D., Johnson, G., Scholes, K.: Exploring Strategy - 12th edn. Pearson Education Limited, Harlow, UK (2020)
34. Dudjak, M., Martinović, G.: In-depth performance analysis of SMOTE-based oversampling algorithms in binary classification. Int. J. Electr. Comput. Eng. Syst. (Online) **11**(1), 13–23 (2020). https://doi.org/10.32985/ijeces.11.1.2

Chatbot Technology Assessment: 40 Cases from Greece

Yannis Charalabidis[1(✉)], Thanos Anagnou[1], Charalampos Alexopoulos[1], Theodoros Papadopoulos[1], Zoi Lachana[1], Christos Bouras[2], Nikos Karacapilidis[3], Vasileios Kokkinos[2], and Apostolos Gkamas[4]

[1] Department of Information and Communication Systems Engineering, University of Aegean, 83200 Mytilene, Samos, Greece
{yannisx,alexop,t.papadopoulos,zoi}@aegean.gr, icsdm619001@icsd.aegean.gr

[2] Computer Engineering and Informatics Department, University of Patras, 26504 Patras, Greece
{bouras,kokkinos}@upatras.gr

[3] Department of Mechanical Engineering, University of Patras, 26504 Patras, Greece
karacap@upatras.gr

[4] University Ecclesiastical Academy of Athens, 14561 Athens, Greece
gkamas@aeavellas.gr

Abstract. In recent years, the field of Artificial Intelligence has seen significant progress, particularly in the development of chatbots via Natural Language Processing (NLP) technology. Recently, however, there has been a real race in this sector with major technology companies constantly presenting new improved solutions. However, the Greek reality presents several peculiarities and difficulties in adopting modern solutions, both due to the idiosyncrasies and rarity of the language and the limited funding capabilities of the Greek economy. The purpose of this research is to evaluate the performance of chatbots in terms of the quality of their responses regarding relevance, naturalness, cohesion, accuracy, vocabulary, as well as to assess the user experience and satisfaction. Another goal is to gain a comprehensive comparative picture of chatbot operation in Greece, both per question and in comparison, between relevant questions. A guided interview with closed-type questions was chosen as the method of evaluation. The aim is to obtain structured and quantified data in an area where the average internet user is not fully familiarized and does not have previous relevant evaluation experience. Conclusions were drawn per question in order to evaluate the level of solutions in a focused and comparative way to identify possible trends and to confirm the consistency of the responses.

Keywords: Artificial Intelligence · Natural Language Processing · Virtual Assistants · Greek Virtual Assistant

M. Papadaki et al. (Eds.): EMCIS 2023, LNBIP 501, pp. 167–181, 2024.
https://doi.org/10.1007/978-3-031-56478-9_12

1 Introduction

The global use of chatbots has surged dramatically in recent years, as businesses and organizations recognize their potential to enhance efficiency and customer satisfaction. One of the key factors for the success of chatbots is their ability to understand and respond to customer queries in a natural and human-like manner. This requires advanced Natural Language Processing (NLP) technology and a deep understanding of the language and communication patterns [1] used by the target audience.

1.1 Chatbots in Greece

Despite their widespread acceptance globally, the number of chatbots in the Greek language remains limited. There is, however, a growing interest in using chatbots in Greek, especially in customer service and government sector. Some businesses in Greece have adopted chatbots to provide 24/7 customer support, aiming to improve efficiency and reduce the reliance on human resources. The results of these implementations have been mixed, with some companies reporting significant improvements in customer satisfaction and efficiency [2], while others have struggled to convince their clients to use them.

Developing chatbots for the Greek language poses several technological challenges [3]. The lack of linguistic standardization, given Greek's ancient, intricate, and multifaceted nature with its many dialects and regional variations [4], complicates Natural Language Processing (NLP) algorithms. This complexity makes it difficult for chatbots to consistently understand and respond to customer queries. Another obstacle is the limited availability of high-quality training datasets, which hampers the ability of chatbots to provide accurate and effective responses. Nevertheless, initiatives are underway to promote the use of chatbots in the Greek language, such as the Pythia project and the development of language models specifically designed for Greek, like Greek-Bert [5].

1.2 Subject and Objectives of the Research

Chatbots are a contemporary tool for communication between humans and machines. Their purpose is to provide answers to user queries and to interact with them in a manner resembling human conversation. However, the quality of their responses depends on the precision and effectiveness of the algorithms used for their training, the accuracy of the training datasets, and the diligence with which they are maintained and improved by their developers.

An effort was made to assess the quality and performance of chatbots through the subjective judgment of individuals. These chatbots are employed by companies and organizations primarily targeting the Greek population.

The goal of the research is to evaluate the chatbots' performance in terms of the relevance, naturalness, coherence, accuracy, and vocabulary of their responses. Additionally, the study aims to assess the user experience and overall satisfaction with the chatbots' functionality. Another objective is to obtain a comprehensive comparative view of the chatbots' operation in Greece, both on a per-query basis and in comparison, between related queries.

2 Background

2.1 Evaluation of Chatbots

Evaluating chatbots is a critical step for their effective development. This assessment encompasses the evaluation of their quality and performance, particularly in understanding user requests and generating appropriate responses in the form of natural and engaging conversations. The aim is to ensure that they meet user needs, are efficient in achieving their intended purpose, and offer a positive user experience [6]. The challenges in evaluation stem from the inherent complexity of human communication itself.

Evaluation models typically involve a range of tasks that assess different aspects of a chatbot's performance, such as intent recognition, entity extraction, and response generation [7]. They can be applied to different data sets and tasks, offering a holistic view of the chatbot's overall quality. Datasets are also vital for the training and testing of chatbot models and evaluating their performance. These datasets should be diverse, representative of the language and environment of the target users, and annotated with relevant information such as intent, entities, and conversational turns.

Evaluation metrics can have both objective and subjective criteria [8]. Objective measures include metrics like perplexity, accuracy, and the F1 score, which are based on quantitative analysis of the chatbot's performance on a specific task or dataset. On the other hand, subjective measures rely on human evaluation of its performance, which can be conducted through surveys, interviews, or user studies. Some of these metrics are:

Perplexity. Is a metric commonly used to evaluate the performance of language models. It measures the uncertainty in predicting the next word in a sentence, with lower values indicating better performance [9]. It is calculated by taking the inverse of the geometric mean of the probabilities assigned to each word in the sentence. For instance, if a chatbot assigns a probability of 0.8 to the word "hello" and a probability of 0.2 to the word "world" in the sentence "hello world", the perplexity score would be $1/\surd(0{,}8 * 0{,}2) =$ 2.24, which equals 2.24. Lower scores can result in more coherent and natural responses.

Accuracy is another metric frequently used to evaluate the overall performance of chatbots [10]. It is calculated by dividing the number of correct responses by the total number of responses. For instance, if a chatbot responds correctly to 80 out of 100 user inputs, its accuracy score would be 80%. Chatbots with higher accuracy scores can execute their designated tasks more effectively.

F1 Score. This is a metric commonly used to assess the performance of chatbots in natural language processing tasks, such as question answering and sentiment analysis. It measures the trade-off between the evaluation metrics of precision and recall [11]. Precision refers to the percentage of identified positive instances that are genuinely positive, meaning the proportion of predictions that are correct. On the other hand, recall pertains to the percentage of actual positive instances that were recognized by the algorithm, that is, the proportion of positive cases that were identified accurately. The balance between them highlights the fact that it's usually not possible to achieve a high level for both simultaneously. If we increase the threshold used to predict positive outcomes, precision will rise, but recall will decrease. Conversely, if we decrease the threshold, recall will increase, but precision will decrease. This happens because as the threshold rises, the number of correct predictions increases, but the total number of predictions made by the algorithm decreases. Inversely, as the threshold drops, the total number of predictions made by the algorithm increases, but it negatively impacts the accuracy of the predictions. The F1 score is especially useful for evaluating the bot's ability to handle complex conversations and to understand the essence of the conversation.

Human Evaluation. Evaluating chatbots is a significant process that involves assessing their quality and performance based on human subjective judgment. This assessment typically pertains to the quality of the chatbot's responses in terms of their relevance, naturalness, and coherence. Moreover, it evaluates the overall user experience and satisfaction derived from the chatbot's functionality [12]. Various methods can be employed for this evaluation, including surveys, interviews, user studies, and expert evaluations. Surveys and interviews usually gather user feedback about their experiences with the chatbot, such as their level of satisfaction, their perceptions of its usefulness, and their opinions on its performance [13]. User studies involve observing users interacting with the chatbot in a controlled environment and collecting data on their behavior and feedback. These studies may employ a random group of users with diverse experiences and backgrounds or focus on expert groups knowledgeable in the subject matter. Several reasons underscore the importance of this evaluation. Firstly, it provides a measure of the chatbot's performance, especially in terms of its ability to attract users and offer a positive user experience. This is crucial as chatbots are designed to interact in a natural and appealing manner, making the quality of user experience paramount to their success. Secondly, human evaluation can reveal the strengths and weaknesses of a chatbot, pointing out areas for improvement. This feedback is invaluable for developers aiming to enhance the performance of their chatbots and better cater to user needs. However, human evaluation has its limitations. It can be time-consuming and costly, especially if many evaluators are required. Additionally, it might be subjective and influenced by factors like personal preferences and biases. Finally, it might not always reflect the chatbot's real-world performance since users might interact differently in various settings or with different objectives. To address these challenges, researchers utilize various techniques such as focused interviews, focus groups, crowdsourcing, and evaluations based on machine learning.

3 Methodology

The chosen method of evaluation was the guided interview using closed-ended questions. This is a widely accepted assessment technique for interactive systems, which falls within both the realms of usability testing and predictive evaluation reviews [14]. The aim is to provide structured and quantified data in an area where the average internet user is not fully familiar with and lacks prior evaluation experience.

Closed-ended questionnaires are a popular research method utilized in guided interviews to collect data systematically and structurally from participants [15]. This method offers several advantages that make it a valuable tool in research. Firstly, it allows for efficient data collection. By employing a standardized set of questions, researchers can swiftly and effortlessly gather data from many participants, which is especially useful in research where time and resources are limited. Secondly, the data can be easily analyzed. Closed-ended questions produce quantitative data that can be readily coded and statistically analyzed, enabling patterns and trends to be quickly and effectively identified. Thirdly, they reduce bias in the data collection process. Closed-ended questions eliminate the need for participants to process their answers, thus reducing the chance of misrepresentation of their response. Lastly, they can be used to gather data from a broad range of participants, irrespective of their background, experience, or expertise.

3.1 Selected of Questions

The following questions were selected based on international literature as being critical criteria, as well as with the input from the supervising professors:

Solution Type: Voice vs. Text. Chatbots can be either voice or text-based, each having its unique characteristics and advantages. Voice-based chatbots provide the benefit of Natural Language Processing (NLP), which allows for more natural and user-friendly interactions. Users can communicate with the chatbot using their voice, akin to conversing with a human, making the experience more engaging and pleasant. On the other hand, text-based chatbots offer the advantage of easy accessibility across various devices such as smartphones, computers, and tablets. Users can interact with the chatbot by typing messages, a familiar and convenient mode of communication.

Economic Sector: Telecommunications, Financial, Commercial, Governmental Entities, Public Benefit, Local Governance, Education. Chatbot systems are widely utilized across various economic sectors, serving diverse areas of business activities. These systems are designed to streamline interactions with customers, enhance customer satisfaction, and provide cost savings for businesses.

Exclusive Mode of Communication: Yes vs. No. The idea of using chatbot systems as the sole point of contact between businesses and customers is relatively new, but it has gained popularity in recent years. This approach involves relying solely on chatbots for customer service, instead of offering multiple channels such as email or phone support. The advantages include reduced costs, 24/7 availability, and improved efficiency. The downsides are limited understanding, lack of human touch, and technical issues.

User Experience with the Interface: Excellent, Good, Average, Poor. User Experience (UX) pertains to the overall experience a user has with a product or service. It encompasses factors such as ease of use, response speed, clarity of content, responsiveness of menus, among others.

User Experience with the Outcome: Excellent, Good, Average, Poor. Users are asked to evaluate whether they achieved their intended goal. This is a crucial metric as it ultimately determines if the customer gets what they need or if they abandon the effort.

Overall Performance: Excellent, Good, Average, Poor. The overall performance from the user's perspective gauges the impression the entire experience left. Even if individual scores were lower or higher, at the end of the day, was the user serviced based on the scenario they had in mind?

3.2 Presentation of the Interview Methodology

The interview, as a focal point of the research, was conducted in a consistent manner for all participants. A neutral location was chosen for this purpose, which was quiet, devoid of decorations, and equipped with common amenities. This included a contemporary computer with a large 65-inch high-resolution (4K) screen and fast internet (1Gbps). The interviewer strived to remain impartial and refrained from suggesting any answers. Websites and phone numbers had been pre-saved to enable faster connections with the least amount of disruption. The duration was set at 2 h per interview to avoid excessive fatigue and irritation that could emotionally influence the responses. The time distribution was as follows: 20 min for introduction, 45 min for solution analysis, 10 min break, and another 45 mins for solution analysis.

The interviews were conducted based on the following scenario: Entry of the interviewee and familiarization with the space and equipment. Collection of the participant's profile data: Gender, Age, Education level, whether they use a computer, and if they own a smartphone. Demonstration and testing of ChatGPT as a benchmark for other solutions. Review of the text evaluating the questions. Commencement of evaluation.

The selection of participants aimed to represent, as closely as possible, potential Greek users who might choose the chatbot as a means of communication instead of abandoning it. For this reason, candidates who have no familiarity with the internet or have a negative attitude towards technology were excluded. Other criteria included availability and willingness to participate, considering the required time commitment (300 min). The characteristics of the group are listed in the table below (Table 1):

Table 1. Characteristics

No	Gender	Age	Educational level	Computer user	Smartphone owner
1	F	42	University	Yes, often	Yes
2	M	48	University	Yes, often	Yes
3	M	50	Elementary	No	Yes
4	M	52	High school	No	Yes
5	F	80	Postgraduate	Yes, often	Yes
6	F	51	Postgraduate	Yes, often	Yes
7	F	17	High school	Yes, often	Yes
8	M	23	University	Yes, often	Yes
9	M	37	Postgraduate	Yes, often	Yes
10	M	29	University	Yes, often	Yes

4 Results

Based on the questionnaires, results emerge for each solution. The table below presents the solutions that were explored, whether they belong to the private or public sector, and the corresponding economic sector they pertain to (Table 2).

Table 2. Chatbot Solutions

Company – Organization	Solution type	Application Domain	Economic sector
2103288000 - Piraeus	Voice	Private	Financial
Winbank - Piraeus	Text	Private	Financial
Vodafone.gr - tobi	Text	Private	Telecommunications
13888 - Cosmote	Voice	Private	Telecommunications
Alpha Bank	Voice	Private	Financial
Eurobank	Voice	Private	Financial
National bank of Greece	Voice	Private	Financial
Attika bank	Text	Private	Financial
Hellenic Development Bank	Text	Private	Financial
leroymerlin	Text	Private	Commercial
ikea	Text	Private	Commercial
eco-mat	Text	Private	Commercial
pennie	Text	Private	Commercial
ledison	Text	Private	Commercial
xtr	Text	Private	Commercial
acs	Text	Private	Commercial
coca-cola	Text	Private	Commercial
goldmall	Text	Private	Commercial
Market4you	Text	Private	Commercial
ReBrain Greece	Text	Public	State entities
oasa	Text	Public	Public utility
deddie	Text	Public	Public utility
dei	Text	Public	Public utility
eydap	Text	Public	Public utility
eopyy	Text	Public	State entities
dypa	Text	Public	State entities

(continued)

Table 2. (*continued*)

Company – Organization	Solution type	Application Domain	Economic sector
Region of Attika	Text	Public	State entities
Region of Stereas Elladas	Text	Public	State entities
Municipality of Papagou-Hollargou	Text	Public	Municipalities
Municipality of Kalamaria	Text	Public	Municipalities
Municipality of Patmos	Text	Public	Municipalities
Municipality of Moschato-Tavros	Text	Public	Municipalities
Municipality of Filis	Text	Public	Municipalities
Municipality of Kastellorizo	Text	Public	Municipalities
Municipality of West Lesvos	Text	Public	Municipalities
Municipality of Platanias	Text	Public	Municipalities
Municipality of Agia	Text	Public	Municipalities
Municipality of Visaltia	Text	Public	Municipalities
Elecectrical & Computer, Engineering Dept - UOP	Text	Public	Education
University of West Attika	Text	Public	Education

For the open-ended questions, the most frequent response given by the interviewees is presented, which we believe represents the majority opinion.

Conclusions will be drawn on a per-question basis to evaluate the level of the solutions in a focused and comparative manner, aiming to identify possible trends.

4.1 Analysis Presentation of the Interview Methodology

Based on the responses given by the participants in the study the functionality and user experience were evaluated.

Sector of Economy. The economic sector of the country plays a pivotal role. The economic profile of each entity implementing a chatbot solution gives us an insight into the penetration these solutions have in the country's economy. However, in relation to the number of sites accessed to find them, they constitute a very small percentage, around 15% (Fig. 1).

Most implementations (10) were found in the commercial sector, which is expected due to the abundance of online stores in the post-Covid era. Surprisingly, local government also had 10 implementations, which seem to be part of a pre-existing software

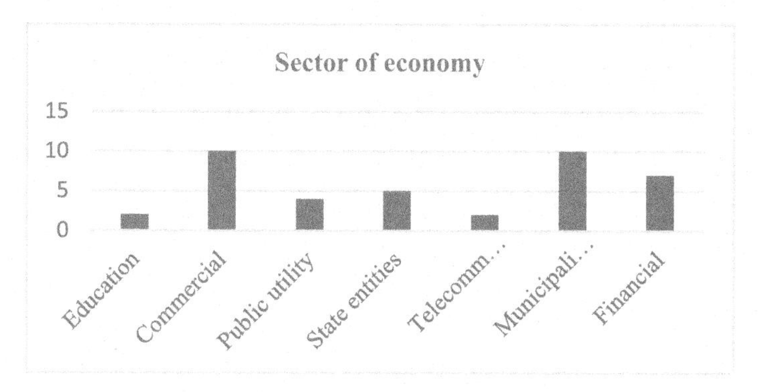

Fig. 1. Sector of economy

package. Other entities (financial, telecommunications & public utilities) have a significant presence relative to their smaller number. Surprisingly low percentages were found in the education sector and central government.

Exclusive Communication Channel. This is the most significant indication of how much a company or organization has relied on chatbots for customer service, and consequently, how much they have invested in the development of this technology (Fig. 2).

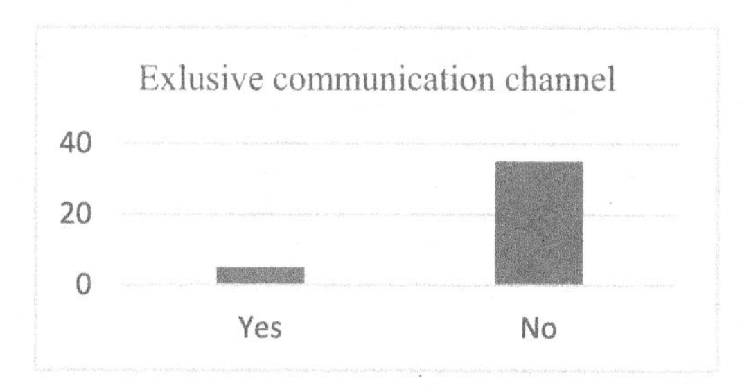

Fig. 2. Exclusive Communication Channel

In the vast majority (35/40) of cases, the implementations we observe serve as a supplementary tool. As a result, most haven't made substantial investments in the advantages that a chatbot offers. The exception is large organizations that have carefully weighed the potential benefits this communication protocol can bring to them.

User Experience with the Interface. Regardless of how advanced the technology of a chatbot is, if the interface with which the user interacts isn't user-friendly, fast, and clear, then it's challenging to evaluate it positively (Fig. 3).

We found that in most cases (34/40), there's an acceptable (ranging from good to average) reception to the way of interaction with the chatbot. The average user's familiarity with messaging applications aids in understanding its functionality. On the

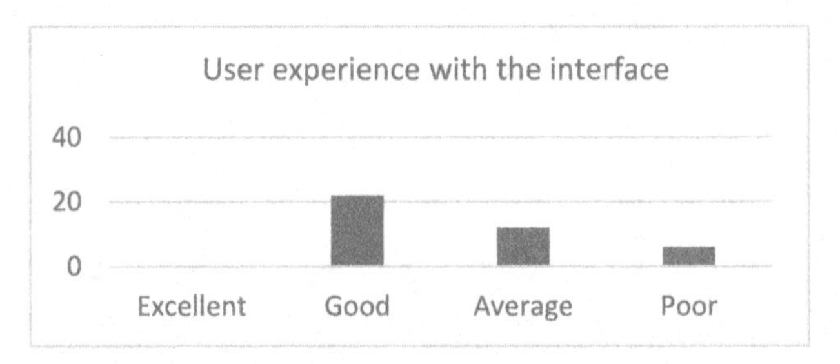

Fig. 3. User Experience with the Interface

other hand, the neutral tone of the responses and the lack of intuition prevented any outstanding evaluations.

User Experience with the Outcome. The primary expectation of a user when engaging with a chatbot is to receive comprehensive assistance with the least effort. The type and importance of the issues one aims to address through the application also play a significant role. For instance, the service quality expected from a bank or telecommunications provider differs from that of a store or municipality.

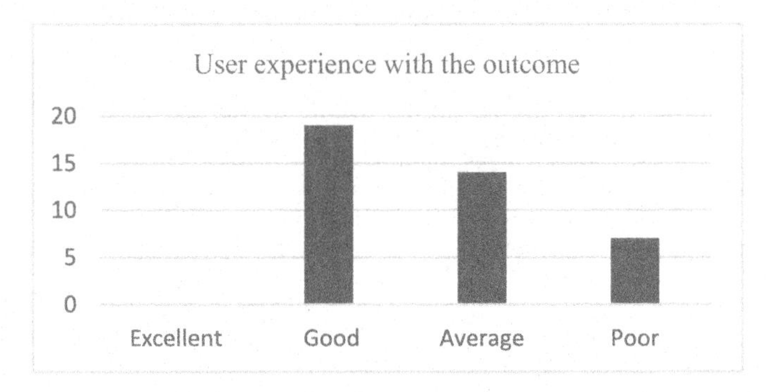

Fig. 4. User Experience with the Outcome

From the results, we observe a moderate to low satisfaction level (33/40) with the outcomes. The bot is mainly used as a search tool within the site rather than a solution-generating engine. The absence of top-rated outcomes isn't surprising since there was no instance of the "magic" of intuitive results typical of a well-functioning AI.

Overall Performance. The interviewee is asked to make an overall assessment of the solution they tested. The main criterion remains the extent and the personal effort and discomfort required to be served. However, it is important to distinguish this from the previous evaluation of the outcome. If the experience was poor, the user might not have continued to that point unless they were participating in a study. On the other hand, they might not have achieved the expected result, but the overall experience might not have been bad (Fig. 5).

Fig. 5. Overall Performance

The pattern observed in the previous questions persists here. Good to moderate performances (33/40) are predominant as users obtained some results, even if it required considerable effort. High performance was not achieved since the goal of a comprehensive intuitive system wasn't even remotely approached. On the contrary, it appears there was an informal compromise and leniency in judgments when the assessed organization or business seemed smaller.

4.2 Comparative Analysis

Based on the responses given by the participants in the survey, and after evaluating the functionality and user experience, a comparative analysis was conducted between the results to ascertain emerging trends and to verify the consistency of the answers.

Solution Category with Overall Performance. We make this comparison to see the satisfaction rate per solution category and to evaluate it (Fig. 6).

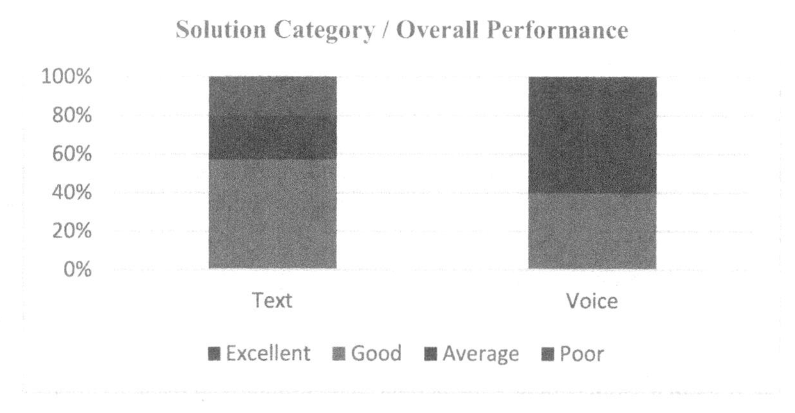

Fig. 6. Solution Category with Overall Performance

Voice solutions, although they only have good and average performance, the good performance vote is lower than the average vote. In contrast, text solutions, even though they included poor overall experiences, still had a satisfaction rate of over 50%. Considering that the voice portals came from large companies, we could infer that user had higher expectations and evaluated them more strictly. Another factor might be the stress of answering a question as in natural speech quickly and fluently.

Overall Performance by Economic Sector. We conduct this comparison to understand the final perception of the interviewees in relation to the economic sector each solution serves (Fig. 7).

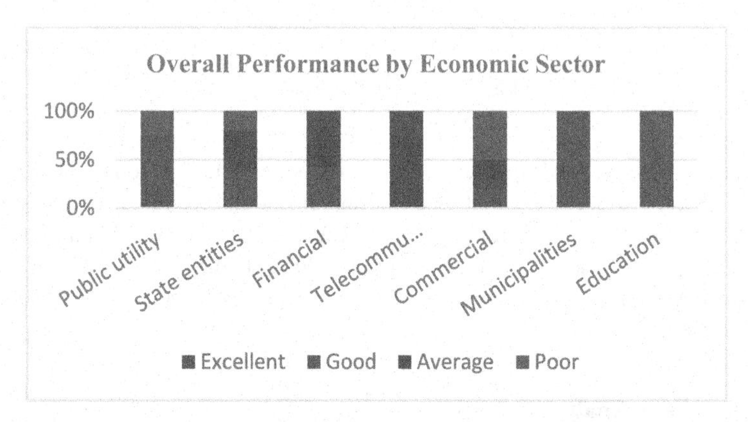

Fig. 7. Overall Performance by Economic Sector

Local governance and education enjoy a positive review. Although they predominantly provide only information, there aren't higher demands placed on them. In contrast, the commercial sector is rated quite low, and rightly so, as the adopted solutions appear to be more complementary to their websites and lack any significant investment in their dynamics.

User Experience in Relation to Overall Performance. We aim to examine whether the obtained results were consistent, given that the user interface invariably has a significant impact on the overall user experience (Fig. 8).

We observe that, with minor deviations, a good interface leads to an effective end performance. This is entirely logical, as it would be improbable for substantial resources to have been invested in the technological foundation without corresponding effort in presenting the outcome.

User Experience in Relation to Overall Performance. We aim to see if the results obtained were consistent, as the interface always has a significant impact on the user's overall experience (Fig. 9).

We observe that, with minimal deviations, a good interface leads to good overall performance. This is perfectly logical, as it wouldn't be possible for a large amount to

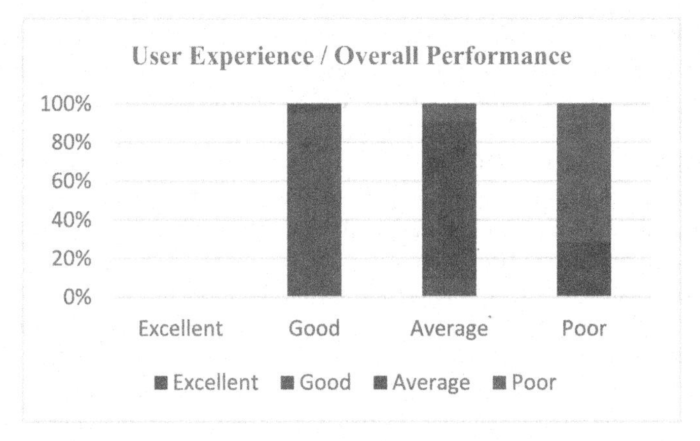

Fig. 8. User Experience. Interface to Overall Performance

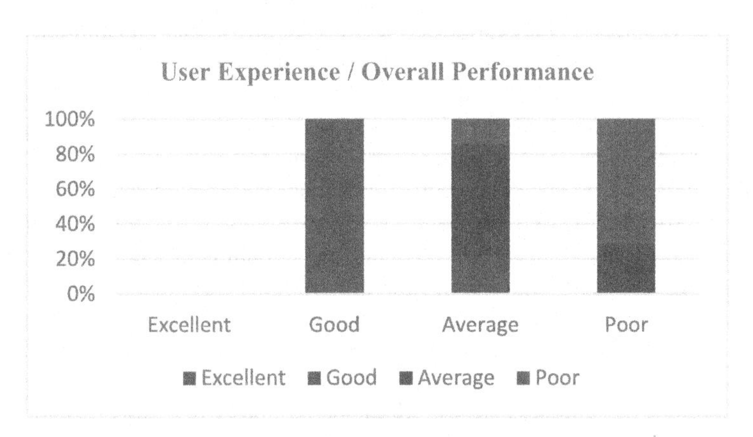

Fig. 9. User Experience. Outcome to Overall Performance

be invested in the technological backbone without corresponding work on presenting the result.

5 Conclusions

As we look to the future, it's clear that chatbots will play an increasingly significant role in our everyday lives, and their impact on society and our economy will continue to grow. The ongoing advancement of artificial intelligence and natural language processing technologies will enable chatbots to understand and respond to customer inquiries in a more human-like manner, providing more precise and personalized responses. As a result, people will feel more comfortable using them, and their adoption will become more widespread.

In Greece, the average to low level of satisfaction reflects the challenges posed by the Greek language, as well as the need for greater investments. Despite these challenges,

Greece should not lag in technological advancements. The country possesses a skilled human workforce with the appropriate educational background, and there is potential for direct application of various solutions in the tourism sector.

In the short term, emphasis should be placed on improving NLP technologies and on increasing specialized staff for training chatbots. In the medium term, investments in new technologies that are more compatible with our linguistic idiom are essential. It is also crucial to promote collaborations between universities, tech companies, the government, and private entities.

From the research results for the private sector, there is a discernible need for more comprehensive electronic services. These services should be available 24/7, allowing an increase in workflow without an uptick in payroll costs. Additionally, merely redirecting to ready-made product websites without the use of AI, which would offer advice and solutions, is unsatisfactory. Such an approach does not make chatbots appealing or attractive for usage. While digital customer service in large companies is satisfactory, it hasn't excelled, indicating that there is a demand for further emphasis on its improvement. It's suggested to initiate collaborations between academic/research institutions and the private sector to develop and train the first "Greek digital sales assistant" leveraging emerging AI platforms, like ChatGPT. Another intriguing development would be the creation of a voice-text hybrid to facilitate complex processes, such as the signing of contracts and agreements, with a higher degree of satisfaction.

For the public sector, chatbots' primary informational role, their complete disconnection from providing real services, and the low expectations set by the research team suggest that the average citizen will resort to more traditional methods if they wish to be served, thereby losing the 24/7 availability advantage. Immediate funding is reccommended for integrating all administrative processes into the National Registry of Administrative Procedures. There should also be an obligation for the legislator to model each new process in stages before it's submitted for approval by the Parliament. Lastly, creating a unified "digital assistant for administrative processes" for the entire Greek public sector could be beneficial. This assistant would support citizens in completing applications, direct them to the appropriate service, and keep them informed about the results. This could pave the way for the complete digitization of the Greek state.

Acknowledgements. This research has been co-financed by the European Union and Greek national funds through the Operational Program "Competitiveness, Entrepreneurship and Innovation", under the call "RESEARCH - CREATE - INNOVATE (2nd Cycle)" (project code: T2EDK-01921).

References

1. Knight, S.: NLP at Work. Hachette Book Group, London (2020)
2. vodafone.gr. 28 June 2022. https://www.vodafone.gr/vodafone-ellados/digital-press-office/deltia-typou/20220628-tovi-o-psifiakos-voithos-tis-vodafone-pio-exypnos-kai-apotelesmatikos-apo-pote/
3. Mageira, K., Pittou, D., Papasalouros, A., Kotis, K., Zangogianni, P., Daradoumis, A.: mpdi.com. 22 March 2022. https://doi.org/10.3390/app12073239

4. Lachana, Z., Loutsaris, M.A., Alexopoulos, C., Charalabidis, Y.: Automated analysis and interrelation of legal elements based on text mining. Int. J. E-Serv. Mob. Appl. (IJESMA) **12**(2), 79–96 (2020)
5. Koutsikakis, J., Chalkidis, I., Malakasiotis, P., Androutsopoulos, I.: Cornell University. arxiv.org: https://arxiv.org/pdf/2008.12014v2.pdf. 03 September 2020
6. Tullis, T., Albert, B.: Measuring the User Experience: Collecting, Analyzing, and Presenting Usability Metrics. Morgan Kaufmann, Burlington (2008)
7. Shawar, A., Atwell, E.: Different measurements metrics to evaluate a chatbot system. In: Proceedings of the Workshop on Bridging the Gap: Academic and Industrial Research in Dialog, pp. 89–96. Seattle: NAACL (2007)
8. Stent, A., Marge, M., Mohit, S.: Evaluating evaluation methods for generation in the presence of variation. In: International Conference on Intelligent Text Processing and Computational Linguistics, pp. 351–354. Springer, Berlin (2009)
9. Jelinek, F., Mercer, Salim: Principles of lexical language modeling for speech recognition. In: Advances in Speech Signal Processing. Dekker Publishers, New York (1991)
10. Liu, B.: Sentiment Analysis and Opinion Mining. In: Synthesis Lectures on Human Language Technologies. Morgan & Claypool Publishers, San Rafael (2012)
11. Manning, C., Raghavan, P., Schütze, H.: Introduction to Information Retrieval. Cambridge University Press, Cambridge (2008)
12. Luger, E., Sellen, A.: Like having a really bad PA: the gulf between user expectation and experience of conversational agents. In: Proceedings of the 2016 CHI Conference on Human Factors in Computing Systems (pp. 5286–5297). Association for Computing Machinery, New York (2016)
13. Radziwill, N., Benton, M.: Cornell University. arxiv.org: https://arxiv.org/ftp/arxiv/papers/1704/1704.04579.pdf. (2017)
14. Koutsampasis, P.: University of Aegean, Department of Product and systems design engineering. eclass.aegean.gr (2015). https://eclass.aegean.gr/modules/document/file.php/511265/merged_document5.pdf
15. Oppenheim, A.: Questionnaire Design, Interviewing and Attitude Measurement. Continuum International Publishing, London (1992)

The Effects of Economic Crisis on the Digitalization of the Greek Social Security

Kavallari Chryso(✉) and Euripidis Loukis

University of Aegean, Samos, Greece
{kavallari,eloukis}@aegean.gr

Abstract. Economic crises repeatedly appear in market-based economies and have serious consequences on them. They are causing serious decreases of the sales revenue and therefor the financial resources of organizations, as well as of their operations and investments. However, they can have some positive effects as well, leading to processes' rationalizations in important functions of them, improvements of their efficiency and better exploitation of their resources. In this paper we investigate the effects of the strong economic crisis that hit Greece between 2010–2018 on the digitization of one of the most important and costly domains of government activity: the social security. It has been concluded that during the economic crisis period there has been a large decrease of the ICT-related investment and operating expenses; however due to some improvement and rationalization of ICT-related processes and practices, as well as a better utilization of knowledgeable ICT personnel (mainly of the central 'Electronic Government Center for Social Security'), a significant increase in the digitalization of the Greek Social Security has been achieved during this difficult period.

Keywords: digitalization · social security · economic crisis · e-government · digital government

1 Introduction

Economic crises repeatedly appear in market-based economies and have serious consequences on them [1–4]. Recently we had the severe 2007 Global Financial Crisis, which had quite negative impact on the economies and the societies of most countries; furthermore, some years ago we had another economic crisis that was initiated by the COVID-19 pandemic [5, 37]. They are causing serious decreases of the financial resources of organizations, as well as of their operations and investments. However, they can have some positive effects as well, leading to processes' rationalizations in important functions of them, improvements of their efficiency and better exploitation of their resources.

However, though we tend to have less and shorter economically 'normal' periods, and more and longer economic crisis periods, with quite serious consequences for the economy and the society, limited empirical research has been conducted on the effects of the economic crises on various aspects of the functioning of organizations; a brief review of this research is provided in Sect. 2.2. Furthermore, this limited empirical

M. Papadaki et al. (Eds.): EMCIS 2023, LNBIP 501, pp. 182–191, 2024.
https://doi.org/10.1007/978-3-031-56478-9_13

research has focused mainly on the effects of economic crises on the financial aspects of organizations. However, only a very small part of it is dealing with the effects of economic crises on the digitalization of organizations, despite the high importance of the latter for modern organizations. Another gap of this limited empirical research on the effects of the economic crises on aspects of organizations is that is has dealt exclusively with the private sector, but not with the public sector.

This paper contributes to filling these important research gaps. It investigates empirically the effects of the strong economic crisis that hit Greece between 2010 and 2018 [6] on the digitalization of one of the most critical and costly domains of government activity: the social security. Social security is of critical importance for modern societies, due to their aging population (a significant and continuously increasing proportion of the population of most countries are elderly individuals who receive pensions and need extensive health care), the increase of chronic diseases as well as the growing inequalities, unemployment and social exclusion (that necessitate large transfers of financial resources to the weakest individuals); furthermore, social security constitutes one of the largest areas of public expenditure in OECD countries [7]. For the above reasons the use of ICT is of critical importance for social security, and for welfare in general, in order to manage efficiently the huge number of beneficiary citizens, the complex administrative processes and rules for the provision of many different kinds of benefits and pensions, as well as the enormous financial resources, and there is a rapid advancement toward 'digital social security' and 'digital welfare' [8–11].

The research objectives of this paper are to:

- investigate the effects of the strong economic crisis that hit Greece between 2010 and 2018 on the digitalization of its social security (meant as development of IS for the support and digital transformation of social security organizations);
- develop an appropriate research model for this, which enables the identification of both the negative effects and also the possible positive ones, based on sound theoretical foundations.

The paper consists of five Sections. The following Sect. 2 presents the background of our study, and then in Sect. 3 its method and data are described. In Sect. 4 the results are presented and discussed in Sect. 4, while in the final Sect. 5 the conclusions are summarized, and future research directions are proposed.

2 Background

2.1 The Greek Social Security Landscape

The Greek social security has been traditionally problematic [12, 13]. There has been a high fragmentation into a large number of individual social security organizations (with each of them being responsible for a specific professional group of citizens, such as lawyers, engineers, journalists, etc.), and this caused many problems: they could not have 'economies of scale', and also as sufficient specialized personnel for some important sophisticated tasks, such as actuarial analyses, funds' investments and digitization; especially for the latter they had small ICT units, with small numbers of ICT personnel, and lack of specialized ICT personnel for some specific important digital technologies.

Also, these numerous social security organizations offered different amounts of benefits and pensions and had different rules for granting them (determined mainly by the political power of the corresponding professional groups and their trade unions). Furthermore, the Greek social security traditionally had high levels of deficits, since the benefits and pensions offered were higher than the incomes (coming mainly from the contributions of employees and employers, as well as the ones from self-employed citizens); these deficits were financed by government, resulting in significant increase of government debt.

For these reasons in the beginning of 2017 these numerous social security organizations were merged into the 'Single Social Security Entity' (EFKA) (https://www.efka.gov.gr). So, in the last two years of the time period covered by our study (2010–2018) the Greek social security had this additional burden (beyond the multiple burdens that the economic crisis had generated) of merging these numerous social security organizations, with respect to their processes, rules for granting various kinds of benefits and pensions, as well as their levels/amounts, registries of beneficiaries and also their information systems.

Therefore, in order to investigate the effects of the economic crisis that hit Greece between 2010 and 2018 on the digitalization of its social security it is necessary to collect relevant data both from the IDIKA and the ICT unit of EFKA.

2.2 Effects of Economic Crises on Organizations

Limited empirical research has been conducted concerning the effects of the economic crises on various aspects of the functioning of organizations, despite the huge economic and social importance of this topic, as well as their increasing frequency in the new century. Most of this empirical research has a financial perspective: it investigates the effects of the 2008 global economic crisis on financial aspects of private sector firms, such as their corporate investment and its finance, in various countries [14–19].

Only the study presented in [20] investigates empirically the effects of economic crisis on firms' digitization; it examines the ICT-related behavior of the five main 'system-relevant' Greek banks in the strong economic crisis that Greece experienced. It has concluded that initially the Greek banks reacted defensively to the crisis by reducing their reducing ICT-related expenses, however later they proceeded to a substantial rationalization of their ICT-related processes and therefore improvement of their relevant ICT capabilities. So, we can distinguish both negative and positive effects of the strong Greek economic crisis on the main banks with respect to digitization.

So, the paper makes a contribution to the limited research stream on this highly important topic, the effects of economic crises on organizations; focusing on the effects on digitization, which are quite under-researched, despite the importance of the use of digital technologies for modern organizations. Furthermore, since this research stream has dealt exclusively with private sector firms, we are focusing on the public sector, examining the effects of economic crisis on the digitization of an important class of government organizations, which are responsible for one of the most critical and costly domains of government activity: the social security. For this purpose, we will be based on the 'Resource-based View' (RBV) of the firm as theoretical foundation, which is outlined in the following Sect 2.3.

2.3 Resource-Based View of the Firm

One of the most widely recognized and used theories in management science if the 'Resource-Based View' (RBV) of the firm [21–23]. According to this theory the main determinants of a firm's performance are on one hand its 'resources' (meant as all kinds of firm's assets, such as equipment, buildings, personnel, etc.) and on the other fact its 'capabilities' (meant as its abilities to select, deploy, exploit and manage these resources in order to perform the main functions and tasks of the firm efficiently and effectively); so performance differences among firms operating in the same external environment are mainly created by differences among them with respect to available resources and capabilities. The RBV theory, though initially developed for private sector firms, has been subsequently used successfully for public sector organizations as well [24–26].

The RBV theory holds not only for the whole firm, but also for each of its functions and tasks separately. So, for the case of the digitization of a firm (meant as the integration of digital technologies in the performance of firm's functions and tasks) there has been extensive relevant research [27–35], which has revealed that its performance and value is determined:

- on one hand by its ICT-related resources, such as ICT hardware, ICT software, ICT personnel, ICT-related external services, etc.,
- and on the other hand, by its ICT-related capabilities, meant as firm's abilities to select, deploy, exploit and manage ICT-related resources in order to increase the efficiency and the effectiveness of the firm.

3 Method and Data

In order to develop our research model we have taken into account that, as mentioned in the Introduction, according to relevant economic research [1–4], economic crises have on one hand negative effects on organizations, leading to decrease of the available financial resources for performing their functions and tasks (i.e. for financing their operational and investment activities); however, on the other hand economic crises have positive effects as well, putting pressure on the organizations to improve the processes they follow for performing their functions and tasks, and exploit better their resources, improving their relevant capabilities. Therefore, economic crises affect both the resources and the capabilities of organizations, so the RBV is an appropriate theoretical foundation for examining the effects of economic crises on organizations (both the overall effects, and also the effects on specific functions or tasks of them), which enables identifying both the negative effects and also the possible positive ones.

So, focusing on the effects of the economic crisis on the digitization of the Greek social security, we examine its effects on the three main elements of the RBV (resources, capabilities, performance):

a) on the financial resources spent for ICT, on one hand for ICT-related investments, and on the other hand for ICT-related operating expenses;
b) on the main ICT-related processes and therefore the relevant firm's ICT-related capabilities;

c) and on the performance of the digitization, meant as development of information systems (IS) for the support and digital transformation, of the Greek social security organizations.

The above multi-dimensional research model is shown in Fig. 1.

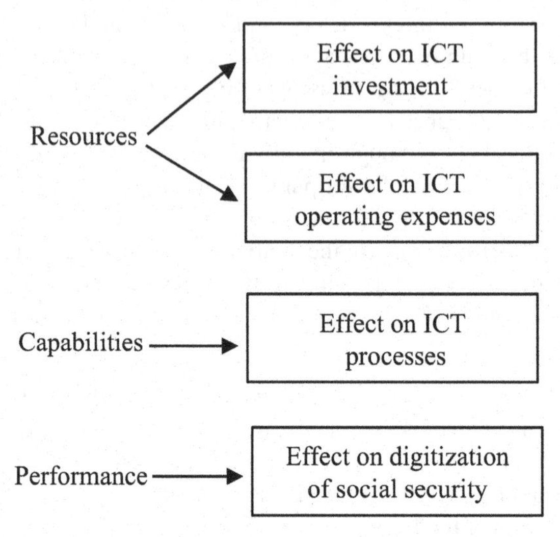

Fig. 1. Research model

The above are examined for each of the two main actors of the digitization of the Greek social security described in Sect. 2.1: the IDIKA, and the ICT units of the individual social security organizations, which have now been consolidated and merged into the ICT unit of the EFKA.

For collecting data about the above components and sub-components of our research model we used a combination of qualitative and quantitative techniques [36]. In particular, we conducted one focus group in the IDIKA, in which participated the President of it, as well as three highly experienced Directors, who had a good knowledge of the development activities over the crisis period 2010–2018; and another focus group in the EFKA, in which participated the ICT Director of it, as well as four experienced former Directors of social security organizations (that were merged into EFKA). In each of these two focus groups the participants initially answered-filled collaboratively a questionnaire, which included one question for each of the sub-components of the four components of our abovementioned research model; for each question each focus group through some discussion arrived at one consensus response'. Then followed an in-depth discussion about the answers than had been provided to these questions, in order to provide further clarifications and explanations about them.

4 Results

In Table 1 are shown the responses of the two focus groups (the IDIKA and the EFKA ones) to the four questions, which concern the general/overall effect of the economic crisis on the ICT-related investments, the ICT-related operating expenses, the improvement and rationalization of the ICT-related processes and the digitalization performance (development of IS for the support and digital transformation of social security organizations).

Table 1. General/overall effects of the economic crisis

	IDIKA	EFKA
on ICT-related investment	7	6 in 2010–2016 3 in 2017–2018
on ICT-related operating expenses	7	6
on the improvement/rationalization of ICT-related processes	3	4
on the digitization (development of IS for social security)	1	6 in 2010–2016 3 in 2017–2018

The first, second and fourth questions are assessed in a 7-points Lickert scale (1 = large increase, 2 = moderate increase, increase, 3 = small increase, 4 = no effect, 5 = small decrease, 6 = moderate decrease, 7 = large decrease); the third question is assessed in a 5-points Lickert scale (1 = not at all, 2 = to a small extent, 3 = to a moderate extent, 4 = to a large extent, 5 = to a very large extent).

We can see in Table 1 that in IDIKA there has been a large decrease of ICT-related investment in the crisis period 2010–2018, and also overall a moderate decrease in the individual social security organizations before their merging (in the 2010–2016 period). In the qualitative focus group discussions, it was mentioned that in most social security organizations there was a large decrease of ICT investment, and only in the 'Social Security Foundation' (IKA), which is the biggest social security organization, there was a stability in the ICT investments, in order to continue the development of its integrated IS. However, after their merging into the EFKA (in the 2017–2018 period) there was a small increase of ICT-related investment of the EFKA, in order: a) to address the new needs posed by this merging, such as the merging of the digital resisters of insured persons of the numerous merged social security organizations (which had problems of data quality, as well as different structure and coding); and b) the need for substantial digital support for overcoming some long-standing problems of the Greek social security, e.g. for the improvement of the collection of insurance contributions, as well as of the control of their high expenditures (both had been traditionally problematic, resulting in high deficits, which were one of the main causes of the crisis). So, overall, there has been a severe decrease of ICT investment during the period of the crisis (2010–2018); the main reason for this had been the reduction of the funding from the central government for social security, due to the strict austerity programs that had been put into effect for overcoming the crisis.

For the same reason, as we can see in Table 1, we had large decrease of the ICT-related operating expenses in the IDIKA, and also decrease of them in the social security organizations, and later in the EFKA. In the qualitative focus group discussions, it was mentioned that this was mainly due to the reduction of the ICT personnel, as there was not recruitment of new ICT personnel for replacing the older ones who retired, and also the contract temporary ICT personnel were terminated.

However, in Table 1 we can see that the crisis has lead to a large extent to the improvement and rationalization of the ICT-related processes and corresponding capabilities of the individual social security organizations and later in the EKFA, which as mentioned in Sect. 2.1, and also in the qualitative focus group discussions, before the crisis had small and immature ICT units that lacked efficient processes and practices, so there was a significant margin for improvement. Furthermore, the crisis led to a moderate extent to the improvement and rationalization of the ICT-related processes and corresponding capabilities of the IDIKA; in the qualitative focus group discussions it was mentioned that this is quite important, since IDIKA is a large, experienced and mature ICT organization, which already had efficient ICT-related processes and practices in place before the crisis.

Furthermore, in the last line of Table 1 we can see that there was a very large increase of the digitization performance of the HDIKA, concerning the development of IS for the social security organizations, though in the individual social security organizations there was overall a moderate decrease until they merged into EFKA (during the 2010–2016 period), followed by a small increase after the merging (for the reasons discussed in 4.1). In the qualitative focus group discussion we had with IDIKA they emphasized that during the crisis period though there was a large decrease of the financial resources for ICT-related investments and operating costs, less ICT personnel as well as dramatic decrease of ICT-related training, there was an 'explosion' of new critical IS for social security, which transformed critical functions and tasks of it. The most important of them was the 'e-prescription' one, which enables a seamless flow of data between doctors, pharmacies, medical examination centers/laboratories, hospitals and social security organizations, and increases dramatically the efficiencies of all these actors, while it also provides large amounts of data to be used for the analysis, control and rationalization of the huge social security expenditures. Also, new IS were developed, which: a) on hand enabled citizens to apply digitally through the Internet for the new benefits that were introduced for supporting citizens who had been severely hit by the crisis, and also for the pre-existing benefits as well as, and b) on the other hand enabled the automated processing of these applications (which is usually quite complex, requiring the application of multiple rules). These IS increased dramatically the efficiency and productivity of the social security organizations, enabling them to handle a much bigger workload (because of the increase of the benefits due to the crisis, as well as of the citizens eligible for them) with less personnel, and also to provide higher quality services to the citizens. Furthermore, IDIKA made a substantial contribution to the abovementioned merging of the digital resisters of insured persons of the numerous merged social security organizations, which was carried out by EFKA.

According to the participants in the IDIKA focus group the main reasons for this high performance of the IDIKA concerning the development of new IS for social security during the crisis period were the following:

- Greece had signed binding agreements with the EU for receiving financial support from it, in order to overcome the crisis, which included strict obligations to rationalize its social security, and reduce its high deficits; this necessitated a substantial increase of its digitalization, and the development of some critical IS that were necessary for overcoming long-standing problems of the Greek social security. So, there was a big pressure from government to the IDIKA to proceed as soon as possible to the development of these critical IS with the available significantly reduced resources.
- Though IDIKA was a large and experienced ICT organization, with mature and efficient ICT-related processes and capabilities, during the crisis it made further improvements of them, in order to meet the abovementioned digitization requirements with significantly reduced resources. For this purpose, they also proceeded to a better utilization of existing highly knowledgeable and experienced ICT personnel, who were under-utilized previously (as they were assigned less development tasks, of lower complexity, that did not fully leverage their true abilities). Furthermore, they made better utilization of the existing equipment, with most of it being obsolete, since, as mentioned above, there has been a large decrease in the ICT hardware investments in ICT during the economic crisis.

5 Conclusions

As mentioned in the Introduction and in Sect. 2.2 there has been quite limited empirical investigation of the effects of the most severe and disruptive event that often appears in firms' external environment, the economic crises, on their digitization; furthermore, this limited empirical investigation id dealing exclusively with private sector firms. This paper contributes to filling these important research gaps. It describes an empirical investigation of the effects of the strong economic crisis that hit Greece between 2010–2018 on the digitization of one of the most important for the society and costly domains of government activity: the social security. For this purpose, we have constructed a research framework, which is based on the RBV theory of the firm, as our theoretical foundation, examining ICT-related resources, capabilities and performance; data have been collected through a combination of qualitative and quantitative techniques.

We have reached a quite interesting conclusion, which is to some extent counter-intuitive and different from our initial expectations: through during the economic crisis period there has been a large decrease of the ICT-related investment and operating expenses, due to some improvement and rationalization of ICT-related processes, as well as a better utilization of knowledgeable ICT personnel (mainly of the central 'Electronic Government Center for Social Security'), a significant increase in the digitalization of the Greek Social Security has been achieved. Important IS have been developed for the support and digital transformation of the Greek social security organizations, with respect both to their internal functions and processes, and also their transaction with citizens and firms.

Further empirical research is required for investigating the effects of economic crisis on the digitization of other types of government organizations, with different levels of importance (especially during crisis) as well as different initial levels of digitalization (before the crisis), and possibly in different national contexts, with different types and intensities of economic crises. For this research the research framework we have developed, based on the RBV theory of the firm, can be useful. Furthermore, this research framework can be further developed and enhanced, using additional theoretical foundations.

References

1. Keeley, B., Love, P.: From Crisis to Recovery - The Causes, Course and Consequences of the Great Recession. OECD Publishing, Paris (2012)
2. Knoop, T.A.: Recessions and Depressions: Understanding Business Cycles, 2nd edn. Praeger Santa Barbara, California (2015)
3. Allen, R.E.: Financial Crises and Recession in the Global Economy, 4th edn. Edward Elgar Publications, Cheltenham (2017)
4. Vartanian, T.P.: 200 Years of American Financial Panics: Crashes, Recessions, Depressions, and the Technology that Will Change It All. Prometheus Books, Guilford, Connecticut (2021)
5. Baldwin, R., Di Mauro, B.W.: Mitigating the COVID Economic Crisis: Act Fast and do Whatever it Takes. Center of Economic Policy Research Press, London (2020)
6. Gourinchas, P.O., Philippon, T., Vayianos, D.: The analytics of the Greek crisis. Nat. Bur. Econ. Res. (NBER) Macroecon. Ann. **31**(1), 1–81 (2016)
7. OECD (2016). Digital Government Strategies for Transforming Public Services in the Welfare Areas. OECD Publishing. Paris
8. McKinnon, R.: Introduction - social security and the digital economy – managing transformation. Int. Soc. Secur. Rev. **72**(3), 5–16 (2019)
9. Larsson, A., Teigland, R.: An introduction to digital welfare: a way forward? In: Larsson, A., Teigland, R. (eds.), Digital Transformation and Public Services - Societal Impacts in Sweden and Beyond, Routledge: Oxon, UK (2020)
10. Coles-Kemp, L., Ashenden, D., Morris, A., Yuille, J.: Digital welfare: designing for more nuanced forms of access. Policy Des. Pract. **3**(2), 177–188 (2020)
11. Larsson, K.K., Haldar, M.: Can computers automate welfare? - Norwegian efforts to make welfare policy more effective. J. Extreme Anthropol. **5**(1), 56–77 (2021)
12. Zambarloukou, S.: Greece after the crisis: still a south European welfare model? Eur. Soc. **17**(5), 653–673 (2015)
13. Expanatory Report of Law 4387/2016 "Unified social security system - Reform of the social security-pension system". https://www.parliament.gr. Accessed
14. Duchin, R., Ozbas, O., Sensoy, A.: Costly External Finance, Corporate Investment, and the Subprime Mortgage Credit Crisis. J. Financ. Econ. **97**(3), 418–435 (2010)
15. Akbar, S., Ur Rehman, S., Ormrod, P.: The impact of recent financial shocks on the financing and investment policies of UK private firms. Int. Rev. Financ. Anal. **26**(C), 59–70 (2013)
16. Kahle, K., Stulz, R.M.: Access to capital, investment, and the financial crisis. J. Financ. Econ. **110**(2), 280–299 (2013)
17. Bo, H., Driver, C., Lin, H.-C.M.: Corporate investment during the financial crisis: evidence from China. Int. Rev. Financ. Anal. **35**, 1–12 (2014)
18. Nguyen, T., Nguyen, H.G.L., Yin, X.: Corporate governance and corporate financing and investment during the 2007–200 financial crisis. Financ. Manage. **44**(1), 115–146 (2015)
19. Zubair, S., Kabir, R., Huang, X.: Does the financial crisis change the effect on financing on investment? Evidence from private SMEs. J. Bus. Res. **110**, 456–463 (2020)

20. Loukis, E., Arvanitis, S., Myrtidis, D.: ICT-related behaviour of Greek banks in the economic crisis. Inf. Syst. Manage. **38**(1), 79–91 (2021)
21. Barney, J.B.: Firm resources and sustained competitive advantage. J. Manag. **17**(1), 99–120 (1991)
22. Barney, J.B.: Resource-Based Theory: Creating and Sustaining Competitive Advantage. Oxford University Press, New York (2007)
23. Nason, R.S., Wiklund, J.: An assessment of resource-based theorizing on firm growth and suggestions for the future. J. Manage. **44**(1), 32–60 (2018)
24. Klein, P., Mahoney, J., McGahan, A., Pitelis, C.: Resources, capabilities, and routines in public organizations. SSRN Electr. J. (2011). http://ssrn.com/abstract=1550028
25. Melián-González, A., Batista-Canino, R.M., Sánchez-Medina, A.: Identifying and assessing valuable resources and core capabilities in public organizations. Int. Rev. Adm. Sci. **76**(1) (2010)
26. Pang, M.S., Lee, G., DeLone, W.H.: IT resources, organizational capabilities, and value creation in public-sector organizations: a public-value management perspective. J. Inf. Technol. **29**, 187–205 (2014)
27. Mata, F.J., Fuerst, W.L., Barney, J.B.: Information technology and sustained competitive advantage: a resource-based analysis. MIS Q. **19**(4), 487–505 (1995)
28. Feeny, D.F., Willcocks, L.P.: Core IT capabilities for exploiting information technology. Sloan Manage. Rev. **39**(3), 9–21 (1998)
29. Bharadwaj, A.: A resource-based perspective on information technology capability and firm performance: an empirical investigation. MIS Q. **24**(1), 169–196 (2000)
30. Wade, M., Hulland, J.: Review: the resource-based view and information systems research: review, extension, and suggestions for future research. MIS Q. **28**(1), 107–142 (2004)
31. Ravichandran, T., Lertwongsatien, C.: Effect of information system resources and capabilities on firm performance: a resource-based perspective. J. Manag. Inf. Syst. **21**(4), 237–276 (2005)
32. Liang, T.P., You, J.J., Liu, C.C.: A resource-based perspective on information technology and firm performance: a meta-analysis. Ind. Manag. Data Syst. **110**(8), 1138–1158 (2010)
33. Gu, J.W., Jung, H.W.: The effects of IS resources, capabilities, and qualities on organizational performance: an integrated approach. Inf. Manage. **50**(2–3), 87–97 (2013)
34. Garrison, G., Wakefield, R.L., Kim, S.: The effects of IT capabilities and delivery model on cloud computing success and firm performance for cloud supported processes and operations. Int. J. Inf. Manage. **35**(4), 377–393 (2015)
35. Raymond, L., Uwizeyemungu, S., Fabi, B., St-Pierre, H.: IT capabilities for product innovation in SMEs: a configurational approach. Inf. Technol. Manage. **19**, 75–87 (2018)
36. Maylor, H., Blackmon, K., Huemann, M.: Researching Business and Management, 2nd edn. Red Globe Press, UK (2017)
37. Papadaki, M., Karamitsos, I., Themistocleous, M.: Covid-19 digital test certificates and blockchain. J. Enterp. Inf. Manage. **34**, 993–1003 (2021). https://www.researchgate.net/publication/353272635_ViewpointCovid-19_digital_test_certificates_and_blockchain

Design, Implementation, and Evaluation of a Food Price Monitoring Tool for Supporting Data Journalists

Papageorgiou Georgios(✉), Lamprinidis Anastasios, and Loukis Euripides

Department of Information and Communications Systems Engineering, University of the Aegean, Mytilene, Greece
{gpapag,icsd18117,eloukis}@aegean.gr

Abstract. Data journalism is a valuable emergent and growing trend, which can lead to higher quality journalism, based on real-life data and not on prejudice and pre-existing stereotypes. However, the effective realization (i.e., the 'real-life' implementation) of this concept requires the development of high-quality software tools that enable journalists to access useful data sources, are easy to use, and provide clear, understandable and intuitive presentations of the data, as well as practices and processes for using them. In this direction our paper is making a contribution. It describes the design, implementation and evaluation of a software tool that utilizes open data sources concerning food prices in Greece, in order to support data journalism on this topic, which constitutes one of the most critical topics that the societies of many countries face. It is a fully automated solution that can generate new visual reports whenever the data provider updates the food prices data and requires minimal intervention from the journalists; this makes the above data usable for journalistic purposes. The results of the evaluation were positive, indicating a high degree of journalists-users satisfaction, and at the same time revealed difficulties in using these open data for journalism purpose due to quality problems, and also revealing directions for future development of the tool.

Keywords: Open Data · Data Journalism · Journalism · Food Price · Information Systems Success Models

1 Introduction

Food prices hold immense significance as food insecurity can substantially result in social unrest and threaten the legitimacy of governments [1, 2]; therefore, close monitoring of food prices can mitigate such evolutions, as it can allow addressing the problem in an early stage. In the contemporary socio-political landscape dominated by the Russo–Ukrainian conflict [3], rising energy prices and increased inflation, especially in Europe, monitoring and communicating the rising food prices to the public, as well as putting pressure on governments to take actions for addressing this problem when required, is of pivotal importance for maintaining social and political coherence. Journalism, often seen as the 'Fourth Estate', can play an elevated role not only in raising public awareness

M. Papadaki et al. (Eds.): EMCIS 2023, LNBIP 501, pp. 192–206, 2024.
https://doi.org/10.1007/978-3-031-56478-9_14

of this crucial subject, but also in acting as a watchdog and applying pressure on the government by holding them accountable for their actions or lack thereof concerning the control of food prices.

In our increasingly interconnected world driven by information, the open data movement promoted the release of data that is freely available, accessible and reusable by anyone. In 2009, the USA government launched Data.gov. After that, other countries joined the open data movement and started to release their data to increase innovation and transparency. Embracing open data can revolutionize business operations, drive innovation, and create new opportunities in the digital landscape [4]. Most importantly, it holds immense promise for fostering transparency, accountability and informed decision-making across various sectors, including improving public services [5]. Among a variety of stakeholders from the public and private domains that can significantly benefit from this explosion of available data, journalists can particularly reap its benefits. This abundance of data led in part to the formation of a new trend in journalism, the data journalism. Veglis and Bratsas (2017) [6] define this new form of journalism as the use of data in all the stages of the journalistic process, the extraction of information, the compilation and the visualization, in a comprehensive way. By utilizing open data, high quality journalism can be developed, which is not based on prejudice and pre-existing stereotypes, but on data from real life, and can go into more detail in analyzing effectively the important societal problems, such as the increase in food prices; this will improve the trustworthiness of journalism, build trust with the public and enable policymakers to make more informed decisions to address the social problems. However, this still abstract concept has to evolve into a set of specific easily applicable ICT-based practices: the effective realization (i.e., the 'real-life' implementation) of this highly valuable data journalism concept requires the development of:

a) High-quality software tools that enable journalists to access useful data sources, are easy to use even by low digital skills journalists, and provide clear, understandable and intuitive presentations of the data; this is very important, as the data are usually provided in tabular forms, which make it difficult for the journalists to understand the data and make sense from them, so they have to be converted into more understandable visualized firms
b) Appropriate practices and processes of using these software tools and data, as well as integrating them in the wider journalism processes.

Our paper makes a contribution to the former. In particular, it presents our research, which was conducted at the news portal HuffPost (Greece), and included the design, implementation and evaluation of a software tool that utilizes open data sources concerning food prices in Greece, in order to support data journalism on this topic, which constitutes one of the most critical topics that the societies of many countries face. The big increase of the prices of food that has taken place recently has undermined substantially the quality of life of millions of citizens, which – if not properly addressed by governments – might give rise to social unrest and political extremism, with quite negative consequences. In particular, our main objective was to provide a fully automated solution that requires minimal intervention from the journalists, which uses as input data provided (as Excel files and also through an API) and can generate new visual reports whenever the data provider updates the data; therefore, these visualizations will be immediately ready to be used by the journalists, enabling them to understand the data easily and make sense of them; this makes these data immediately usable for journalistic purposes. Our research included also an evaluation of the tool by gathering feedback from its users to assess its usefulness and potential impact in supporting journalists in their work. With our findings, we aim to take a step forward in exploring the open data journalism domain.

Our paper consists of six sections. The following Sect. 2 describes the background of our study, and in Sect. 3, we outline our methodology. In Sect. 4, the results are presented and then discussed in Sect. 5, while in the final Sect. 6, the conclusions are summarized, and future research directions are proposed.

2 Background

In this section we outline the background of our study; it includes: a) the research that has been done concerning data journalism tools that use open data (in Sect. 2.1); and b) the widely recognized and used Delone and McLean model of Information Systems Success, which has been used as a basis for evaluating the abovementioned data journalism software tool (in Sect. 2.2).

2.1 Open Data Journalism Tools

Previous research on the intersection of open data and data journalism, referred to as open data journalism, is limited according to the literature review [7]. Most of the publications on open data journalism revolve around creating software tools that can assist journalists with data collection and analysis, as well as visualization of their findings, aiming to provide technological solutions to the lack of technical skills on the journalistic side. However, most of these tools were not focused on specific problems; they could enhance journalists' capabilities in the fields of the journalistic process: data collection, data analysis, visualization, and presentation. Furthermore, only some of these technical solutions were accompanied by real-world case studies [8–10], or even limited evaluation scenarios [11, 12], and in other cases, they only provide a demonstration of the tool [13, 14].

For instance, Gupta et al. [8] contributed a paper about the state elections in India that focuses on communicating electoral insights by employing interactive visual techniques and promoting engaging data communication. Petricek [11] describes a user-friendly low-coding tool, named Gamma, for data exploration and analysis, and demonstrates its capabilities by evaluating it with a user study involving 13 participants; however, this experiment was conducted in a controlled environment. A paper by Bozsik et al. [9] showcases the creation of an affordable housing dashboard with the involvement of local stakeholders in the town of Charlottesville; however, there is no evaluation of the tool and the impact it can have on the journalists. Shehu et al. [10] present a tool for gathering, analysing, and visualising findings based on data for public procurements in North Macedonia, utilising data organisation techniques and the application of data mining algorithms. Another notable publication by Evequoz et al. [12] presents a tool for visualising and analysing parliamentary voting. The study also includes an evaluation of the tool's usefulness, although it was conducted with limited exposure of the users to the platform. The paper by Cao et al. [14] presents a technological solution that can assist journalists in enhancing fact-checking functionalities by extracting linked open data from Excel sheets. In the Belink [13] paper, a tool for fact-checking is presented for use in data journalism; the authors use complex SPARQL 1.1 queries to extract timed facts, statements, and beliefs.

The majority of these publications lacked an evaluation of the tools and while they seem to hold value for the journalistic community, no concrete evidence supports this claim. In the case of Gamma, an evaluation of its usability was included in the publication, but it was conducted in a controlled environment. In the publication by Evequoz et al. [12], analyzing parliamentary voting in Sweden, users had limited exposure to the tool. Therefore, it becomes evident from examining these studies that more research on the usefulness of these tools is required to accurately measure their impact and, if necessary, guide their development to better suit the needs of journalists based on solid evidence. So, this paper describes another useful software that supports data journalism concerning a critical topic, which affects negatively millions of citizens undermining their lives: the food prices.

2.2 The DeLone and McLean Information Systems Success Model

The DeLone and McLean information systems success model [15] was based on the previous work on Mason [16] concerning the measurement of the output of an information system. DeLone and McLean identified six critical dimensions of information systems success: System Quality (technical level), Information Quality (semantic level) and Use, User Satisfaction, Individual Impact, and Organizational Impact (influence level) (Fig 1).

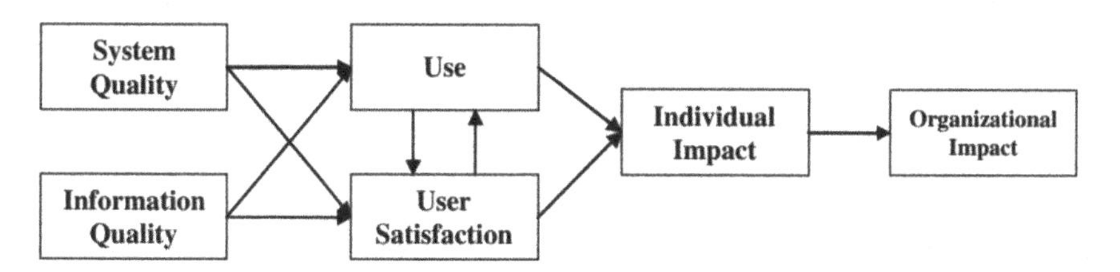

Fig. 1. DeLone and McLean IS Success Model (1992)

This model was designed to be complete and parsimonious, aiming to provide a holistic view of information systems success while maintaining simplicity. Since its initial formulation in 1992, researchers have widely adopted and cited the DeLone and McLean model. However, the model did not remain static. Researchers in the field of information systems began proposing enhancements and modifications shortly after its publication. Recognising the dynamic nature of information systems, feedback loops were incorporated into the model to account for system evolution and maintenance. These adjustments, based on both empirical studies and constructive feedback, led to an updated version of the DeLone and McLean Model in 2003 [17]. So, we have used this model as a basis (with some adaptations) for evaluating the data journalism software tool we developed (see Sect. 3.3 and then Sect. 4.2).

3 Methodology

This section outlines the methodology employed for the design, development, and evaluation of the open data journalism tool we developed. Our methodology consisted of three interrelated phases, each contributing to the creation of an automated tool for transforming raw open data about food prices in Greece into compelling visual narratives.

3.1 Understanding the Needs

The first phase involved a comprehensive literature review of Open Data, Data Journalism, and Open Data Journalism, combined with discussions with journalists from HuffPost Greece concerning their data needs. This investigation revealed the need for an automated solution that could translate open data into visual representations to boost their data-driven storytelling with minimal intervention of them.

3.2 Software Development

The second phase focused on designing the tool, where we prioritized robustness, simplicity, and automation, ensuring it met the needs identified in the abovementioned first step. The adoption of Scrum for agile and iterative development assisted us through weekly sprint meetings with the journalists and other stakeholders to align priorities, focus on the user's needs, and resolve uncertainties proactively. Emphasis was placed on refining the visual aspects of the tool, ensuring accuracy and user engagement. The tool was designed for automation, enabling hands-off operation and efficient data processing.

3.3 Evaluation

For the evaluation of the tool, we used as basis an elaboration of the DeLone and McLean, which has been proposed for digital government information systems evaluation [18], and made an adaptation of it to the data journalism context. The structure of our evaluation model is shown in Fig. 1. We can see that in its first layer it includes assessment of the five first layer dimensions proposed by the abovementioned elaboration of the DeLone and McLean Model [18]: information quality, service quality as well as three aspects of system quality (ease of use, capabilities (meant as functionalities provided to the user) and technical quality). In the second layer it includes an adaptation of the 'impact' evaluation dimension proposed by the model proposed in [18]: the business-level influence of the tool. Furthermore, given that our tool is designed for journalists and intended for daily use by them, ensuring high user satisfaction is a dimension that greatly concerns us; therefore, the second layer of our evaluation model includes also an assessment of users; satisfaction.

So, our evaluation method comprises the above seven core evaluation dimensions, which are shown in Fig. 2, forming the foundation of our assessment. The participants

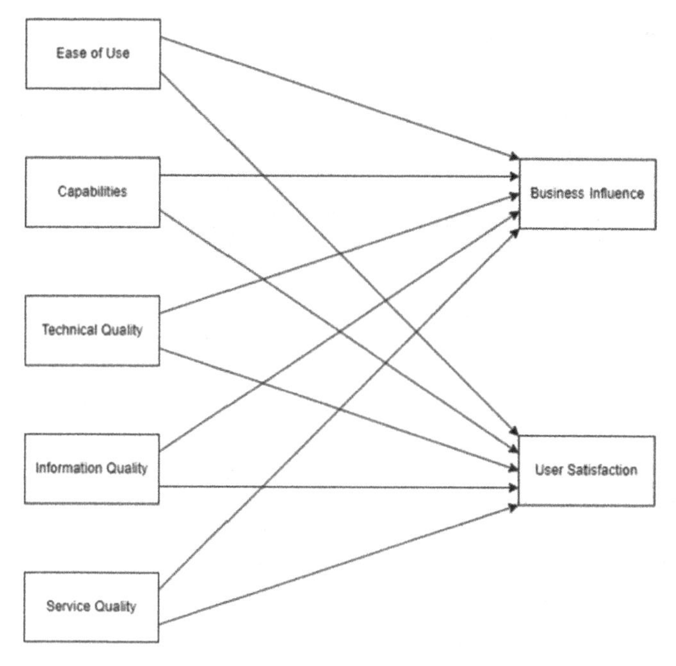

Fig. 2. Structure of our evaluation model

in our evaluation were users, including ICT professionals and journalists, who provided feedback tailored to their roles. We have used a combination of quantitative and qualitative evaluation methods in order to collect data about these four dimensions/aspects of our tool. In particular, we used a questionnaire, which included questions concerning these evaluation dimensions seeking nuanced perspectives; it is shown in the Appendix. The evaluation participants initially filled this questionnaire, and then we conducted direct interviews of them, in which we had in-depth discussions about the abovementioned dimensions/aspects of the tool.

4 Results

Our approach to creating a helpful software tool for journalists revolved around automation and simplicity. The dataset we used was provided by the Greek Organization of Central Markets and Fisheries (OKAA) and included the prices of the main food products. OKAA is the leading wholesale distributor of food in Greece and the most significant food park management organization in the Balkans. Therefore, the provided prices would reflect the overall market situation and could be used to uncover potential problems (e.g., big price increases) in the local food retail markets. We also expected the system to be robust and the errors from the data source to be minimal. This dataset, covering food supply information in Greece, also met our openness and regulatory requirements.

4.1 Software Development: Challenges and Solutions

OKAA offered data access through both Excel files and an API. Given our preference for automation, we chose the API to acquire the data. Python was selected to implement the data fetching and transformation process due to its simplicity and widespread adoption. A script of 120 lines was developed to search the API URL for meat, fruit and vegetable products (which, according to interviewed journalists in initial needs elicitation phase, are the most important for the consumers, so there is high journalistic interest in them), and then process the open data into three JSON files, ready for use in the next steps of our pipeline. For the visualisations of these prices, we used the widely recognized and used Power BI tool, a user-friendly visualizations tool that allowed us to create complex visuals without intricate coding. The publication of the visualisations was simplified with the creation of HTML IFrame tags that can be easily embedded in any online media article. Automation was one of the most essential features of our tool. To facilitate this, we converted the Python script into an executable file and scheduled it for automated execution via Windows 10's Task Scheduler. The Power BI visualisations also received automated data refresh capabilities, ensuring that our visuals stayed up-to-date and the tool operated fully automated.

Challenges during development included discrepancies in historical data from OKAA and data formatting for Power BI. However, our commitment to day-to-day operations drove us to create a system for swift data rectification. Deploying our open data journalism tool presented new challenges. Differences in computing environments required careful adjustments for seamless automation. Software installations on deployment machines posed another challenge, demanding technical expertise and troubleshooting to resolve potential conflicts. Deployment underscored the importance of adaptability, attention to detail, and technical understanding. Comprehensive testing across diverse environments and robust documentation were essential. The deployment phase marked a crucial transition from development to practical utilisation, offering valuable insights for tool enhancement.

We scheduled the execution/activation of our tool three times a day with a four-hour window between them, considering that the maximum timeframe of the tool completing its processes is 20 min. This method has been proven highly effective since our tool has been up-to-date every day that new open data arrives from the source, emphasising its punctuality and dependable performance.

The simple and innovative approach we took benefits the tool in many ways; one of them is that the only requirements are a Power BI subscription and a typical PC with medium specs to be operational every day, since the only processes that take place are the script execution and the Power BI online environment checking for new data on our PC. Leveraging state-of-the-art software as a service, like Power BI, has been an amazing opportunity to focus on the business aspects of our project rather than spending valuable time with technical difficulties.

In Fig. 3 we can see a typical visualization provided by the tool, which shows the prices of some important food products at a specific date.

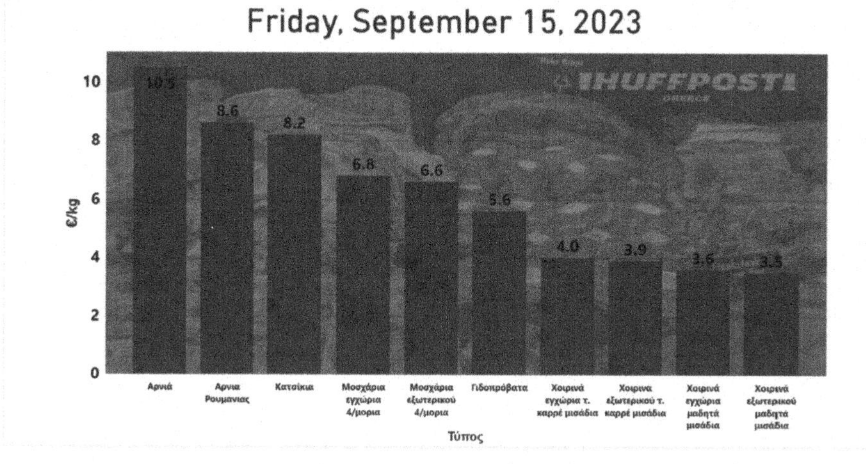

Fig. 3. A typical visualization provided by the tool

4.2 Evaluation

In the evaluation phase we focused on assessing the effectiveness and usability of this open data journalism tool we meticulously developed. The evaluation encompasses two distinct yet complementary approaches. First, we employed a quantitative method by distributing a questionnaire (shown in the Appendix) to one journalist-user and one ICT employee to assess their satisfaction. Subsequently, we delved into the qualitative aspect of our research by conducting in-depth interviews with the above two users who participated in the quantitative evaluation. Through these interviews, we aimed to extract invaluable insights, gaining a deeper understanding of the tool's real-world usability, challenges, and potential for further refinement. This two-fold evaluation process, which combines quantitative and qualitative techniques, provides a comprehensive overview of the tool's performance and facilitates the refinement of our open data journalism solution in alignment with user needs and expectations.

4.2.1 Quantitative Evaluation

In Table 1 we can see the responses to the questionnaire. We can see that with respect to the with the ease of use and usability, the technical quality and performance, the ease of the routine required to update and renew open data and the level of service and support/documentation provided both users are satisfied to a very large extent; also, both are overall satisfied with the open data journalism tool to a very large extent. With respect to the other questions one user expressed satisfaction to a very large extent and the other to a large extent. These indicate a high level of satisfaction with this tool.

Table 1. Results of the quantitative evaluation

	not at all	to a small extent	to a moderate extent	to a large extent	to a verylarge extent
How satisfied are you with the ease of use and usability of the open data journalism tool?	0	0	0	0	**2**
How Satisfied are you with the capabilities and functionalities provided by the open data journalism tool?	0	0	0	**1**	**1**
How Satisfied are you with the technical quality and performance of the open data Journalism tool	0	0	0	0	**2**
How would you rate your satisfaction with the ease or difficulty of the routine required to update and renew open data?	0	0	0	0	**2**
How satisfied are you with the clarity and intuitiveness of information provided by the open data journalism tool?	0	0	0	**1**	**1**
How satisfied are you with the level of service and support/documentation provided by the open data journalism tool?	0	0	0	0	**2**
How satisfied are you with the business benefits the open data journalism tool has on your journalistic work?	0	0	0	**1**	**1**
What is the anticipated frequency of usage projected for this tool?	0	0	0	**1**	**1**
How satisfied are you overall with the open data journalism tool as a user?	0	0	0	0	**2**

4.2.2 Qualitative Evaluation

With respect to the Ease of Use dimension our findings from the qualitative interviews align with the positive trends observed in our quantitative data. Users comment positively the user-friendly interface of the tool, highlighting its accessibility, whch eliminate the need for extensive training. The cloud-based server on Power BI ensures 24/7 availability, further enhancing usability. From a technical standpoint, users reported smooth setup processes with minimal challenges, underscoring the tool's reliability and ease of implementation. These qualitative insights reinforce our commitment to a user-centric design, emphasising accessibility and robustness, paving the way for the tool's continued success in our data-driven journalistic endeavours. The Editor-In-Chief of the news portal HuffPost mentioned.

> *"It is evident that our tool aligns excellently with our primary objective: to be user-friendly and require minimal training, particularly for individuals in our diverse newsroom with varying backgrounds."*

In the Capabilities dimension, the interviews that we conducted gave us positive feedback about the tool's technical capabilities and highlighted some data-related issues that come from the source of the data. The Editor-In-Chief mentioned in the interview that naming convention variations and occasional grammatical errors from the data set are areas for improvement.

Regarding technical quality, our interview responses strongly endorse the tool's efficiency and reliability. Both users consistently reported that the tool completes updates within the appropriate time frame, aligning perfectly with our goal of delivering swift and effective performance. There have been no reported errors until the end of our research, underscoring the tool's remarkable robustness.

Regarding satisfaction with the routine for updating and renewing open data, user feedback was remarkably positive. The routine is refreshingly straightforward, primarily involving data upload validation. These insights highlight the tool's user-friendly design and emphasise its ability to streamline complex data processes, ensuring that users can focus on journalism's insights and storytelling aspects rather than grappling with technical intricacies.

In Information Quality, dedicated to assessing the clarity and intuitiveness of our open data journalism tool's information, the Editor-In-Chief's "good" satisfaction rating reflects the tool's strong performance and opportunities for refinement. Challenges linked to data quality, such as spelling errors and naming conventions misalignment, underscore the critical importance of data quality within our tool and the broader open data sources. Conversely, our discussions regarding colour schemes and data visualisation have yielded positive outcomes.

In the domain of service quality, feedback highlighted the substantial value of the installation process, described as "extra helpful", providing a robust foundation for further exploration and learning. This positive experience not only eased the initial setup but also empowered users to delve deeper into automation intricacies, unlocking a spectrum of possibilities, including data and source fusion for richer insights.

In the domain of business influence and the advantages our tool brings to journalistic work, enthusiasm and excitement were palpable. Our innovation centred around constant updates from open data sources, marked a significant milestone. It was commended for providing reliable, up-to-date information that builds trust with readers and fosters a unique and enduring relationship, setting the whole journalistic company apart. The Editor-In-Chief's perspective added an intriguing dimension, describing our endeavour as a "low-budget attempt" that had surpassed expectations, showcasing our ability to achieve automation and reliability aligned with our vision within budget constraints.

Regarding the tool's projected frequency of usage, it's essential to consider the unique rhythm of change of the food price data it processes. Unlike industries with constant data updates, food prices fluctuate based on market dynamics, seasons, and global events. Our tool's adaptability and relevance, and while it may not be in constant use, its effectiveness lies in its ability to respond promptly and accurately when required, demonstrating the versatility and practicality of our tool in the dynamic field of open data journalism.

In the User Satisfaction dimension, the final segment of our evaluation, both respondents expressed a high degree of satisfaction. They consistently lauded our tool for its exceptional user-friendliness and its immediate comprehensibility to viewers without the need for a learning curve. This innate simplicity and intuitiveness emerged as prominent strengths. Furthermore, our users emphasised the tool's effectiveness in aiding individuals to assess product affordability within their budget constraints.

5 Discussion

By developing and evaluating an open data journalism tool in a news media organisation, we had the opportunity to explore the boundaries of open data provided by OKAA and understand the journalists' daily needs. Furthermore, we had the opportunity to evaluate the use of the tool in a real-world setting. Our research targeted the day-to-day aspect of open data journalism due to the limited timeframe we had for its completion; tackling the investigative aspect of data journalism [19] would have required significantly more time and resources from the media organisation, and it would have also increased its technological complexity. Therefore, we removed it from the scope of the research.

Our approach was to create a robust, easy-to-use and fully automated tool that will be easily integrated into the work of journalists and will require minimal training and maintenance. According to the evaluation, this fully automated tool was what the company required to tackle the critical social issues of rising food prices due to the contemporary geopolitical situation. However, we encountered several limitations, and we have identified topics for feature research and improvements.

The most severe impediments we faced revolved around the quality of data provided by OKAA. We often encountered inconsistencies in the titles of the products, misspellings, and formatting errors in the data. Handling formatting errors were addressed with the use of data-cleaning techniques. However, dealing with the inconsistencies and misspellings posed challenges for the consistency of the visualisations. Additionally, the seasonal nature of the products created inconsistencies in the generation of historical charts and severely impacted their usefulness. Therefore, it became evident that the tool's functionality is inextricably linked to the data quality ensured by the provider.

To address these limitations, from our end, the tool has to be expanded in the feature cycle of development with more advanced data cleaning techniques for the normalization and standardization of the product names and formats to enhance data consistency. Another useful feature is an automated quality checker that could validate the updated data against predefined standards; this could enhance the overall data integrity of the tool.

Furthermore, in future research, we need to reevaluate the impact this tool has on the daily work of journalists and the business advantages it can provide to organisations. It is also important to expand our evaluation framework by incorporating technology acceptance and innovation diffusion models [20–22].

6 Conclusion

In the previous sections of this paper we have presented our research findings concerning the utilization of open data for supporting data journalism concerning the crucial societal problem, which undermines the quality of life of millions of citizens: the rising food prices. We embarked on this endeavour along with journalists from the news portal HuffPost (Greece) and used the open data provided by OKAA to acquire the food prices in Greece. We created a Python script to fetch and process the data so that they could be used by Power BI, our visualisation tool; we automated the data fetching process and generated the visualisations. So, we converted the above open data provided by OKAA to a visualized form, which can be easily and immediately understood by journalists, enabling them to make sense of the data, and draw conclusions from them, leading to high-quality 'evidence-based' journalism on this highly important topic. Despite the room for improvement, the evaluation of the tool's usefulness was ranked highly by the journalists.

Our study has important implications for research and practice. With respect to research, it contributes to the evolution of this abstract concept of data journalism towards a set of specific and easily applicable ICT-based practices; it creates new knowledge concerning the effective realization (i.e., 'real-life' implementation) of this valuable concept for improving the quality and trustworthiness of modern journalism through high-quality software tools that provide useful information to journalists and are easy to understand and draw conclusions from even by low digital skills journalists. With respect to practice, we have developed a useful ICT-based practice, which enables media organizations (both paper-based and electronic ones) to make their first steps towards data journalism easily and at low cost, which can be extended to other types of open data concerning various social problems. Furthermore, our research has revealed various issues concerning the quality of the open data that restrict their usefulness for journalistic purposes.

Further work can be done concerning the enhancement of the present capabilities of our data journalism support tool with new features, such as historical price charts, as well as its evaluation by larger numbers of users. Moreover, we have to research more systematically to what extent the tool can be adopted by the rest of the journalists in the newsroom using technology acceptance and innovation diffusion theories [20–22].

Acknowledgements. This project has received funding from the European Union's Horizon 2020 research and innovation programme under the Marie Skłodowska-Curie grant agreement No 955569.

The opinions expressed in this document reflect only the author's view and in no way reflect the European Commission's opinions. The European Commission is not responsible for any use that may be made of the information it contains.

Appendix

The questionnaire included the following questions, which had to be answered in a five levels Lickert scale (1= not at all, 2 = to a small extent, 3 = to a moderate extent, 4 = to a large extent, 5 = to a very large extent).

Ease of Use:
How satisfied are you with the ease of use and usability of the open data journalism tool?

Capabilities:
How satisfied are you with the capabilities and functionalities provided by the open data journalism tool?

Technical Quality:
How satisfied are you with the technical quality and performance of the open data journalism tool?

How would you rate your satisfaction with the ease or difficulty of the routine required to update and renew open data?

Information Quality:
How satisfied are you with the clarity and intuitiveness of information provided by the open data journalism tool?

Service Quality:
How satisfied are you with the level of service and support/documentation provided by the open data journalism tool?

Business Influence:
How satisfied are you with the business benefits the open data journalism tool has on your journalistic work?

What is the anticipated frequency of usage projected for this tool?

User Satisfaction:
How satisfied are you overall with the open data journalism tool as a user?

References

1. Weinberg, J., Bakker, R.: Let them eat cake. Confl. Manag. Peace Sci. **32**(3), 309–326 (2015)
2. Rudolfsen, I.: Food price increase and urban unrest: the role of societal organizations. J. Peace Res. **58**(2), 215–230 (2021). https://doi.org/10.1177/0022343319899705
3. Saâdaoui, F., Ben Jabeur, S., Goodell, J.W.: Causality of geopolitical risk on food prices: considering the Russo–Ukrainian conflict. Financ. Res. Lett., **49**, 103103 (2022). https://doi.org/10.1016/j.frl.2022.103103.
4. Brolcháin, N.Ó., Porwol, L., Ojo, A., Wagner, T., Lopez, E.T., Karstens, E.: Extending open data platforms with storytelling features. In: Proceedings of the 18th Annual International Conference on Digital Government Research, Staten Island, pp. 48–53. ACM, NY, USA (2017). https://doi.org/10.1145/3085228.3085283.
5. Charalabidis, Y., Alexopoulos, C., Ferro, E., Janssen, M., Lampoltshammer, T., Zuiderwijk, A.: The world of open data: concepts, methods, tools and experiences, 1st ed. In: Public Administration and Information Technology, no. 28. Springer, Cham (2018).https://doi.org/10.1007/978-3-319-90850-2
6. Veglis, A., Bratsas, C.: Reporters in the age of data journalism. J. Appl. Journal. MEDIA Stud. **6**(2), 225–244 (2017). https://doi.org/10.1386/ajms.6.2.225_1
7. Papageorgiou, G., Loukis, E., Magnussen, R., Charalabidis, Y.: Open data journalism: a domain mapping review (2023). https://doi.org/10.1145/3614321.3614340 [Unpublished manuscript]
8. Gupta, K., Sampat, S., Sharma, M., Rajamanickam, V.: Visualization of election data: using interaction design and visual discovery for communicating complex insights. EJournal EDemocracy Open Gov. **8**(2), 59–86 (2016). https://doi.org/10.29379/jedem.v8i2.422
9. Bozsik, S., Cheng, X., Kuncham, M., Mitchell, E.: Democratizing housing affordability data: open data and data journalism in charlottesville, VA. In: 2022 Systems and Information Engineering Design Symposium (SIEDS), pp. 178–183. Charlottesville, VA, USA: IEEE (2022). https://doi.org/10.1109/SIEDS55548.2022.9799410.
10. . Shehu, V., Mijushkovic, A., Besimi, A.: Empowering data driven journalism in Macedonia. In: ACM International Conference Proceeding Series. Association for Computing Machinery (2016).https://doi.org/10.1145/2955129.2955187
11. Petricek, T.: The gamma: programmatic data exploration for non-programmers. In: 2022 IEEE Symposium on Visual Languages and Human-Centric Computing (VL/HCC), pp. 1–7. IEEE, Roma, Italy (2022).https://doi.org/10.1109/VL/HCC53370.2022.9833134
12. Evéquoz, F., Castanheiro, H.: Which lobby won the vote? Visualizing influence of interest groups in swiss parliament. In: Lindgren, I., Janssen, M., Lee, H., Polini, A., Rodríguez Bolívar, M.P., Scholl, H.J., Tambouris, E. (eds.) EGOV 2019. LNCS, vol. 11685, pp. 155–167. Springer, Cham (2019). https://doi.org/10.1007/978-3-030-27325-5_12
13. Cao, T.-D., Duroyon, L., Goasdoué, F., Manolescu, I., Tannier, X.: BeLink: querying networks of facts, statements and beliefs. In: International Conference on Information Knowledge Manage, Association for Computing Machinery, pp. 2941–2944 (2019). https://doi.org/10.1145/3357384.3357851
14. Cao, T.D., Manolescu, I., Tannier, X.: Extracting linked data from statistic spreadsheets. In: Gruenwald, L., Groppe, S., (eds.) Proceedings of International Workshop on Semantic Big Data, SBD - Conjunction ACM SIGMOD/PODS Conference, Association for Computing Machinery, Inc (2017). https://doi.org/10.1145/3066911.3066914.
15. DeLone, W.H., McLean, E.R.: Information Systems Success Measurement (2016). https://ieeexplore.ieee.org/document/8187232
16. Mason, R.O.: Measuring information output: a communication systems approach. Inf. Manage. **1**(4), 219–234 (1978). https://doi.org/10.1016/0378-7206(78)90028-9

17. The DeLone and McLean model of information systems success: a ten-year update. J. Manag. Inf. Syst. **19**(4), pp. 9–30 (2003). https://doi.org/10.1080/07421222.2003.11045748.
18. Loukis, E.N.: A methodology for evaluating and improving digital governance systems based on information systems success models and public value theory. In: Charalabidis, Y., Flak, L.S., Viale Pereira, G. (eds.) Scientific Foundations of Digital Governance and Transformation. PAIT, vol. 38, pp. 245–273. Springer, Cham (2022). https://doi.org/10.1007/978-3-030-92945-9_10
19. Lück, J., Schultz, T.: Investigative data journalism in a globalized world. Journal. Res. **2**(2), 93–114 (2019). https://doi.org/10.1453/2569-152X-22019-9858-en
20. Rogers, E.M.: Diffusion of Innovations, 5th edn. Free Press, New York (2003)
21. Venkatesh, V., Davis, F.D.: A theoretical extension of the technology acceptance model: four longitudinal field studies. Manag. Sci. **46**(2), 186–204 (2000). https://doi.org/10.1287/mnsc.46.2.186.11926
22. Momani, A.M., Jamous, M.M.: The Evolution of Technology Acceptance Theories (2017)

Healthcare Information Systems

Smartphone Apps for Parents of Preterm Infants from NICU to Home: A Quality, Evidence-Based Content and Data Protection Assessment

Roxane Coquoz[1(✉)], Camille Pellaton[2], Leo Bettelini[2], Laura Rio[1], and Alessio De Santo[2]

[1] HE Arc Santé, HES-SO, University of Applied Sciences Western Switzerland, Neuchâtel, Switzerland
{roxane.coquoz,laura.rio}@he-arc.ch
[2] HEG Arc, HES-SO, University of Applied Sciences Western Switzerland, Neuchâtel, Switzerland
{camille.pellaton,leo.bettelini,alessio.desanto}@he-arc.ch

Abstract. Over one in ten babies are born preterm annually, presenting challenges beyond the neonatology intensive care unit (NICU), including depression and post-traumatic stress for parents. Research has demonstrated that tailored interventions supporting parents transitioning from NICU to home can decrease these adverse outcomes. Parents often seek online answers for children's health before consulting medical professionals. Smartphone applications (apps) supporting parents are being increasingly developed, however, the literature suggests that current apps lack quality and credibility. This study offers a methodical assessment of the quality, level of evidence-based content, and data protection of apps aiming to support parents of premature infants in their transition from NICU to home. The web-based application aggregator Appagg was used to list free and paid Android and iOS apps using keywords such as NICU, preterm, discharge and parenting support in English and in French. The apps were evaluated between March and July 2023. Quality was evaluated using the Mobile Application Rating Scale (MARS) to measure engagement, functionality, aesthetics and information. Then, this study suggests an evidence-based content (EBC) assessment based on the most recent recommendations, guidelines, and scientific literature regarding NICU discharge for premature infants. Finally, the apps' data protection was evaluated in regards to compliance with the European General Data Protection Regulation (GDPR) or GDPR-like regulation. The search yielded a total of 896 unique apps. Screening for title and abstract selected 22 apps. 12 remained for final analysis as 9 were not accessible to Switzerland or needed patient ID for access and one did not work (video content was not working at all). The results showed that three apps (3/12, 25%) received a good MARS score on overall quality (>4.0 out of 5.0), five apps (5/12, 42%) presented good levels of EBC assessment (≥4 out of 5) and four apps (4/12, 33%) were explicitly compliant to at least one data protection standard (≥4 out of 5).

Keywords: parenting · smartphone apps · NICU · premature infants · data protection · evidence-based content analysis

M. Papadaki et al. (Eds.): EMCIS 2023, LNBIP 501, pp. 209–224, 2024.
https://doi.org/10.1007/978-3-031-56478-9_15

1 Introduction

Parents of preterm neonates represent a growing population (one in ten births) worldwide [15, 39]. Compared to term infants, hospitalized preemies have a higher risk of mortality, morbidity and handicap [5, 31]. Parents of these infants are more susceptible of experiencing emotional distress, anxiety and depression [19] that can result in altered parenting patterns upon discharge home, and may be a major contributor to adverse family and infant outcomes [8, 17]. A recent position paper published by the French Neonatal Society reviewing literature from 2000 to 2018 for scientific evidence on how to help families throughout NICU discharge compiles high-level evidence of clinically tested interventions [32]. Interventions and programs have demonstrated improved family outcomes, with decreases in hospital stay and readmission rates and decreased rates of parental depression, anxiety and stress [24, 54], as well as persistent benefits in parents-child interactions, cognitive and behavioural development [43].

The use of smartphone applications (apps) is extensive across every aspect of daily life. Smartphones have been globally adopted faster than any other technological innovation in history. The number of smartphone users worldwide rose to a record-breaking 3.6 billion in 2020 and is expected to reach 4.5 billion by 2024 [28]. On average, people spend over 3 hours on their phones per day [2, 30] and despite some health concerns associated with excessive screen time consumption, there is a strong interest in developing apps to promote health. This has been amplified during the COVID-19 pandemic with a rise in health and fitness apps [21] and also clinical-grade apps as therapeutic interventions [38] showing efficiency, economic benefits and increased patient empowerment and accessibility to healthcare. Apps are being used more and more, even in situations when clinical care is provided in person, as they can be used to help tailor patient-provider communication and support patient self-management and care engagement [51].

The French Neonatal Society evidenced that information is key for NICU families but there are still problems with how the information is communicated and how parents adhere to it. A standardized, progressive, family-centered, and coordinated approach should be used to deliver information that is personalized, repeated, and given to both parents. Tailored apps have been developed as a way to support and improve this transition home and make sure that families are more involved and better informed. These apps possess many advantages, as they can provide specialized information that is easily accessible at any time, and personalized education tailored to anticipated parents' needs. They can be incorporated in a family-centered approach, support care continuity [9] and improve communication with healthcare professionals [49]. In 2019, Richardson et al., published a review of apps for parents of infants in the NICUs and found that they were functional but had low quality and credibility [34]. Since this review, several apps targeting NICU parents have been removed, others added, and new legislation on data protection has been issued. In Europe, the European General Data Protection Regulation (GDPR) has been effective since May 25th, 2018 in all member states to harmonize data protection laws across Europe [18, 53]. In Switzerland, the New Federal Act on Data Protection (nFADP), compatible with the GDPR, has been issued and will be enforced from September 1, 2023 [47]. This implementation of new legislation to better protect citizens' data shows the increasing importance of data protection and the need for trust and reliability in the solutions offered to the public, especially in regards with their personal and health data.

In this review we assessed the quality and reliability of apps aimed at NICU parents and available in Switzerland. We included new apps released since the 2019 review from Richardson et al. [34] and added two dimensions of importance in the review, which are the levels of evidence-based content and data protection. In this article, we present an overview of existing guidelines and interventions for supporting premature newborns and their parents as well as the use of digital solutions with their promising benefits and their current limitations. The process and methodology for selecting and assessing the eligible apps is then detailed, including validated tools and the additional items specifically developed for the study. Results and conclusions are then presented with their limits and perspectives in the field of neonatal discharge and beyond.

2 Supporting Parents Through Discharge – What Works and What Research Says

2.1 Overview of Existing Guidelines and Interventions

In the US, best practices and recommendations for NICU discharge preparation and transition planning of the American Academy of Pediatrics (AAP) were published in 2008, reaffirmed in 2018, and in 2022 [10, 40]. They include parental education, identification and involvement of support services, care planning in the hospital as well as a follow-up plan for care at home. In Europe, the position paper from the French neonatal society cited earlier, reviewed 939 articles and also concluded that an early family-centered, structured, coordinated and individualised approach to discharge improves transition and outcomes. Support should be targeted to both parents and also consider optimal timing and delivery modes tailored to parents' situation and needs [52]. Moreover, it has been shown that young parents and those with very low birth-weight premature infants may need additional support and resources [33].

Based on these guidelines, several programs, strategies, and interventions to support parents have been developed. They have demonstrated positive and clinically meaningful effects on parental health outcomes, lowering stress, anxiety and depressive symptoms and increasing self-efficacy, confidence, satisfaction, bonding and short-term outcomes of preterm infants (breast-feeding rate, decreased retinopathy of prematurity, decreased hospital length of stay, increased weight gain velocity) [54].

2.2 NICU Parents and Digital Solutions

Parents frequently use the Internet to search for general health issues affecting their child's health and well-being before consulting a healthcare professional [25, 26, 46]. New parents are increasingly using digital supports during the perinatal period, reporting an interest in monitoring their infants and overall family health [14, 16, 27]. Notably, parents with an infant in the NICU reported to prefer accessing the Internet via their mobile phone, particularly the younger ones favoring smartphone apps [7, 29].

The literature targeting existing digital interventions in the NICU setting and the various established guidelines make it possible to classify possible functions of these interventions into categories: information, communication and education, monitoring and tracking, neonatal care, skills training, emotional and peer support, as well as goal setting [34, 40]. The topics covered by digital interventions are diverse and may concern: NICU and newborn care, prematurity, health and development, pain, medication, breastfeeding and feeding, pregnancy and childbirth, discharge, parenting, relationships, emotions, mental health, support, and entertainment [11, 34, 42]. Some authors also recommend paying specific attention to family-centered needs assessment [35], transfer and care coordination [36] or protection [22]. Apps geared at NICU parents have shown positive impact on parent's self-efficacy, competence, and confidence [9] and parent–infant attachment [42].

Apps are becoming part of routine care and thus should meet practice standards to provide information that is understandable, reliable, current, evidence-based and independent of commercial associations. Since the NICU experience can expose parents to great insecurity, vulnerability, and stress, ensuring the quality, credibility and evidence-based nature of the information provided by existing apps are of the utmost importance, and their evaluation is therefore highly warranted [9, 29, 49].

However, there are currently no recognized quality standards for the development of such solutions [4] and among thousands of mobile apps aimed at NICU parents, it is unclear how much of their design and content is evidence-based and how effective their use is [9, 34, 49]. The Mobile App Rating Scale (MARS) is one of the most widely used tools for apps quality evaluation but some limitations have been pointed out [34]. To overcome these limitations, the literature recommends that the MARS tool be adjusted to place greater emphasis on credibility, provision of sources and also data safety and protection [34, 44, 50]. To address these issues, this paper provides a critical analysis of apps targeting NICU parents of premature infants returning home while focusing on the ones available in the french part of Switzerland. A mapping of existing applications and an assessment of their quality, evidence-based content level and data protection have been carried out using the MARS tool, the EBC assesment tool and a data protection likert scale.

3 Switzerland (French Speaking) App Landscape – What Currently Exists

Switzerland has four language regions: German, French, Italian and Romansh. The number of German, Italian and Romansh speakers is decreasing, while French speakers are increasing [48]. Non-national languages are also gaining in importance. The two most widely spoken non-national languages are English and Portuguese. Multilingualism is an essential part of Switzerland's identity. French part of Switzerland includes the cantons of Geneva, Vaud, Neuchâtel, Jura, Valais as well as the French speaking parts of the cantons of Fribourg and Berne and host seven Swiss level III and level IIB neonatal units [1]. In this review, we used systematic methods to identify, select and evaluate apps supporting NICU parents of preterm infants going home, that were available in the French speaking region of Switzerland. The collection and assessment of the apps took part between March and July 2023. We evaluated their quality using the validated Mobile App Rating Scale (MARS), their level of evidence-based content with a table based on the latest published recommendations and guidelines and the apps' data protection according to the level of adherence to GDPR or GDPR like guidelines.

3.1 Step 1: App Selection

The app collection was conducted through the search engine of the app aggregator Appagg (https://appagg.com/). The following search terms were used against the database: *nicu, preemie, preterm, premature baby, neonatal, infant care, parenting support, child health, post discharge*, inspired by a previous research [34]. The same search was performed with French keywords but retrieved no additional results.

Table 1. Inclusion and exclusion criteria

Inclusion criteria
Apple iOS and Google Android app
only apps with available descriptions in the English or French language
only apps published in the Apple App Store or the Google Play Store
only apps under Education, Health & Fitness, Medical or Parenting categories
Exclusion criteria
apps unrelated to facilitating the transition of parents of premature infants from hospital to home
apps designed for healthcare professionals rather than parents
exact duplicates between different app stores

Inclusion and exclusion criteria, as presented in Table 1, were applied to exclusively select relevant apps. Figure 1 depicts the complete process.

At the subset stage containing 602 apps, ChatGPT API was used to classify the relevance of the applications based on their descriptions. The following prompt was utilized with the GPT-3 API: *"Does the following app description mention that the app contains features to support parents of a PRETERM NEWBORN before, during and after discharge from the neonatal intensive care unit (NICU). This digital intervention should increase parents' self-efficacy and decrease health services overuse?"*. 10% of the result set was manually labeled in order to assess the classification results. Based on the outcomes presented in Table 2, the calculated metrics revealed an accuracy of 0.68, indicating that the model was able to label correctly approximately 68% of the apps. The model exhibited a sensitivity of 1 and a specificity of 0.672, meaning that all positive values are correctly identified whilst 67.2% of negative instances are. With a negative predicted value of 1, the model identified every negative instance. Conversely, the positive predicted value of 0.095, highlighted that only 9.5% of instances predicted as positive were true positives. The model's predictions lean toward leniency, however, this aligns with the scope of our research, since this cautious approach guarantees the inclusion of all relevant apps. Thus, from 602 apps, 444 were removed automatically and a manual verification was performed on the 158 remaining applications to rectify instances where the AI model's screening was insufficient. 137 apps were manually removed, resulting in a set composed of 21 apps.

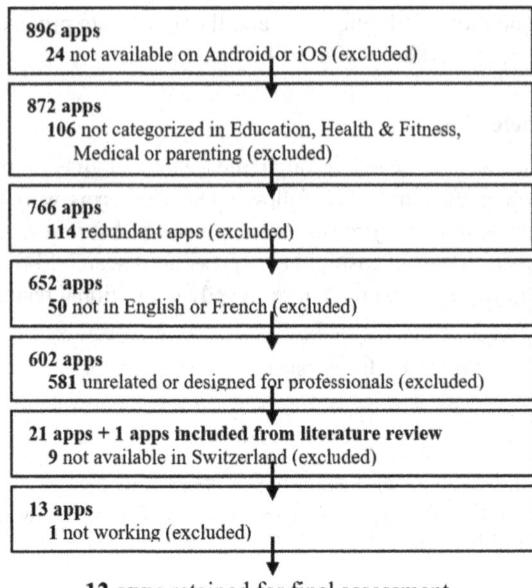

Fig. 1. Procedure for our app sample selection

Table 2. Outcomes of the prediction algorithm

	Prediction	
True label	1	0
1	2 (TP)	19 (FP)
0	0 (FN)	39 (TN)

3.2 Step 2: App Evaluation

In this step, selected apps were evaluated for quality, level of evidence-based content and data protection.

Quality. App quality was evaluated using the MARS, which is a simple, objective, reliable and validated tool that is widely used for classifying and assessing the quality of mobile health apps [50]. It has excellent internal consistency (alpha = .90) and interrater reliability intraclass correlation coefficient (ICC = .97). MARS assesses app quality on four objective scales: *engagement: 5 items on interesting, fun or interactive content, functionality: 4 items on app navigation and logical usability, aesthetics: 3 items on graphic design and visual appeal, and information quality: 7 items on credibility of source.* The 23-items are rated on a 5-point Likert scale: "1. Inadequate", "2. Poor", "3. Acceptable", "4. Good", and "5. Excellent". The objective MARS score is calculated as a 5-star rating using the mean from the engagement, functionality, aesthetics and information quality scales scores. The MARS also has one *subjective quality* scale for assessing user's judgment on their likelihood of recommending, using and purchasing the app on a personal 5-star rating, and is reported separately as the subjective MARS score. Here, we only evaluated the objective score as we aimed for an objective ranking.

Evidence-Based Content. The authors of the MARS scale point out that it is a general tool and suggest adding items for a more specific and contextualised assessment [45]. This has been done in the reproductive health, infant feeding or post-partum depression fields. Indeed, several researchers have proposed revised versions of the MARS scale [6, 23, 44]. For the neonatal discharge field, no existing MARS scale version was specifically adapted. In addition, we noted that the original MARS scale does not accord substantial weight to the level of evidence-based nature of the app. It only evaluates if the app itself has been tested in a clinical setting with the following question: *"Has the app been trialled/tested; must be verified by evidence (in published scientific literature)"*. This does not ensure that the content of the app (information, recommendations) is evidence-based. As a result, we modified the original MARS scale by adding specific items based on recent published guidelines that compile EBC recommendations [10, 40, 49]. There are 30 items covering six main domains: *Newborn care and the NICU setting, Preemie's health, growth and development, Breastfeeding and feeding, Discharge preparation, Discharge follow-up and, Family support and mental health.* Each domain contains five items that should be addressed according to guidelines to provide a wholesome support. Each item is verified and if present obtains 1 point. Each domain obtains a score from 15, from "inadequate" to "excellent". The total score is calculated by summing the points of the six domains, and are categorized as follow: "0–6 points. Inadequate", "7–12 points. Poor", "13–18 points. Acceptable", "19–24 points. Good" and "25–30 points. Excellent". The details of the 30 items can be obtained from the authors upon request.

Data Protection Assessment. In order to evaluate the level of data protection proposed by the app, we added one item with a 5-point Likert scale: "NA. no privacy policy or not found", "1. Privacy policy does not detail data privacy nor security", "2. Privacy policy mentions adherence to specific rules in relation to data security and privacy", "3. Explicit compliance to either a security or privacy standard", "4. Explicit compliance to at least one privacy and security standard", and "5. Explicit GDPR compliance".

Cumulative Score. The cumulative score is obtained by adding the Engagement, Functionality, Aesthetics, Information, EBC and Data protection scores of each app. It will be used for the ranking of the apps to have a global approach.

4 Results

4.1 Step 1: App Selection

App search was performed from 10th of March to 9th of July 2023. A total of 896 apps were identified, using Appagg (Fig. 1). After removing in an automated, verified way, non-iOS or Android apps (n = 24), off-topic (n = 550), duplicates (n = 114) and non-English (n = 50), 158 apps were considered for manual evaluation leading to the exclusion of 137 apps. One supplementary app was added to the initial output of Appagg, based on extensive literature search and finally 22 apps were selected to be downloaded for full evaluation. When downloading these apps, nine were either not accessible to Switzerland (CH), or required an invitation or code certifying that you have a baby in neonatal care and one could be downloaded but its content, consisting in short videos, was not working. This resulted in 12 apps fully reviewed for quality, level of evidence-based content and data protection.

Out of the 12 apps, five were found in peer-reviewed scientific articles. Two articles were about the apps development that were either based on semi-structured interviews with parents, testing and user feedback [42] or following a participant centered design approach with a focus groups method and co-design sessions [20]. One article described the translation and cultural adaptation of an app from English to Portuguese [41]. Finally, two articles were about the apps being tested with families and outcomes measured [3, 13]. The Babble app was tested using a between-subjects post-test and a quasi-experimental design compared participants grouped by Babble app use (yes/no). It showed no significant impact on parents self-evaluated distress or self-efficacy [13]. The IFDC app was tested in an interventional study following a pre-post design and showed a significantly reduced length of stay in the NICU [3].

Overview information about the selected applications is provided in Table 3. Out of the 12 apps, eight apps (8/12, 67%) were developed by universities, hospitals or foundations and four were commercial. Overall, eleven apps (11/12, 92%) were free to download and use and one was only accessible by purchase. All apps were available in English with five proposing a second language (5/12, 42%), one app was available in French, another in Indian and three could be used in Spanish.

Among the 12 apps, "MyPreemie app" has the highest download rate with more than 10 K (1/12, 8%). "Peekaboo ICU Preemie Baby" is second with more than 5 K (1/12, 8%). Three apps have a download rate of 1 K + (3/12, 25%). One app has more than 500 downloads (1/12, 8%) and five apps have a download rate of 100 + or below (5/12, 42%). "Quantum Caring for Parents" and "PretermConnect" have the lowest download rates. For one app, download rate information was not available (1/12, 8%).

4.2 Step 2: App Evaluation

The content of 12 apps was evaluated for quality (engagement, functionality, aesthetics, information), level of evidence-based content and data protection. Table 3 presents the main results of our study, with the 12 apps ranked based on their cumulative score. In order to better see the distribution of scores, Fig. 2 shows the distribution of the scores by items measured. Detailed attribution of evidence-based content scoring is further detailed in Table 4. The maximum achievable total score is 30. The best score of our sample is 25.1 and was obtained by only one app (1/12, 8%) while the average total score was 19.8.

As shown in Table 3 below, the best MARS score is 4.4 and was reached by only one app (1/12, 8%), while the average MARS score is 3.6. The average score for user engagement is 3.4. The average score for functionality is 3.9. The average score for aesthetics is 3.6. The average score for information is 3.5. The average score for evidence-based content assessment is 3.2. The average score for data protection is 2.5. For the engagement dimension, the highest score is 4.6 and was obtained by only one app (1/12, 8%). For functionality, the highest score is 4.8 and was obtained by three apps (3/12, 25%). The highest score for aesthetics is 4.7 and was reached by three apps (3/12, 25%).

The highest score for information is 4.4 and was obtained by one apps (1/12, 8%). Among the top scoring apps on MARS, none obtained the highest scores for every dimension.

Regarding the added dimension of evidence-based content, the highest score is 4 out of 5 and was obtained by 5 apps (5/12, 42%). More details about this item results are provided in Table 4. For the added dimension of data protection, the best score is 5 and was reached by only 2 apps (2/12, 17%). Among the 12 rated apps, 5 obtained the lowest score of 1 (5/12, 42%).

Table 3. Overview, MARS, EBC and data protection scores of the reviewed apps

id	App name (editor)	Overview					MARS							
		iOS	Android	Free	Language	Downloads	Engagement	Functionality	Aesthetics	Information	*MARS Score*	Evidence-based	Data protection	*Cumulative Score*
1	BABBLE NZ Neonatal Family App (MidCentral District Health Board)	✓	✓	✓	en	1K+	3.6	4.8	4.3	4.4	4.3	4	4	25.1
2	Hand to Hold (Hand to Hold)	✓	✓	✓	en, es	100+	3.6	4.3	4.7	3.0	3.9	3	5	23.6
3	My Prem Baby - by Tommy's (Tommy's)	✓	✓	✓	en	1K+	4	4.8	4.7	4.0	4.4	4	2	23.5
4	IFDC (Imperial College Healthcare NHS Trust)	✓	✓	✓	en	1K+	4	4.8	4.7	3.9	4.3	4	2	23.4
5	Quantum Caring for Parents (Caring Essentials Collaborative, LLC)	✓	✓		en, fr	1+	4.6	4	3.3	4.0	4.0	2	5	22.9
6	MyPreemie app (Graham's Foundation)	✓	✓	✓	en	10K+	3.8	3.8	3	3.4	3.5	3	3	20.0
7	Peekaboo ICU Preemie Baby (Peekaboo ICU, LLC)	✓	✓	✓	en	5K+	3.8	3	4.3	3.7	3.7	3	1	18.8
8	NICU Companion (Indiana University)	✓	✓	✓	en, es	NA	3.6	2.8	3.3	3.7	3.4	1	4	18.4
9	Life's Little Love (Courtney Larocque)	✓	✓	✓	en	500+	2.6	4.5	3	3.1	3.3	4	1	18.2
10	PretermConnect (CHIH H WANG)	✓	✓	✓	en	50+	2.6	3	3.3	3.6	3.1	4	1	17.5
11	Our Journey in the NICU (Phoenix Children's Hospital, Inc)	✓	✓	✓	en, es	100+	2.6	4.5	1.3	2.7	2.8	3	1	15.1
12	NICUFIC (Watch Puls)		✓	✓	en, ind	100+	1.8	2.5	3	2.3	2.4	NA	1	10.6
	Average						3.4	3.9	3.6	3.5	3.6	3.2	2.5	**19.8**

Notes: Apps are sorted based on their cumulative score. en=English, es=Spanish, fr=French, ind=Indian. NA=App content partially unavailable in English.

Table 4 below provides details of the level of evidence-based assessment for each application. The "IFDC" app obtained the best score of our sample with a total score of 24. It was rated as "Good" with a Likert score of 4, which was also obtained by 4 other apps (5/12, 42%). Four apps were rated "Adequate" with a Likert score of 3 (4/12, 33%), one app was rated "Poor" with a Likert score of 2 (1/12, 8%) and one app was rated "Inadequate" with a Likert score of 1 (1/12, 8%). The NICUFIC app could not be evaluated for its entire content, as some videos were not available in English and was therefore rated NA (1/12, 8%).

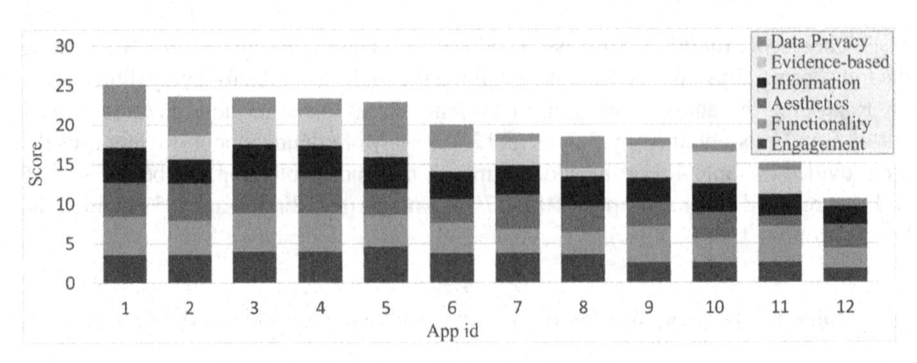

Fig. 2. Overview of the reviewed apps' scores

The main domains addressed in the apps were "Newborn care in the NICU setting" and "Breastfeeding and feeding" with a cumulative score of 39. Then come "Preemie's health, growth and development", "Discharge preparation" and "Family mental health and support" with cumulative scores of 34, 34 and 31 respectively. The domain that was the less frequently addressed was the "Postdischarge preparation and follow-up" with a cumulative score of 24. Out of the 30 items, 17 were found in six apps or more (17/30, 57%) and the other in less than six apps (13/30, 43%). Out of the 17, four are found in all apps (12/12, 100%): "Medical equipment and NICU setting information (1a)", "Premature newborn care (1d)", "Breastfeeding and feeding education (3a)", "Establishing, expressing and maintaining breastfeeding (3c)". Out of the 13, three items are only covered in 3 apps (3/12, 25%): "Discharge planning tools (5a)", "Mother physical health and post-partum follow-up (5e)", "Support and information related to emotional and intimate relationships after the birth of a child (6d)" and one item is found in only one app (1/12, 8%) and is "Both parents involvement in feeding (3b)".

Table 4. Details of the level of evidence-based assessment

id	Newborn care in NICU set.						Preemie's health and development						Breastfeeding and feeding						Discharge preparation						Post-discharge preparation						Family mental health and support							
	1a	1b	1c	1d	1e	*Total 1*	2a	2b	2c	2d	2e	*Total 2*	3a	3b	3c	3d	3e	*Total 3*	4a	4b	4c	4d	4e	*Total 4*	5a	5b	5c	5d	5e	*Total 5*	6a	6b	6c	6d	6e	*Total 6*	***Grand Total***	***Likert Score***
1	1	1	1	1	1	**5**	1	1	0	1	1	**4**	1	0	1	0	1	**3**	1	1	0	0	1	**3**	0	1	1	1	0	**3**	1	1	1	0	0	**3**	**21**	**4**
2	1	0	0	1	1	**3**	1	0	0	1	1	**3**	1	0	1	0	1	**3**	1	0	1	1	0	**3**	0	0	0	0	1	**1**	1	1	1	1	0	**4**	**17**	**3**
3	1	0	0	1	0	**2**	1	1	1	1	1	**5**	1	0	1	0	1	**3**	1	1	1	1	1	**5**	0	0	1	1	1	**3**	1	1	1	1	1	**5**	**23**	**4**
4	1	0	1	1	1	**4**	1	1	1	0	1	**4**	1	0	1	1	1	**4**	1	1	1	1	1	**5**	0	1	1	1	0	**3**	1	1	1	0	1	**4**	**24**	**4**
5	1	1	1	1	0	**4**	1	1	0	0	0	**2**	1	0	1	0	0	**2**	0	0	1	0	0	**1**	0	0	0	0	0	**0**	1	1	1	0	0	**3**	**12**	**2**
6	1	1	1	1	0	**4**	1	0	1	1	1	**4**	1	0	1	1	1	**4**	1	0	0	0	0	**1**	0	0	1	1	0	**2**	1	1	0	0	0	**2**	**17**	**3**
7	1	0	0	1	0	**2**	1	1	1	0	1	**4**	1	0	1	0	1	**3**	1	1	0	1	1	**4**	1	0	0	1	0	**2**	1	1	1	0	0	**3**	**18**	**3**
8	1	0	0	1	0	**2**	0	0	1	0	0	**1**	1	0	1	0	0	**2**	0	0	0	0	0	**0**	0	0	0	0	0	**0**	0	0	1	0	0	**1**	**6**	**1**
9	1	1	1	1	0	**4**	0	1	0	1	1	**3**	1	1	1	1	1	**5**	1	1	1	0	1	**4**	1	1	1	1	0	**4**	0	0	0	0	0	**0**	**20**	**4**
10	1	0	0	1	0	**2**	1	0	0	1	1	**3**	1	0	1	1	1	**4**	0	1	1	1	1	**4**	0	1	1	0	1	**3**	1	1	1	1	1	**5**	**21**	**4**
11	1	1	1	1	0	**4**	0	0	0	0	0	**0**	1	0	1	1	1	**4**	0	1	1	0	1	**3**	1	1	1	0	0	**3**	0	0	0	0	1	**1**	**15**	**3**
12	1	0	0	1	1	**3**	0	0	0	0	1	**1**	1	0	1	0	0	**2**	N.	N.	1	N.	N.	**1**	N.	N.	N.	N.	N.	**0**	0	0	0	0	0	**0**	**7**	**NA**
Σ	**12**	**5**	**6**	**12**	**4**	**39**	**8**	**6**	**5**	**6**	**9**	**34**	**12**	**1**	**12**	**5**	**9**	**39**	**7**	**7**	**8**	**5**	**7**	**34**	**3**	**5**	**7**	**6**	**3**	**24**	**8**	**8**	**8**	**3**	**4**	**31**		

Notes: Column 1 represents the apps' id as presented in Table 3. Then, each numbered column represents an item of our novel evidence-based content assessment scale.

5 Discussion

This review of apps for NICU parents of premature neonates going home followed the Richardson et al. [34] review with an added evaluation of the level of evidence based content of the apps, as well as an evaluation of their data protection.

5.1 Quality Assessment

The comparison we can make with Richardson is of the app selection and assessment of quality using the MARS. In 2019, 18 apps were selected and analysed. In this study, we ended up re-analysing nine same apps and three new ones. It is worth noting, that in our initial search they were eight new more apps found and corresponding to the selection criteria, but unfortunately, they were not accessible to Swiss users. Comparable to the results of the initial study, none of the apps reached the "excellent" level with MARS. In our study, five were evaluated as good, five as acceptable and two were evaluated as poor. In Richardson study, they concluded that MARS did not reflect exactly the quality of the apps as, based on expert evaluation, the information within one app was found to likely enhance parents learning and feelings to support their NICU experience, despite receiving lower scores in the MARS core sub-scales.

The only MARS item evaluating the evidence-based content is item 19 "The app has not been trialed/tested". This item only assesses if the app has been trialed in a study and if outcomes have been measured. The fact that the app has been developed following a scientific process, or translated or adapted would not be captured by MARS. Thus, we suggest that grade 1 of item 19 could be, "the development or adaptation of the app followed a scientific approach and has been published", because, such a step should ensure a certain level of app quality.

Also, there is no item in MARS that evaluates if the app content is based on evidence and published results, despite the fact that this is crucial when developing any intervention, including apps or eHealth.

Thus, in order to analyse the app content in a systematic and objective way, we proceeded to an in-depth analysis of the apps and verified if their content was compliant to the latest guidelines published in US and Europe to best support families and improve their outcomes.

5.2 Evidence-Based Content (EBC) Assessment

When evaluating the level of evidence-based content, none of the apps reviewed reached the "excellent" grade. This result is in line with the conclusions of authors who have proposed an assessment of the evidence-based content of apps in related fields [6, 23, 44], demonstrating a recurring trend in healthcare towards poor adherence of apps developed to guidelines and results of scientific literature. Interestingly, the EBC classification did not paralleled with MARS results, with some apps obtaining a "good" grade for MARS but only an "acceptable" for the EBC assessment (e.g."Hand to Hold") or inversely some apps obtaining an "acceptable" grade for MARS but a "good" for the EBC assessment (e.g. "Preterm Connect app"). These discrepancies could be due to the fact that the EBC tool provides a more in-depth assessment of the information dimension, complementing the one performed by the conventional MARS, which focuses essentially on the way in which information is transmitted. Also of interest, is to observe that it would be possible to create an app that would reach practically a full score (29/30) on the EBC if various domains were extracted from different apps. My Preterm Baby obtained the highest score for the EBC domains 2, 4 and 6, Babble for the domain 1 and Life's Little Love for domains 3 and 5. This in-depth review addressing 30 items that are recommended, allowed us to do a precise cartography of items frequency. Thus topics around medical information and premature neonatal care are systematically covered, however, items addressed to both parents are mostly not present. This underlines a recurrent issue in the intervention for NICU parents, where the father is often forgotten, despite the fact that his involvement in the NICU, discharge and beyond is of the utmost importance for all family members.

5.3 Data Protection Assessment

To our knowledge, data protection has not been evaluated systematically in other NICU apps reviews. However, in this day and age, it seems that apps should respond to the highest data protection policies. In the present study, two apps reached the maximum score and complied with GDPR and for nine apps, the data protection was either not present at all or in a vague manner.

5.4 Limitations and Future Research Directions

The findings of this study should be interpreted in the context of certain limitations. One is our strategy to retrieve and screen the apps with Appagg. Overall it provided a relevant list of apps but it was too inclusive, and thus we had to exclude over 500 apps. We also had to add an app that escaped the Appagg listing. Another limitation is that we performed the app collection in English and French, but not in German, Italian and Romansh, which are the official languages of Switzerland and would have therefore enabled us to extend this review to all Swiss regions and not only the French one. However, the app collection in French did not yield to any supplementary results than in English. Based on the literature review we strongly believe, that search in German, Italian and Romansh would have lead to a similar observation. Despite demonstrating the interest and high relevance of our new EBC and data protection tools, there are limitations to their development and implementation. Although they are both based on the most recent guidelines and recommendations, we did not proceed to their psychometric evaluation. To strengthen the repeatability and objectivity of our new assessment tools, future research directions could be their revision and validation by a committee of experts using a Delphi-like approach or measurement of their psychometric validity in a dedicated study.

6 Conclusion and Perspectives

This study provides an updated review of the apps that support NICU parents of premature neonates returning home and currently available for the french speaking part of Switzerland in French or English. Quality, evidence-based content level, and data protection/security assessment were investigated and demonstrated that none of the apps in the market reached the highest level in all three domains. Based on this review, we were able to identify a number of avenues to develop a new app for NICU parents that would respond to the highest standard in the future. We believe this app should be free, co-designed with healthcare professionals and parents, and should address the 30 items listed in the EBC assessment to answer the latest guidelines. This app should be available in a multitude of languages in order to reach a maximum of parents, and more specifically those who cannot communicate easily in the country where they are having their children. Indeed, it is known that most of the preterm neonates hospitalized in the Swiss NICUs are from parents who are not Swiss. This is due to multiple factors, including a less frequent pregnancy monitoring. These parents who are more vulnerable do not always speak the national language or English. Thus, an app that could be available in their own language will be of great support.

Finally, we also believe that the app should be scientifically built and its impact tested in a clinical setting within a randomized controlled trial or quasi experimental study, such as the NICU2HOME app developed by Garfield and its team [11]. This app was not included in the current study because it needs a NICU security code to enter. In regards to evaluation of eHealth apps, we believe that MARS is a relevant tool for the evaluation of quality but that it could be modified or completed as we did to evaluate the app level of evidence-based content and data protection.

As more and more apps are being developed in healthcare, it is warranted to establish systematic protocols for their screening, evaluation and development. Such protocols have recently been developed and published [37], and among them the TECH framework [12]. Although this review has not been designed according to the TECH framework, its methodology follows a close rationale. The distinctive feature of the TECH framework is that the formulation of the research question and eligibility criteria during an app review should be systematically guided through 4 dimensions: target user, evaluation focus, apps connectedness and the health domain explored [12]. We strongly believe that future work aimed at analysing apps in our field and in healthcare more generally should embrace such a systematic approach, as the one provided by the TECH framework.

References

1. Adams, M., Natalucci, G., Bassler, D.: An overview of the Swiss Neonatal Network & follow-up group (SwissNeoNet). Pediatr. Med. **6** (2023)
2. Atas, A.H., Çelik, B.: Smartphone use of university students: patterns, purposes, and situations. Malays. Online J. Educ. Technol. **7**(2), 59–70 (2019)
3. Banerjee, J., et al.: Improving infant outcomes through implementation of a family integrated care bundle including a parent supporting mobile application. Arch. Dis. Child. Fetal Neonatal Ed. **105**(2), 172–177 (2020)
4. Boulos, M.N.K., Brewer, A.C., Karimkhani, C., Buller, D.B., Dellavalle, R.P.: Mobile medical and health apps: state of the art, concerns, regulatory control and certification. Online J. Public Health Inform. **5**(3), 229 (2014)
5. Chen, F., Bajwa, N.M., Rimensberger, P.C., Posfay-Barbe, K.M., Pfister, R.E.: Thirteen-year mortality and morbidity in preterm infants in Switzerland. Arch. Dis. Child. Fetal Neonatal Ed. **101**(5), F377–F383 (2016)
6. Cheng, H., et al.: Content and quality of infant feeding smartphone apps: five-year update on a systematic search and evaluation. JMIR mHealth uHealth **8**(5), e17300 (2020)
7. De Rouck, S., Leys, M.: Information needs of parents of children admitted to a neonatal intensive care unit: a review of the literature (1990–2008). Patient Educ. Couns. **76**(2), 159–173 (2009)
8. van Dokkum, N.H., de Kroon, M.L., Reijneveld, S.A., Bos, A.F.: Neonatal stress, health, and development in preterms: a systematic review. Pediatrics **148**(4) (2021)
9. Dol, J., Delahunty-Pike, A., Siani, S.A., Campbell-Yeo, M.: EHealth interventions for parents in neonatal intensive care units: a systematic review. JBI Evid. Synth. **15**(12), 2981–3005 (2017)
10. Pediatrics, A.A.O.: Committee on fetus and newborn. Hospital discharge of the high risk neonate. Pediatrics **122**, 1119–1126 (2008)
11. Garfield, C.F., Kerrigan, E., Christie, R., Jackson, K.L., Lee, Y.S.: A mobile health intervention to support parenting self-efficacy in the neonatal intensive care unit from admission to home. J. Pediatr. **244**, 92–100 (2022)
12. Gasteiger, N., et al.: Conducting a systematic review and evaluation of commercially available mobile applications (apps) on a health-related topic: the tech approach and a step-by-step methodological guide. BMJ Open **13**(6), e073283 (2023)
13. Gibson, C., Williams, M., Ross, K., de Vries, N.: Distress, self-efficacy, feeling informed and the babble app: a New Zealand neonatal parent sample. J. Neonatal Nurs. **29**(2), 273–277 (2023)
14. Goetz, M., et al.: Perceptions of patient engagement applications during pregnancy: a qualitative assessment of the patient's perspective. JMIR mHealth uHealth **5**(5), e7040 (2017)
15. Grunberg, V.A., Geller, P.A., Hoffman, C., Patterson, C.A.: A biopsychosocial model of nicu family adjustment and child development. J. Perinatol. **43**(4), 510–517 (2023)
16. Guerra-Reyes, L., Christie, V.M., Prabhakar, A., Harris, A.L., Siek, K.A.: Postpartum health information seeking using mobile phones: experiences of low-income mothers. Matern. Child Health J. **20**, 13–21 (2016)

17. Harris, R., Gibbs, D., Mangin-Heimos, K., Pineda, R.: Maternal mental health during the neonatal period: relationships to the occupation of parenting. Early Hum. Dev. **120**, 31–39 (2018)
18. Hoofnagle, C.J., Van Der Sloot, B., Borgesius, F.Z.: The European union general data protection regulation: what it is and what it means. Inf. Commun. Technol. Law **28**(1), 65–98 (2019)
19. Ionio, C., et al.: Mothers and fathers in NICU: the impact of preterm birth on parental distress. Eur. J. Psychol. **12**(4), 604 (2016)
20. Jani, S.G., et al.: PretermConnect: leveraging mobile technology to mitigate social disadvantage in the NICU and beyond. In: Seminars in Perinatology, vol. 45, p. 151413. Elsevier (2021)
21. Kalgotra, P., Raja, U., Sharda, R.: Growth in the development of health and fitness mobile apps amid Covid-19 pandemic. Digit. Health **8**, 20552076221129070 (2022)
22. Kim, H.N., Garfield, C., Lee, Y.S.: Paternal and maternal information and communication technology usage as their very low birth weight infants transition home from the NICU. Int. J. Hum.-Comput. Interact. **31**(1), 44–54 (2015)
23. Li, Y., Zhao, Q., Cross, W.M., Chen, J., Qin, C., Sun, M.: Assessing the quality of mobile applications targeting postpartum depression in China. Int. J. Ment. Health Nurs. **29**(5), 772–785 (2020)
24. Liu, Y., McGowan, E., Tucker, R., Glasgow, L., Kluckman, M., Vohr, B.: Transition home plus program reduces medicaid spending and health care use for high-risk infants admitted to the neonatal intensive care unit for 5 or more days. J. Pediatr. **200**, 91–97 (2018)
25. Logsdon, M.C., et al.: Preferred health resources and use of social media to obtain health and depression information by adolescent mothers. J. Child Adolesc. Psychiatr. Nurs. **27**(4), 163–168 (2014)
26. Logsdon, M.C., Mittelberg, M., Myers, J.: Use of social media and internet to obtain health information by rural adolescent mothers. Appl. Nurs. Res. **28**(1), 55–56 (2015)
27. Lupton, D., Pedersen, S.: An Australian survey of women's use of pregnancy and parenting apps. Women Birth **29**(4), 368–375 (2016)
28. Newzoo: Global mobile market report (2018)
29. Orr, T., et al.: Smartphone and internet preferences of parents. Adv. Neonatal Care **17**(2), 131–138 (2017)
30. Oulasvirta, A., Rattenbury, T., Ma, L., Raita, E.: Habits make smartphone use more pervasive. Pers. Ubiquitous Comput. **16**, 105–114 (2012)
31. Patrick, S.W., Burke, J.F., Biel, T.J., Auger, K.A., Goyal, N.K., Cooper, W.O.: Risk of hospital readmission among infants with neonatal abstinence syndrome. Hosp. Pediatr. **5**(10), 513–519 (2015)
32. Pladys, P., et al.: French neonatal society position paper stresses the importance of an early family-centred approach to discharging preterm infants from hospital. Acta Paediatr. **109**(7), 1302–1309 (2020)
33. Prouhet, P.M., Gregory, M.R., Russell, C.L., Yaeger, L.H.: Fathers' stress in the neonatal intensive care unit: a systematic review. Adv. Neonatal Care **18**(2), 105–120 (2018)
34. Richardson, B., et al.: Evaluation of mobile apps targeted to parents of infants in the neonatal intensive care unit: systematic app review. JMIR mHealth uHealth **7**(4), e11620 (2019)
35. Rio, L., Fadda, M.D., Lambert, S., Ramelet, A.S.: Beliefs and needs of fathers of newborns hospitalised in a neonatal unit: a descriptive correlational study. Aust. Crit. Care **35**(2), 167–173 (2022)
36. Rio, L., Tenthorey, C., Ramelet, A.S.: Unplanned postdischarge healthcare utilisation, discharge readiness, and perceived quality of teaching in mothers of neonates hospitalized in a neonatal intensive care unit: a descriptive and correlational study. Aust. Crit. Care **34**(1), 9–14 (2021)

37. Roberts, A.E., Davenport, T.A., Wong, T., Moon, H.W., Hickie, I.B., LaMonica, H.M.: Evaluating the quality and safety of health-related apps and e-tools: adapting the mobile app rating scale and developing a quality assurance protocol. Internet Interv. **24**, 100379 (2021)
38. Sapanel, Y., et al.: Economic evaluation associated with clinical-grade mobile app–based digital therapeutic interventions: systematic review. J. Med. Internet Res. **25**, e47094 (2023)
39. Siva, N., et al.: Stress and stressors experienced by the parents of high-risk neonates admitted in neonatal intensive care unit: systematic review and meta-analysis evidence available from India. Stress Health (2023)
40. Smith, V.C., Love, K., Goyer, E.: NICU discharge preparation and transition planning: guidelines and recommendations. J. Perinatol. **42**(Suppl. 1), 7–21 (2022)
41. Souto, N., Curado, M.A., Henriques, M.A., Garcia, A., Coughin, M., Vasconcellos, T.: Quantum caring for parents: translation and adaptation of a mobile application into Portuguese. Enfermería Global **21**(1), 532–544 (2022)
42. Spargo, P., de Vries, N.K.: 'Babble': a smartphone app for parents who have a baby in the neonatal unit. J. Paediatr. Child Health **54**(2), 121–123 (2018)
43. Spittle, A., Orton, J., Anderson, P.J., Boyd, R., Doyle, L.W.: Early developmental intervention programmes provided post hospital discharge to prevent motor and cognitive impairment in preterm infants. Cochrane Database Syst. Rev. (11) (2015)
44. Stifani, B.M., Peters, M., French, K., Gill, R.K.: There's an app for it: a systematic review of mobile apps providing information about abortion using a revised mars scale. PLOS Digit. Health **2**(7), e0000277 (2023)
45. Stoyanov, S.R., Hides, L., Kavanagh, D.J., Zelenko, O., Tjondronegoro, D., Mani, M.: Mobile app rating scale: a new tool for assessing the quality of health mobile apps. JMIR mHealth uHealth **3**(1), e3422 (2015)
46. Sundstrom, B.: Mothers "Google It Up:" extending communication channel behavior in diffusion of innovations theory. Health Commun. **31**(1), 91–101 (2016)
47. Swiss Confederation: Data protection (2023). https://www.kmu.admin.ch/kmu/en/home/facts-and-trends/digitization/data-protection.html
48. Swiss Confederation: Language – facts and figures (2023). https://www.eda.admin.ch/aboutswitzerland/en/home/gesellschaft/sprachen/die-sprachen---fakten-und-zahlen.html
49. Tenfelde, K., Antheunis, M., Krahmer, E., Bunt, J.E.: Using digital communication technology to improve neonatal care: two-part explorative needs assessment. JMIR Pediatr. Parent. **6**, e38435 (2023)
50. Terhorst, Y., et al.: Validation of the mobile application rating scale (MARS). PLoS ONE **15**(11), e0241480 (2020)
51. Tofighi, B., Abrantes, A., Stein, M.D.: The role of technology-based interventions for substance use disorders in primary care: a review of the literature. Med. Clinics **102**(4), 715–731 (2018)
52. Treyvaud, K., Spittle, A., Anderson, P.J., O'Brien, K.: A multilayered approach is needed in the NICU to support parents after the preterm birth of their infant. Early Hum. Dev. **139**, 104838 (2019)
53. Voigt, P., Von dem Bussche, A.: The EU general data protection regulation (GDPR). A practical guide, 1st edn, Springer, Cham (2017)
54. Zhang, X., Kurtz, M., Lee, S.Y., Liu, H.: Early intervention for preterm infants and their mothers: a systematic review. J. Perinat. Neonatal Nurs. **35**(4), E69–E82 (2021)

Assessing the Progress of Portuguese Hospitals' Online Services

Demetrios Sarantis(✉), Delfina Soares, and Joana Carvalho

The United Nations University Operating Unit on Policy-Driven Electronic Governance (UNU-EGOV), Rua de Vila Flor 166, 4810-445 Guimarães, Portugal
{sarantis,soares,joana.carvalho}@unu.edu

Abstract. Online health services provision is the future of heath sector globally. Hospitals that do not make the transition from paper-based systems to electronic may swiftly undermine their chances of sustaining competitiveness in the market and efficiency towards their patients. This paper applies Hospital Online Service Index (HOSPI) assessment methodology to analyse progress in Portuguese hospitals' migration to a digital state. It employs data from an assessment conducted in 2021 to examine online health services provision. The article compares the results of that assessment to a previous one in 2019 and finds that Portuguese hospitals' websites remain in a stable status with slight progress in administration procedures, navigability, usability and readability aspects. The paper also suggests possible steps that could be considered, by Portuguese hospitals, towards covering the existing gaps.

Keywords: health · hospital · assessment · evaluation · website · service · patient · Portugal

1 Introduction

In the digital age we live today, it is important for the healthcare industry to have an online presence. Most (92%) US residents demand quick and easy access to their medical records. However, nearly half of them have trouble accessing medical records from their patient portal (45%) or physician's office (42%). At the same time, most health executives (77%) are investing in enhanced portals and mobile apps [1].

A hospital website acts as a digital gateway, providing potential patients and their families' easy access to important information about the facility and the services it offers [2]. This information could include the hospital's address, how to reach it, doctors' profiles, departments, and offered services. Patients can also learn more about the hospital's specialties, infrastructure and facilities if it has a website. This information can help patients make better decisions about their health care and make it more likely to choose the hospital as their provider [3]. A website shows how professional and knowledgeable a hospital is. It gives the hospital an opportunity to interact with patients, to show its capacity, accomplishments, awards, and accreditations [4]. By reading about the hospital's history, mission, and vision, potential patients and their families can quickly decide

M. Papadaki et al. (Eds.): EMCIS 2023, LNBIP 501, pp. 225–233, 2024.
https://doi.org/10.1007/978-3-031-56478-9_16

if they can trust it. Also, hospitals can include reviews and testimonials from other patients, which can add to their credibility [5].

Furthermore, online healthcare websites have attracted people's attention as a platform for information exchange between patients and between doctors and patients. Online consultation is a way for doctors to communicate with patients through healthcare websites directly. For policymakers, online consultations can make up for the deficiencies in traditional consultations and achieve a balance of medical resources. For hospitals, online consultation can increase hospitals' visibility, increase profit, and coordinate resources between hospitals. Managers of healthcare websites aim to create value by increasing the number of participating patients and doctors. Patients can save time and reduce medical costs through online consultations, and doctors can increase their own income and grow popularity through online consultations [6].

Patients do not have to go to the hospital in person to make appointments, check test results, pay bills, or look at medical records. A website can also make it easier for hospital staff to do their jobs by giving them online forms, like admission forms, that can be filled out and sent in.

A website is an economical way for a hospital to market its services to a large audience. It costs less than traditional ways to advertise, like on TV, radio, or in print. A hospital website can also be updated quickly and easily, giving patients and people who might become patients the most up-to-date information. A website can make the experience of a patient more efficient and pleasant by giving them information and tools to manage their health. On their websites, hospitals can post health and wellness-related articles, blogs, and videos that can help patients learn more about their conditions and treatments. It can also be used to offer online support groups, which can help people with some health problems find others with the same problems.

Therefore, understanding the dynamics of using online healthcare platforms and assessing the provided services is an important area of research from the perspective of all the stakeholders [7]. In this paper, after presenting the basic elements of the applied methodology, the assessment findings are presented in comparison with 2019. The final sections discuss possible improvements of identified gaps and conclusions.

2 Methodology

The empirical data for this study is extracted from an assessment survey conducted in 2021 by The United Nations University Operating Unit on Policy-Driven Electronic Governance (UNU-EGOV). 146 Portuguese hospitals, from public and private sector, were included in the survey. 144 have been assessed since two websites were down and impossible to access. These hospitals cover the whole geographical (coastal, inland, islands) and administrative areas (large region, medium region, small region according to the resident population) of the country.

HOSPI methodology, grounded in [8], is composed of four criteria, each with a set of indicators and sub indicators that allow hospital website assessment. The assessment was done through direct observation of the hospitals' websites, of the set of criteria, indicators, and sub-indicators described in [9]. During the stage of data collection, value 1 was ascribed to the presence of the considered sub-indicator, value 0 to its absence, and NA if it was not applicable.

The assessors were instructed to assume, during the assessment, the logic and attitude that would typically be of an average user when navigating the website. This means that the effort put in the search for the assessment of the sub-indicators should be like the one performed, on average, by a citizen while using the website, and not an exaggerated and extensive effort. The task was performed by a team of two assessors, under the supervision of a third one (supervisor), who is an expert on the assessment process. This means that for each hospital website there are two observations (one from each assessor) which were validated by the supervisor. In cases where the two assessors assigned different values to a specific sub-indicator, this was signalled to them to be reassessed more thoroughly. In case the assessment discrepancy remained, the supervisor decided which value was assigned to the sub-indicators. Data was then treated to attribute one single value to each sub-indicator, eliminating discrepancies and avoiding misclassifications. Assessors' commentaries were construed to facilitate this task and regarded as complementary information [10].

3 Findings

In this section, the results of the 2021 survey are presented in comparison with the 2019 survey results [11]. In this way, the progress of Portuguese hospitals websites can be monitored, and useful conclusions can be extracted.

3.1 Content

In Content criterion all indicators show a drop. Analysing C1.i1 (Health institution information available on the website) subindicators, only 'provision of phone directory' performs better than 2019. Portuguese hospital websites perform worse regarding 'provision of a welcome message', 'VAT number', 'map of the hospital area', 'complementary services (e.g. press, cafeteria, television, telephone, parking, religious service)', 'institution history', 'area covered by the hospital (population served)', 'quality management certification (e.g. ISO, EFQM)', 'management reports (e.g. activities plan, budget, activities report, account report)', 'public procurement announcements', 'hospitalization information' and 'legislation applied to hospital's context'. The rest of subindicators perform similarly to 2019 (Table 1)

.

- Analysing C1.i2 (Quality Metrics) subindicators, 'number of institution beds disclosed', 'waiting time to be seen in the emergency room', 'date of last monitoring of the waiting list disclosed', 'results of surveys regarding patient satisfaction' and 'number of internships accepted by the hospital' perform worse than 2019. The rest of subindicators perform similarly to 2019.
- In indicator C1.i3 (Organisational Structure and Medical Information), only 'hospital's departments list' perform better than 2019. 'Organisation chart (medical management, nursing management, institution management)', 'contact details of departments' and 'provision of health-care booklets' perform worse than 2019. The rest of subindicators perform similarly to 2019.

Table 1. Content criterion indicators

	2019			2021		
Indicator	Average value	Max value	Min value	Average value	Max value	Min value
C1.i1-Health institution information available on the website	0,626	0,905	0,222	0,586	0,81	0,333
C1.i2-Quality Metrics	0,123	0,533	0	0,085	0,438	0
C1.i3-Organisational Structure and Medical Information	0,399	0,75	0	0,346	0,75	0,063
C1.i4-Patient Information	0,442	0,933	0	0,316	0,867	0
C1.i5-Research and Teaching	0,472	1	0	0,239	0,778	0

- In indicator C1.i4 (Patient Information) only 'information and rules to be followed on admission discharge' perform better than 2019. 'Information regarding patient privacy', 'types of admission', 'information and rules to be followed during the stay at the institution', 'information and rules to be followed by visitors', 'patient care service business hours', 'details of how to pay charges or fees' and 'list of consultations/services with fees available' perform worse than 2019. The rest of subindicators perform similarly to 2019.
- In indicator C1.i5 (Research and Teaching), 'institution's scientific studies', 'institution's publications', 'institution's undergraduate or postgraduate courses', 'schedule of activities (e.g., courses, congresses, conferences)' and 'library related information' perform worse than 2019. The rest of subindicators perform similarly to 2019. This indicator is only analysed for hospitals that are labelled as teaching hospitals.

3.2 Service Provision

In Service Provision criterion, only C2.i1 (Administration Procedures) indicator has improved compared to 2019. More specifically, in this indicator, the 'possibility of forms downloading' shows an improvement. The other subindicators remain at a similar level (Table 2).

Subindicators of C2.i2 (Appointments) indicator do not show a significant change compared to 2019.

In C2.i3 (Patient Care), there are 2 subindicators, 'provision of telemedicine (videoconference system) services' and 'private area access (with login and password)' that perform better than 2019 and 3 subindicators, 'electronic directory with patient's records', 'patient telemonitoring (e.g. specific vital signs, blood glucose, peak flow rate, blood/urine chemistry)' and the 'possibility to require and/or obtain medical prescription' that perform worse than 2019. The rest of the subindicators do not present significant changes.

Table 2. Service Provision criterion indicators (indicators that increased with grey colour)

	2019			2021		
Indicator	Average value	Max value	Min value	Average value	Max value	Min value
C2.i1-Administration Procedures	0,292	0,75	0	0,318	0,5	0
C2.i2-Appointments	0,457	1	0	0,431	0,667	0
C2.i3-Patient Care	0,183	0,667	0	0,119	0,333	0

3.3 Community Interaction

In Community Interaction criterion, Publicity/Marketing indicator scores the same as 2019 while the other two underperform compared to 2019 (Table 3).

Table 3. Community Interaction criterion indicators

	2019			2021		
Indicator	Average value	Max value	Min value	Average value	Max value	Min value
C3.i1-Participation	0,325	0,7	0,1	0,283	0,6	0
C3.i2-Media	0,374	0,889	0	0,245	0,778	0
C3.i3-Publicity/Marketing	0,337	0,704	0	0,337	0,75	0

- In indicator C3.i1 (Participation), 'suggestions via web', 'information request via web' and 'associations that work at the institution (voluntary associations)' perform worse than 2019.
- In indicator C3.i2 (Media), 'institution's presence in the media (e.g., features news that appeared in press, radio, TV, social networks)', 'up-to-dated news/events schedule/newsletter', 'public relations office contact information' and 'institution news' underperform when compared to 2019.
- In indicator C3.i3 (Publicity/Marketing), 'disclosure of sponsors and investors', 'Facebook link', 'Twitter link', 'LinkedIn link' and 'other social networks links (e.g., Instagram)' perform better than 2019. The only subindicator that underperforms compared to 2019 is 'Youtube link'.

3.4 Technology

In the Technology Features criterion, 2 indicators, Navigability and Usability/Readability perform better than 2019. The rest perform worse than 2019 (Table 4).

- In indicator C4.i1 (Navigability), 'website name appears on browser title bar' and 'interwebsite links are distinguished from intrawebsite links' perform better than 2019.

Table 4. Technology Features criterion indicators (indicators that increased with grey colour)

	2019			2021		
Indicator	Average value	Max value	Min value	Average value	Max value	Min value
C4.i1-Navigability	0,646	0,857	0,143	0,772	0,857	0,25
C4.i2-Accessibility	0,393	0,7	0,1	0,356	0,5	0,2
C4.i3-Usability/Readability	0,533	0,786	0	0,590	0,786	0,286
C4.i4-Credibility	0,516	0,714	0,286	0,423	0,571	0,143
C4.i5-Privacy/Security	0,581	0,857	0	0,507	0,857	0

- In indicator C4.i2 (Accessibility), 'compliance with WCAG 2.1' and 'availability of accessibility symbol or declaration' perform worse than 2019.
- In indicator C4.i3 (Usability/Readability), 'illustrations/pictures/photos accompany text to assist description' and 'website technological sophistication (universal services use via web services, APIs, widgets)' underperform when compared to 2019. On the contrary, 'website search engine', 'website layout responsiveness (i.e., does it adapt to varying screen sizes)' perform better than 2019.
- In indicator C4.i4 (Credibility), 'pages have dates associated with them' score lower than 2019 and 'webmaster characteristics' better than 2019.
- In indicator C4.i5 (Privacy/Security), 'provision of disclaimers', 'copyright notice' and 'privacy policy' underperform while 'responsible person for website content' and 'website security (encryption)' perform better than 2019.

4 Discussion

It is noticeable that the change between the two assessments (2019, 2021) is not always positive as it may be expected. We observe a slight drop in all the content criterion indicators. On the one side, this may suggest that Portuguese hospitals did not improve their content during this period as they focused on facing COVID 19 related issues and providing information on the pandemic. In some cases, COVID 19 related sections have been developed in hospital websites to inform and properly guide citizens [12]. Quality metrics indicator is the lowest performing indicator in this criterion. Bearing in mind that quality metrics are used to assess the clinical performance, care quality, and safety measures of healthcare providers, proper quality indicators should be chosen and added to the websites.

Room for improvement also exists in service provision criterion. Even though there is marginal progress in digitizing administrative procedures, hospitals should modernize appointment procedures and use technology to upgrade and simplify doctor-patient interaction. Interaction digitization and data exchange could boost hospital's efficiency and patient's satisfaction [13]. Use of advanced digitized administrative functions lags and a digital divide appears to be emerging. Providing patients with real-time clinical information during their hospitalization may enhance their ability to engage in caregiving tasks, critical to ensuring inpatient care quality and safety [14]. Electronic patient record, information access and tele-services are key contributors to improving hospital performance in this regard.

There is an increasing trend towards using social networks to facilitate communication between hospital and patient. Hospitals seem to focus and investigate the social networks possibilities to come closer to patients, especially regarding information and communication issues [15]. These results of social media adoption and utilization among Portuguese hospitals, provide the framework for future studies investigating the effect of social media on patient outcomes, including links between social media use and the quality of hospital care and services. Patient engagement and interaction mechanisms need to be enhanced as well as media and news web presence.

Portuguese hospitals show an improvement in navigability, and usability/readability aspects, which shows an interest in facilitating the interaction of patients in an easier and more responsive way. On the contrary, accessibility, credibility and privacy/security aspects need to be re-examined. This results in the fact that many disabled citizens are not able to interact with hospital's website at all, thus depending on assistance which they would not otherwise need. Nor do non-disabled citizens have a better chance to get access to information in many cases, as the pages are hard to find. Improvement of accessibility levels will provide the possibility to various special patients' groups to use hospital websites. Updated data provision, privacy and security policies will increase the credibility of their content and the level of patients trust. These problems may be solved by a stricter approach to standards, rules, and legislation, at least as far as accessibility is concerned.

Some methodological differentiations are identified between the 2 assessments (2019–2021) which can to some extent explain some of the results variations. 132 hospital websites were assessed in 2019 compared to 146 in 2021. Different assessors were used for the 2 assessments. There are slight changes in the subindicators from 2019 to 2021 assessment. These changes occurred after the feedback received and modifications decided from the research team to meet the current needs and ameliorate the methodology. It should be also highlighted that assessment took place during the COVID 19 pandemic. It has been noted during the assessment that most of the Portuguese hospitals were focused on facing the pandemic, developing various initiatives in this direction, and modifying their websites to respond to contemporary needs.

Alone or in combination, there are some barriers to the adoption of online health services. For example, if hospital management lacks technology and website administration staff and expertise, it would be hard pressed to implement and manage even basic websites. If the hospital lacks financial resources, it may not be able to afford to develop and support online services. If hospital management has concerns about the security of their systems and data, even after other barriers have been addressed, it may remain cautious about what it places on the website and whether its services are interactive. Another relevant aspect concerns the difference between private and public hospitals in Portugal. The later having administrative limitations and shared information technology services that slow down progress in website service provision.

Portuguese hospitals that are in the early stages of website development need more time (and other scarce resources) to develop them. We would expect that newer hospital websites would be less mature than older ones. Some hospitals have website managers whose sole responsibility is to develop and maintain their website. A substantial infrastructure is needed to support hospital information systems operations and services and the existence of an information technology infrastructure is related to the ability of the hospital to implement innovative online services.

5 Conclusion

Overall Portugal's hospital websites seem to show a stagnation between 2019 and 2021. Content and service provision aspects should be reconsidered and redesigned based on patients' expectations and hospital management's targeting. The information provided by hospital websites are quite dynamic and a user may not need to physically visit or make a phone call to get any further information. As for the transactional services, there seems to be some reluctance both on the public hospitals and on the private ones to fully embrace the idea of digital administrative transactions and online exchange of patient data with the hospital. Digitization of administration procedures and patient information handling are important pillars in this direction. Social media and active interaction with stakeholders are increasingly considered by Portuguese hospitals. This area has also room for systematic improvement. Technological issues are handled adequately but they could also be improved by following standards and rules, established by central government in health sector.

As numerically more hospitals adopt online services, the innovation also penetrates more deeply-into their organizations and permits a wider array of uses and potentially greater payoffs. Hence, we would expect future studies to show an increase in both adoption and sophistication of online services among Portuguese hospitals.

As Portuguese hospitals attempt to move forward, they must not only put their plans into action but also make certain that perceived impacts are positive and deliver promised results. In addition, barriers to more sophisticated offerings must be dealt with if hospitals and their patients can expect to reap the benefits promised from these investments. Because it is still early in the development of online health services, there is still a need for more study. Both case study and assessment research are critical to further understanding online health services adoption, the extent of sophistication of these, and whether the hospitals and the patients are beginning to realize the promised payoffs from them. A future research direction would be to apply the assessment method in other countries and compare the results [16].

References

1. Erum, A.: Patients are using portals more than ever and high adoption is linked to shorter hospital stays, Insider Intelligence (2022). https://www.insiderintelligence.com/content/patients-using-portals-more-than-ever-and-high-adoption-linked-shorter-hospital-stays. Accessed 14 Sept 2023
2. Hong, Y.J., Kim, J.: Analysis of interactive e-health tools on United Arab Emirates patient visited hospital websites. Healthcare Inform. Res. **25**(1), 33–40 (2019)
3. Huang, E., Wu, K., Edwards, K.: Integrated patient education on US hospital Web sites. In: eHealth, pp. 215–222 (2016)

4. Sarantis, D., Soares, D. S.: Literature on website evaluation in health sector (2017)
5. Hulter, P., Pluut, B., Leenen-Brinkhuis, C., de Mul, M., Ahaus, K., Weggelaar-Jansen, A.M.: Adopting patient portals in hospitals: qualitative study. J. Med. Internet Res. **22**(5), e16921 (2020)
6. Almathami, H.K.Y., Win, K.T., Vlahu-Gjorgievska, E.: Barriers and facilitators that influence telemedicine-based, real-time, online consultation at patients' homes: systematic literature review. J. Med. Internet Res. **22**(2), e16407 (2020)
7. Sarantis, D., Soares, D.S.: From a literature review to a conceptual framework for health sector websites' assessment. In: Janssen, M., Axelsson, K., Glassey, O., Klievink, B., Krimmer, R., Lindgren, I., Parycek, P., Scholl, H.J., Trutnev, D. (eds.) EGOV 2017. LNCS, vol. 10428, pp. 128–141. Springer, Cham (2017). https://doi.org/10.1007/978-3-319-64677-0_11
8. Fan, W., Zhou, Q., Qiu, L., Kumar, S.: Should doctors open online consultation services? An empirical investigation of their impact on offline appointments. Inf. Syst. Res. **34**(2), 629–651 (2023)
9. Sarantis, D., Soares, D., Carvalho, J.: HSWAI: a health sector website assessment instrument. In: Proceedings of the 13th International Conference on Theory and Practice of Electronic Governance, pp. 359–368 (2020)
10. Sarantis, D., Soares, D., Carvalho, J.: Developing health sector websites assessment instrument: challenges and pitfalls. Eval. Program Plann. **92**, 102065 (2022)
11. Sarantis, D., Soares, D.S., Carvalho, J.: Assessment of hospitals' websites in Portugal. Front. Public Health **10**, 995153 (2022)
12. Xu, X., et al.: Assessment of internet hospitals in China during the COVID-19 pandemic: national cross-sectional data analysis study. J. Med. Internet Res. **23**(1), e21825 (2021)
13. Moro Visconti, R., Morea, D.: Healthcare digitalization and pay-for-performance incentives in smart hospital project financing. Int. J. Environ. Res. Public Health **17**(7), 2318 (2020)
14. Kelly, M.M., et al.: Parent perceptions of real-time access to their hospitalized child's medical records using an inpatient portal: a qualitative study. Hosp. Pediatr. **9**(4), 273–280 (2019)
15. Gonçalves, G.: Are hospitals our friends? An exploratory study on the role of Facebook in hospital organizations' dialogic communication. Health Mark. Q. **37**(3), 265–279 (2020)
16. Sarantis, D., Papadopoulos, S., Soares, D., Carvalho, J.: Design and implementation of hospital online services: in-depth assessment of Greece. In: 2023 Ninth International Conference on eDemocracy & eGovernment (ICEDEG), pp. 1–6. IEEE (2023)

A Cross-Sector Data Space for Correlating Environmental Risks with Human Health

Athanasios Kiourtis(✉), Argyro Mavrogiorgou, and Dimosthenis Kyriazis

Department of Digital Systems, University of Piraeus, Piraeus, Greece
{kiourtis,margy,dimos}@unipi.gr

Abstract. Data spaces are one of the key technology pillars of the European Strategy for Data, which intends to establish a single market inside the European Union (EU) for the efficient and secure sharing and interchange of data across industries. A data space that could combine and correlate cross-sector knowledge from the healthcare and environmental sectors could play a crucial role in determining the future of healthcare. Given that climate change is the single greatest health threat facing humanity and that health professionals worldwide are already responding to the health harms caused by this developing crisis, such a solution would enable the identification and correlation of environmental influences on human health as well as the extraction of novel biomarkers. However, the difficulties in creating such infrastructure necessitate cutting-edge, multidisciplinary research in numerous fields. This manuscript contributes into providing a visionary approach toward a single-entry point ecosystem to access, share, and trade cross-sector data assets originating from the environmental and healthcare domains through a Cross-sector Data Space (CDS), thereby effectively promoting European technological autonomy in data sharing. This CDS will consider a variety of analytics as ready-to-use solutions to facilitate analysis, prediction, and monitoring of the causality, correlation, reasoning, and practical visualization of real-time environmental settings, as well as to identify the effects of climate change to human health.

Keywords: Data Space · Data Sharing · Climate Risk · Human Health · Environmental Impact · Environmental Risk

1 Introduction

The European Strategy for Data aims at creating a single market for data to be shared and exchanged across sectors efficiently and securely within the European Union (EU) [1]. Data spaces is one of the main technological building blocks towards this target [2]. During the last years, the development of various data spaces has taken place as part of initiatives like GAIA-X [3], including data spaces in sectors like healthcare, finance, and industry. The latter enable organizations to share data in trusted and interoperable ways, while at the same time realizing mechanisms for data generation, curation, trading, and monetization. Nevertheless, there are many cases where data economy stakeholders must simultaneously access, combine, and process with data from multiple ecosystems and different data spaces from various sectors. This is for example the case for applications

M. Papadaki et al. (Eds.): EMCIS 2023, LNBIP 501, pp. 234–247, 2024.
https://doi.org/10.1007/978-3-031-56478-9_17

that need to derive or correlate insights from different, yet related, sectors such as finance and insurance, energy and manufacturing, as well as environment and healthcare. In such cases, state of the art data spaces initiatives fall short, as data spaces from different sectors and thematic areas must be integrated, nested and inter-related. Interconnecting such data spaces is a key enabler for cross-domain use cases. As a prominent example, such inter data space capabilities are required in the healthcare and environment sectors, where several isolated data spaces exist [4], without effective ways for combining correlated knowledge. This combination can play an instrumental role in shaping the future of healthcare, through identifying, anticipating, and factoring environmental influences on human health, while extracting novel biomarkers. However, the challenges of developing such infrastructure require multi-disciplinary, cutting-edge research.

Due to the ongoing climate crisis and its proven impact to human health [5], healthcare providers, environmentalists, regulators, researchers, as well as citizens require more efficient access to cross-domain knowledge, concerning collateral impacts and implications of the environmental domain, and more precisely the impacts of natural hazards (e.g., heatwaves, floods) to human health, with high data availability, accessibility, interoperability, and reusability, along with the support of analytical capabilities. A cross-sector data space can facilitate decision making and planning, increasing its involved parties' awareness, along with the fact that the available information on human health (e.g., wearable sensors, social media) and environment (e.g., air and water sensors, micro-satellites) is explosively growing, holding great promise in providing cross-sectoral critical insights. Hence, what is of critical importance is not only to offer siloed markets of healthcare and environmental data to assess these emerging threats, but to realize a cross-sector data space for collecting, storing, processing, discovering, and analyzing data of both sectors (i.e., healthcare and environment), supporting the correlation and the causal inference of their available data, based on Machine/Deep Learning (ML/DL), and Artificial Intelligence (AI) driven analytical procedures. Despite recent improvements in cutting-edge ML/DL and AI techniques, it is still challenging to clearly identify and correlate the impacts of environmental factors on human health, whereas their effects are not broadly understood and accounted to support effective decision-making. This is because climate-driven health risks are complex, often invisible, and hard to be accurately predicted. Moreover, the identification of causality relationships beyond simple correlations is hardly possible with traditional ML and analytics techniques. Therefore, the need for advanced AI reasoning techniques that put in place all the causes that lead to threatening environmental situations, while also identifying and analyzing the diversity of their factors that affect human health is more than crucial.

What is more, there is a great need for realizing interoperable data spaces [6], by viewing interoperability [50] as the new norm for facilitating mass adoption and scalability, on top of agreed standards, common vocabularies, ontologies, data models, and reference architectures. Therefore, an innovative, robust, and interoperable data space for the entire lifecycle of healthcare and environmental data must be developed. The data space should aim at secure and efficient AI-based data operations, further supporting explainability, interoperability, sharing, reusing, and trading high-quality and high-value fairly priced data assets (i.e., raw data, analysed data, data services, algorithms, etc.), thus expanding common European data spaces. What is also lacking is proper motivations for data providers to be convinced about the value of their data assets, thus assured of sharing them for additional research and exploitation. Hence, data providers must be encouraged to join in data spaces and must be provided with customized incentives to share and trade more data assets. Moreover, data spaces should be established in a coordinated way, combining technological, functional, and legal processes, adopting legislative measures on data governance and reuse, applying security/privacy-by-design, and assuring compliance to applicable laws (e.g., General Data Protection Regulation (GDPR) [7]).

In this context, the aim of this manuscript is to provide a visionary approach towards a single-entry point ecosystem to access, share, and trade cross-sector data assets deriving from the environmental and healthcare domains, through a Cross-sector Data Space (CDS), effectively contributing to the European technological autonomy in data sharing. This CDS will consider various analytics as ready-to-use data assets, exploiting ML/DL/AI/ Explainable AI (XAI) to facilitate the analysis, prediction, and monitoring of the causality, correlation, reasoning, and pragmatic visualization of real-time environmental settings, identifying human health impacts of environmental hazards, and their reverse correlation. Hence, direct and indirect human health impacts could be measured, creating, extending, explaining, and interlinking domain-specific and domain-agnostic knowledge.

The rest of the manuscript has the following structure. In Sect. 2, the ambition behind the envisioned CDS is being provided, referring to the emerging need of technological solutions towards addressing the climate impact on human health. Section 3 contains the visionary architecture of the CDS, while Sect. 4 summarizes the CDS's added value to its stakeholders, concluding with our next plans.

2 Ambition and Related Challenges

2.1 Environmental Impacts on Human Health

The World Meteorological Organization (WMO) highlighted that "2021 was a make-or-break year for climate action, with the window to prevent the worst impacts of climate change closing rapidly" [8]. The modern climate era is mainly characterized by the increase of climate changes, influencing the amount of solar energy that is received in our planet. Planet warming is happening almost ten times faster than the average rate of ice-age-recovery warming, with carbon dioxide produced by human activities exponentially increasing [9]. Currently it is proved that the average surface temperature has risen about 1.18 °C since the late 19th century, due to increased carbon dioxide emissions and other human activities [10]. This heat increase was absorbed by the ocean, with the top 100 m of ocean showing warming of more than 0.33 °C since 1969. At the same time, the amount of snow has decreased over the past five decades and the snow is melting earlier, affecting the overall sea levels. All these phenomena are mostly related to the global warming trend, provoking the "greenhouse effect". Considering such evidence, the World Health Organization (WHO) estimates that climate change will cause at least 250.000 additional deaths per year globally between 2030 and 2050, whereas in the EU environmental factors are estimated to account for almost 20% of all deaths [11].

It is already predicted [12] that increased concentrations of greenhouse gases will lead to increased deaths and illnesses from heat and potential decreases deaths from cold, whereas augmented air pollution or chemicals will negatively impact indoor and outdoor air quality, affecting the human respiratory and cardiovascular systems. In this context, it is estimated that higher pollen concentrations and longer pollen seasons will increase allergic sensitization and asthma episodes, decreasing the overall productivity and public health, while extreme events will create health risks due to damage of property, destruction of assets, loss of infrastructure and public services, social and economic impacts, environmental degradation, and other factors. Additionally, biodiversity anomalies or disruption of ecosystems will transfer viruses, bacteria, and protozoa, and higher concentrations of carbon dioxide in the atmosphere will affect food safety and nutrition, disrupt, or slow the distribution of food, and distress mental health. With these emerging challenges, EU is trying to decrease climate change through policies at home and close cooperation with international partners, where based on these it is already met a reduction of gas emissions [13], whereas it expects that emissions will be furtherly reduced by at least 55% by 2030 [14]. Furthermore, by 2050, EU aims to become the world's first climate-neutral continent and a climate-resilient society [15]. Nevertheless, these are just a few examples of what EU is targeting towards "Living and working in a health-promoting environment", and therefore a combination of technology, policies, and strategies should be provided towards addressing the climate-change impact on human health.

2.2 Cross-Sector Data Space Challenges

Transparency. Despite the well-structured specifications [16] about data spaces development, including standardizations and rules, there are still very few practical and scalable implementation of data spaces. In healthcare, the creation of a European Health Data Space (EHDS) is one of the priorities of the Commission, while on the other hand, the environmental sector is supported by the Green Deal Data Space along with the European Green Deal priority actions and EU programmes [17] (e.g., Galileo, Copernicus). Nevertheless, these initiatives have a specific sectoral focus and are confined to siloed use cases with limited interoperability, domain-specific authorization methods, and common analytics capabilities, providing partial discoverability and usability, and lacking advanced and trusted AI tools that can turn cross-sector data into insights. Data sharing, and trading schemes are not always considered, while not any cross-sector interlinkings are triggered, even though the environmental aspect is a driving factor for humans' health [18]. Hence, the vision is to implement cross-sector data spaces' building blocks, interlinking healthcare and environment, through guidelines, standards, and protocols.

Data Interoperability. The variability of vocabularies refers to generating mismatches (e.g., syntax, logical representation), ontologies (e.g., scope, model coverage, granularity), explications (e.g., paradigm, concept), and terminologies (e.g., synonyms) that lack standardisation initiatives, followed by unstructured legacy data [19, 48]. On top, existing ontologies, vocabularies, models, and data cannot be efficiently accessed or even located, since most of them are not uploaded to trusted repositories, not exploiting data spaces and persistent identifiers (e.g., Digital Object Identifiers (DOIs) [20]). Moreover, data can still be found trapped in silos, making it hardly accessible due to lack of single-entry points identity management mechanisms, or uninterpretable due to lack of explainability tools (e.g., XAI) [21]. Scientific complexity may be another data issue, whereas the different data models and policies of the environment and health domains are creating additional barriers in interpreting the extracted interlinking insights, since no interoperable cross-sector standards exist, while the outcomes importance and priorities differ between the two sectors [22]. Hence, the vision is to enable the ingested data interoperability through advanced semantic modelling, representation, and interlinking of cross-sector ontologies and models, applying data cleaning and interoperability standards-based techniques to detect and correct corrupt or inaccurate records and homogenize the collected data, whilst assigning data usage license and provenance information to make the data nature clear.

Data Assessment. Data scientists devote up to 80% of their time managing data before extracting any insights, delaying its assessment, monetization, and further treatment as an asset, while continuity is still a challenge [23]. These create additional data valuation issues, since there is currently no data monetization law, while data collection is highly decentralized, arising legal and social implications [24]. The treatment of data as separate assets is missing, through recognizing its importance, better organizing metadata, integrating data monetization as a business strategy, and encouraging growth, whilst communicating the data value, through market dynamics. Data monetization comes along with data sharing, including the challenges of data organization, copyright and licensing, multi-variety of repositories, lack of time to deposit data, and increased costs of sharing [25]. Balancing the benefits of data access and sharing with the risks, while considering private, national, and public interests may be an option for promoting data sharing. What is also needed is the reinforcement of trust and empowering users to share data and maximize the data re-use value. To this end, the encouragement of data provision via incentive tools is needed, acknowledging data markets limitations, and addressing data ownership uncertainties, intellectual property rights (IPRs) and ownership-like rights [26]. What is required, is a set of tools to value the assets nature, provenance, maturity, implementation/collection way, attributes, production, quality, completeness, application context, potential use cases, and relevant data assets. For efficient assets' assessment suitable data valuation and market dynamics methodologies should be applied to determine growth opportunities and supply/demand factors, whereas an incentives mechanism should encourage data providers to share/trade more data assets.

Reasoning of Climate Change and Human Health. Interpretability and reasoning over both domain-specific and cross-sector knowledge (i.e., healthcare, environment) through AI systems is to ensure AI operations' transparency and accountability to minimize poor decisions [27]. However, this is still performed based on past patterns and correlations, while since environmental conditions evolve, so does the human health, and thus static environments, closed problems, and lack of "predicting" scenarios may lead to cross-sector correlation inefficiencies [28]. With AI systems focusing on singular/limited tasks, relying solely on "training" data, the further usage of deep reasoning tools cannot adjust networks' weights for admissible solutions as the learning process advances. That is why, complex, often invisible, climate-driven health risks, are becoming extremely difficult to be accurately predicted, considering also that causality beyond simple correlations is hardly possible with traditional ML/analytics techniques [29], followed by increased energy inefficiencies. Also, there are missing reasoning techniques for explainable models, maintaining high searching, learning, and causal performance, while most XAI models deal with siloed non-interlinked sectors. In this context, it should be provided analytical capabilities, for large amounts of transactions. Data-based knowledge graphs extracting causal drivers combined with constraint reasoning could identify/quantify the reasons for climate hazards that threaten human health. Multisectoral causal AI analysis could quantify implications of environmental factors to human health, combined with the reasoning results, identifying enduring causal signals. Thus, the web of causes of an event could be identified, providing insights, enhancing trustworthiness and explainability, enabling humans to reason the outcomes of AI models.

3 Visionary Cross-Sector Data Space

3.1 Overall Concept

Considering current needs and challenges, the overall concept of the envisioned Cross-sector Data Space (CDS) ecosystem is to be accessible through a single-entry point, facilitating healthcare groups, environmental groups, regulatory groups, as well as citizens and researchers to securely access, share, trade, and analyze cross-sector data covering the environmental and healthcare domains towards a data economy capable of ensuring digital autonomy. These cross-sector data will concurrently consider various sources and initiatives, existing healthcare data spaces as well as environmental sources and data spaces, and will follow the Findable, Accessible, Interoperable, and Reusable (FAIR) paradigm being efficiently and fairly priced, and able to be traded in a secure and user-friendly way. This CDS ecosystem will offer analytics services to facilitate inter-domain analysis, prediction, quantification, and monitoring of the causality and reasoning of real-time environmental conditions, identifying the human health impacts of environmental hazards due to climate change and their reverse correlation. All these will be both developed and supported by cutting-edge technologies, supporting environmental sustainability, and an attractive, secure, and dynamic data-agile economy.

3.2 Cross-Sector Data Space Architecture

The envisioned CDS aims on effectively contributing to the EU technological autonomy in data sharing through its two (2) layers (Fig. 1), as explained below:

Data Catalogue and Data Assets Management Suite. This layer will consist of the components related with the *Data Catalogue*, *Multi-sources Data Collection*, *Data FAIRification*, *Data Assets Monetization & Trading*, and *Data Market Dynamics Analysis*, where each component will have its unique role, as indicated below (Fig. 2):

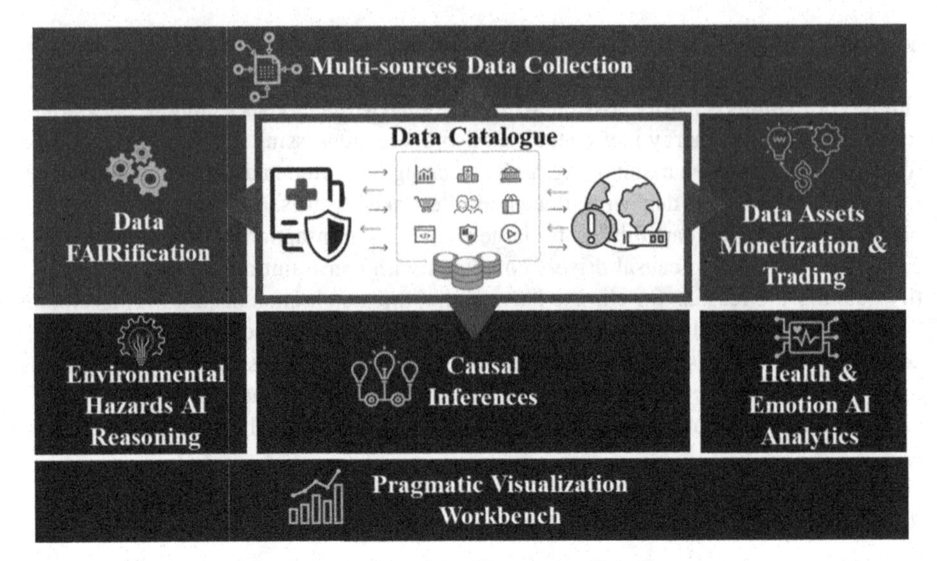

Fig. 1. Overview of the Cross-sector Data Space.

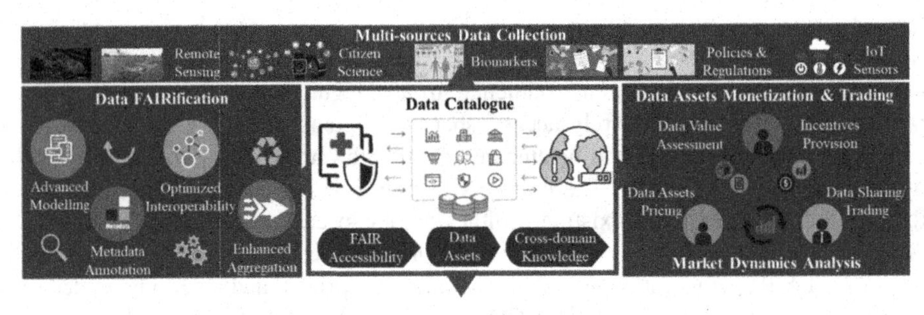

Fig. 2. Overview of the Data Catalogue & Data Assets Management Suite.

Data Catalogue. The Data Catalogue will deliver FAIR, discoverable, efficient, and fairly priced data assets to its various stakeholders for easily finding, accessing, processing, sharing, trading, and reusing the CDS data assets. It will include assets deriving from individuals, sector-specific target groups, as well as cross-sector and currently isolated data spaces in the healthcare and environmental domains. For each entry in the data space, additional information will be stored with respect to the data source nature, the use cases in which the data asset has been used, and its purpose within the context of the CDS.

Multi-sources Data Collection. These techniques will collect scattered healthcare and environmental data in one place securely ingesting: (i) Remote sensing data (satellite imagery (e.g., Copernicus [30]), social listening, data from surveillance tools and climate data sources (e.g., NASA Earth Observatory [31]), Sensors data of smart environments, (ii) Citizen science data collected from citizens' electronic devices (e.g., geolocation, demographic, health-related statistics, crowdsourcing, emotional data), (iii) Individuals' raw data/biomarkers (e.g., vital signs from wearable devices), (iv) Existing regulations (e.g., GDPR), and policies (health and climate), and (v) Data assets from external sector-specific data spaces (e.g., DIAS [32], Green Deal [33]).

Data FAIRification. These techniques will offer cross-sector data modelling, cleaning, interoperability, and aggregation for effectively managing/interlinking the collected data and their FAIR delivery. For data heterogeneity, it will be provided semantic representation via a meta-interpretation layer, enabling modelling and annotation of the ingested data with metadata and identifiers, for efficient discovery (Findability, Accessibility). Diverse data cleaning/interoperability standards-based methods (e.g., OpenRefine [34]) will be applied to detect and correct inaccurate records and homogenize the collected data (Interoperability), whilst a cross-sector data aggregation layer will fuse, link, and aggregate all the incoming data. The latter will also assign data usage license and provenance information to make clear how, why and by whom the data has been created/processed (Reusability). This will result to high transparency standards and wider adoption of the data.

Data Assets Monetization and Trading. These services will rely on the combination of the Big Data Value Chain (BDVC) strategies [35] along with Big Data monetization techniques, offering an XAI-driven multi-phased data value assessment mechanism to capture and harvest the value of the CDS data assets over time. It will automatically value the assets nature (e.g., volume, variety, reliability, sensitivity), provenance (e.g., data from vulnerable groups such as cancer survivors related with pollution levels, may have higher cross-sector added value than data deriving from an activity tracking device), maturity (e.g., raw, analyzed), collection/implementation way (e.g., data I/O), data attributes, production pathway (i.e., low to high ease of production), current application context, and potential future data both individual and cross-domain (healthcare and environmental) use cases, concurrently identifying and inventorying relevant data assets, whilst considering existing valuation models (i.e., Data/Insight/Analytics enabled Platform/Multiple industry Platforms-as-a-Service) for better XAI results. On top of the valuable data assets, these services will include an Adaptive Multi-Agent System for fairly implying the data assets' price, focusing on "how" (i.e., what are the best sales conditions to reach the required objectives) instead of "what" (i.e., which price to propose for the data), supporting, also the possibility for a user to price a data asset by herself. A customized incentives mechanism will also be offered tailored to each population group according to their profile, location, and generic requirements, to encourage data providers to join in the CDS and share/trade more assets. The data consumers will be able to decide on incentives to pay to the data providers, by considering the profit to be made from the collected data. By sharing the consumers' profit with the data provider as incentive, the data provider will get fair prices for providing her data (i.e., data awarding system). This will include a truthful price report mechanism guaranteeing that the data provider takes the optimal profit when she reports honestly her privacy price, while based on this, the consumer can maximize her profit within a potential budget constraint. These will be established via appropriate Service Level Agreement (SLA [36]) auditing among the interacting parties articulating the terms of the monetization and trading agreement, guaranteeing quality, completeness, and the respective provenance guidelines.

Market Dynamics Analysis. These tools will identify the factors that affect the CDS market, and the data monetization models, including the supply, demand, price, and quantity of the offered/traded/shared data assets, also considering what may potentially affect the envisioned business models. Market dynamic tests (e.g., Knowage [37]) will: (i) evaluate overarching market dynamics to determine CDS growth opportunities, (ii) determine the factors affecting the supply and demand of specific assets, and (iii) most properly and fairly assess and value the provided data assets. The toolkit will also identify what is driving the current demand, leading to decisions for growing, shrinking, or leaving the demand flat. Finally, it will focus on the demand drivers and limiters (including relevant macroeconomic indicators impacting this market), by studying fiscal/monetary policies to see the interest rates and prices in similar cross-sector data markets, to observe how stakeholders react to projections about the market's future, and to identify how much the CDS is needed and has the potential to stand out amid competitors. To boost user-friendliness, interactive visualizations will rationalize the market dynamics evolution pathway.

Inter-domain AI Analytics Toolbox. This layer will consist of the components related with the *Environmental Hazards AI Reasoning*, *Health & Emotion AI Analytics*, *Causal Inferences*, and *Pragmatic Visualization Workbench*, where each component will have its unique role, as indicated below (Fig. 3):

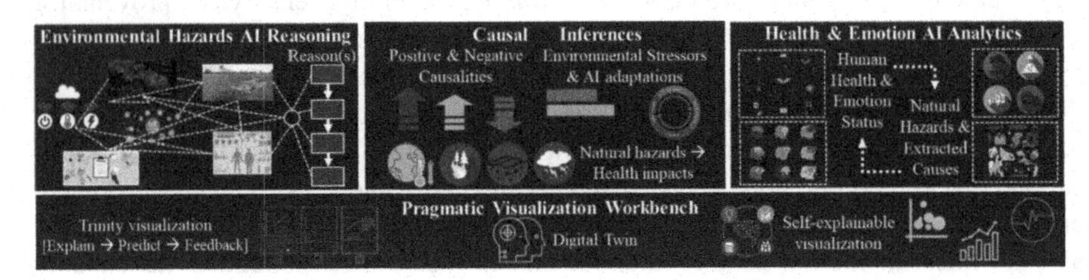

Fig. 3. Overview of the Inter-domain AI Analytics Toolbox.

Environmental Hazards AI Reasoning. These services will identify the reasons that led to environmental hazards due to climate change and their effects (e.g., increased heating levels lead to wildfires, causing air pollution), supporting what-if scenarios prior to analytics, for energy and time efficiency. This will be achieved by training reasoning structures to predict the most logical causes and effects of environmental hazards, by exploiting a data-oriented knowledge graph platform to apply advanced reasoning to the collected data and discover new insights following DRNets [38] and constraint reasoning, understanding why something happened and how current circumstances must change to improve future outcomes. Thus, the cause-and-effect concept will be supported explaining why a hazard happened, extracting non-predictable correlation results, and concluding on how current circumstances must change to improve future outcomes.

Health and Emotion AI Analytics. These services will provide an additional source of information towards interventions for improving the quality of life of individuals, by facilitating the analysis of data towards extracting knowledge (i.e., healthcare biomarkers that factor environmental parameters) regarding their medical/physical and mental health data (i.e., emotions), thus predicting how, when, and in whom a disease will develop, by applying traditional ML (e.g., SciKit Learn Support Vector Machines (SVM)[39]), and DL (e.g., TensorFlow Artificial Neural Networks (ANNs) [40]) techniques. Regarding medical/physical health data, data-driven analytics will be applied, for assessing individuals' health status and predicting the onset of a disease based on their living conditions, identifying the reasons behind such disease (e.g., air pollution leads to respiratory episodes, increasing the need for air ionizers, leading to raised heating levels).

Causal Inferences. These services will monitor the evolution of human health by observing causal inferences based on humans' exposure to specific environmental hazards (e.g., wildfires may lead to respiratory episodes since they cause air pollution, which causes respiratory episodes), whereas also identifying the causal inferences between the individuals' daily habits and environmental hazards (e.g., increased usage of air ionizers leads to increased heating levels, which may cause wildfires). Thus, the services will identify how environmental hazards affect human health and how human activities lead to environmental hazards, and thus will discover the causal inference regarding how humans' activities affect their own health. To support causality inferences the services will employ graph AI and XAI techniques (e.g., SHAP [41]) that explain interrelationships across features, combining Structural Equation Modelling (SEM) [42], and Exploratory/Confirmatory Factor Analysis (EFA/CFA) [43], providing causality relationships beyond conventional (naïve) correlations.

Pragmatic Visualization Workbench. This tool will offer a user-friendly visualization workbench that will visualize and assess the analytics results in different ways (e.g., correlation matrixes, histograms), while visualizations can be modified on-the-fly by the users via parameters experimentation. It will leverage advanced graph frameworks (e.g., Neo4J [44]) and libraries (e.g., NetworkX [45]) for graph analysis, inference, and visualization over CDS data and captured results, also enabling pragmatic visualization capabilities for demonstrating the analytical results and their potential impacts on top of its users' digital twins [46]. Each digital twin, acting as a digital replica of high trustworthiness with realistic behavior, will be constructed following the notion of digital twin data, considering their data gathering, interaction, universality, mining, fusion, iterative optimization, and on-demand usage. The workbench will also prompt the users to draw their own visualization predictions via a trinity visualization capability where: (i) the input data will be portrayed to users, (ii) the users will make their predictions, and (iii) the workbench will annotate the gap between users' prior knowledge and actual predictions.

4 Discussions, Limitations and Next Steps

The envisioned CDS aims to offer a solution for cross-sector data assets trading, monetizing, exchange, and interoperability being supported by mature systems towards technological autonomy in data sharing, by exposing to its stakeholders its entire ecosystem consisting of the layers and components explained in Sect. 3. These will provide the means to develop a cross-sector data space of realistic scope and size, deployable in real-world applications being AI-oriented, user-centric, transparent, trustworthy, FAIR-based, ethical, and legal compliant, offering a variety of added value. *(i) Cross-sector single sign-on Data Space linking environment and health*: The CDS will be a single-entry point hosting environment, offering to its stakeholders a novel data space for enabling the access, sharing, trading, and exchanging capabilities of data assets belonging on two (2) hardly correlated and interlinked sectors (i.e., environment and healthcare). All the available data assets (i.e., collected data, data models, policies, results of analytics mechanisms, etc.) will flow within Europe and across sectors aiming towards the goal of the common European data spaces (e.g., green deal, health, skills). *(ii) Environmental and healthcare FAIR data management and interoperable delivery*: The Data FAIRification mechanisms will enable the findability, accessibility, interoperability, and reusability of the ingested data to all the target groups. It will provide advanced semantic modelling and representation via a meta-interpretation layer, applying diverse data cleaning and interoperability standards-based techniques to detect and correct corrupt or inaccurate records and homogenize the collected data, which in principle rely on diverse data models and policies, whilst assigning data usage license and provenance information to make clear the nature of the data. *(iii) User-centered energy-efficient data assets assessment, monetization, sharing, and trading*: The techniques of AI Data Monetization and Trading will harvest the value of the CDS data assets, by monetizing them, zooming-in to their nature, emphasizing to their CO_2 efficiencies for their value adjustment and CO_2 inefficiencies for carbon-penalties application. For facilitating the assets' assessment, sharing, and trading, market dynamics analysis will be applied and visualized via a "market demand review", to determine growth opportunities, and factors affecting the assets' supply/demand. It will be implied a fairly assets' price, based on their nature, their CO_2 footprint, and the best sales conditions to reach their objectives, supporting their self-pricing. Above all, an incentives mechanism will encourage data providers to join in the CDS and share/trade more assets, via proper SLAs. *(iv) Novel and pragmatic inter-domain analytical insights*: An analytics toolbox including Environmental Hazards AI Reasoning, Health & Emotion AI Analytics, Causal Inferences, and Pragmatic Visualization Workbench will offer to the stakeholders with analytical capabilities, facilitating explainability, being capable of handling large amounts of transactions. The toolbox will be able to identify/quantify the reasons for environmental hazards that threaten human health (physical, mental, well-being), exploiting the data assets residing on the CDS. Also, the toolbox will quantify the implications of diverse environmental factors to human health combined with the reasoning results, identifying enduring causal signals. Thus, the web of causes of a behavior or event will be identified, providing critical insights and correlation results. For increasing the value/explainability of the results, stakeholders will be delivered a visualization workbench supporting the interactive estimation of multisectoral effects of environmental hazards to human health under various conditions, and their parameterization/application on top of users' digital twins for better decision making, what/if analysis, and interactive understanding.

Our next steps include the realization of the CDS envisioned functionalities based on scenario-driven business and technical requirements, considering limitations such as the difficulties into integrating all the different envisioned functionalities, or calculating the carbon footprint of energy-hungry components and algorithms. These functionalities will be put under functional stress tests in the context of external health-related projects (e.g., CrowdHEALTH [47], InteropEHRate [49]) reflecting the needs of diverse EU countries, while the overall architecture will be continuously adapted. Performance evaluations will be also considered, targeting the functionality, the practicality, and the user-friendliness of the envisioned CDS offerings. The extracted results will be continuously monitored and evaluated, whereas they will be communicated to external stakeholders (healthcare stakeholders, environmental stakeholders, regulatory stakeholders, citizens, and researchers), whose services will be facilitated, while their volunteerism will be enhanced for sharing and providing their data assets via the single-entry point of the CDS.

Acknowledgements. The SmartCHANGE project has received funding from the Horizon Europe R&I programme under the GA No. 101080965.

References

1. Bruzzone, G., et al.: As Open as Possible, as Closed as Needed: Challenges of the EU Strategy for Data. Les Nouvelles-Journal of the LESI **56**(1), 41–49 (2021)
2. Mavrogiorgou, A., et al.: FAME: federated decentralized trusted data marketplace for embedded finance. In: International Conference on Smart Applications, Communications and Networking (SmartNets), pp.1–6 (2023)
3. Braud, A., et al.: The road to European digital sovereignty with Gaia-X and IDSA. IEEE Netw. **35**(2), 4–5 (2021)
4. Curry, E.: Future research directions for dataspaces, data ecosystems, and intelligent systems. Real-time Linked Dataspaces, pp. 297–304 (2020)
5. Mitra, S., et al.: Impact of heavy metals on the environment and human health: Novel therapeutic insights to counter the toxicity. J. King Saud Uni. Sci. **34**(3), 101865 (2022)
6. Solmaz, G., et al.: Enabling data spaces: existing developments and challenges. In: Proceedings of the 1st International Workshop on Data Economy, pp. 42–48 (2022)
7. Kiourtis, A., et al.: Identity management standards: a literature review. Comput. Inform. **3**(1), 35–46 (2023)
8. Wise, J.: Climate change: window to act is closing rapidly. BMJ **380**, 674 (2023)
9. Noll, M.: Exponential life-threatening rise of the global temperature (2023)
10. Malla, F.A., et al.: Understanding climate change: scientific opinion and public perspective. Climate Change: The Social and Scientific Construct, pp.1–20 (2022)
11. Climate Change, https://www.who.int/health-topics/climate-change#tab=tab_1. Accessed 29 Sept 2023
12. Naiyer, S., Abbas, S.S.: Effect of greenhouse gases on human health. In: Sonwani, S., Saxena, P. (eds.) Greenhouse Gases: Sources, Sinks and Mitigation, pp. 85–106. Springer, Singapore (2022). https://doi.org/10.1007/978-981-16-4482-5_5
13. Is Europe reducing its greenhouse gas emissions?, https://www.eea.europa.eu/themes/climate/eu-greenhouse-gas-inventory/is-europe-reducing-its-greenhouse, last accessed 2023/09/29
14. Progress made in cutting emissions. https://climate.ec.europa.eu/eu-action/climate-strategies-targets/progress-made-cutting-emissions_en. Accessed 29 Sept 2023
15. 2050 long-term strategy. https://climate.ec.europa.eu/eu-action/climate-strategies-targets/2050-long-term-strategy_en. Accessed 29 Sept 2023

16. Design Principles for Data Spaces, H2020 OPEN-DEI Position Paper. https://design-principles-for-data-spaces.org/. Accessed 29 Sept 2023
17. Copernicus and Galileo: boosting their integration and synergies around the world. https://www.copernicus.eu/en/news/news/observer-copernicus-and-galileo-boosting-their-integration-and-synergies-around-world. Accessed 29 Sept 2023
18. Korançe, F.: The growing relation between environment and public health. SciMedicine J. **3**(2), 100–115 (2021)
19. Ramis Ferrer, B., et al.: Comparing ontologies and databases: a critical review of lifecycle engineering models in manufacturing. Knowl. Inf. Syst. **63**(6), 1271–1304 (2021). https://doi.org/10.1007/s10115-021-01558-4
20. Rosenberg, J. et al.: Leveraging ElasticSearch to improve data discoverability in science gateways. In: Practice & Experience in Advanced Research Computing, pp. 1–5 (2019)
21. Novakovsky, G., et al.: Obtaining genetics insights from deep learning via explainable artificial intelligence. Nat. Rev. Genet. **24**(2), 125–137 (2023)
22. Bokulich, A., Parker, W.: Data models, representation and adequacy-for-purpose. Eur. J. Philos. Sci. **11**(1), 1–26 (2021)
23. Data Scientist: The Dirtiest Job of the 21st Century, https://towardsdatascience.com/data-scientist-the-dirtiest-job-of-the-21st-century-7f0c8215e845, last accessed 2023/09/29
24. Spiekermann, M.: Data marketplaces: trends and monetisation of data goods. Intereconomics **54**(4), 208–216 (2019)
25. Data sharing challenges. https://www.datarepublic.com/resources/resources-guides/the-most-common-challenges-of-data-sharing. Accessed 29 Sept 2023
26. Angela, G.: Enhancing Access to and Sharing of Data: Reconciling Risks and Benefits of Data Re-use Across Societies. OECD Publishing, Tokyo (2019)
27. Lu, J.L.: Correlation of climate change indicators with health and environmental data in the Philippines. Acta Medica Philippina **56**(1) (2021)
28. Anderegg, W.R., et al.: Climate-driven risks to the climate mitigation potential of forests. Science **368**(6497), eaaz7005 (2020)
29. Došilović, F.K., et al.: Explainable artificial intelligence: a survey. In: 41st International Convention on Information & Communication Technology, pp. 210–215 (2018)
30. Copernicus. https://www.copernicus.eu/en. Accessed 29 Sept 2023
31. Nasa Earth Observatory. https://earthobservatory.nasa.gov/. Accessed 29 Sept 2023
32. Copernicus: Data and Information Access Services. https://www.copernicus.eu/en/access-data/dias. Accessed 29 Sept 2023
33. A European Green Deal. https://commission.europa.eu/strategy-and-policy/priorities-2019-2024/european-green-deal_en. Accessed 29 Sept 2023
34. OpenRefine. https://openrefine.org/. Accessed 29 Sept 2023
35. Faroukhi, A.Z., et al.: An adaptable big data value chain framework for end-to-end big data monetization. Big Data Cogn. Comput. **4**(4), 34 (2020)
36. Barcelos, A.M.F.: Researching beliefs about SLA: a critical review. Beliefs about SLA: New research approaches, 7–33 (2003)
37. Knowage. https://www.knowage-suite.com/site/. Accessed 29 Sept 2023
38. Zhang, J., et al.: DRNet: a deep neural network with multi-layer residual blocks improves image denoising. IEEE Access **9**, 79936–79946 (2020)
39. Mavrogiorgou, A., et al.: A comparative study of ML algorithms for scenario-agnostic predictions in healthcare. In: IEEE Symposium on Computers and Communications (ISCC), pp. 1–7 (2022)
40. Kaveh, A., et al.: Efficient training of two ANNs using four meta-heuristic algorithms for predicting the FRP strength. In Structures **52**, 256–272 (2023)

41. Mosca, E., et al.: SHAP-based explanation methods: a review for NLP interpretability. In: Proceedings of the International Conference on Computational Linguistics, pp. 4593–4603 (2022)
42. Zhang, H.: Structural equation modeling. In: Models and Methods for Management Science, pp. 363–381 (2022)
43. Gomez, R., et al.: Confirmatory factor analysis and exploratory structural equation modeling of the factor structure of the questionnaire of Cognitive and Affective Empathy (QCAE). PLoS ONE **17**(2), e0261914 (2022)
44. Saad, M., Zhang, Y., Tian, J., Jia, J.: A graph database for life cycle inventory using Neo4j. J. Clean. Prod. **393**, 136344 (2023)
45. Hadaj, P., Strzałka, D., Nowak, M., Łatka, M., Dymora, P.: The use of PLANS and NetworkX in modeling power grid system failures. Sci. Rep. **12**(1), 17445 (2022)
46. Kleftakis, S., et al.: Digital twin in healthcare through the eyes of the Vitruvian man. In: Innovation in Medicine and Healthcare: Proceedings of 10th KES-InMed, pp. 75–85 (2022)
47. Kyriazis, D., et al.: The CrowdHEALTH project and the hollistic health records: collective wisdom driving public health policies. Acta Informatica Medica **27**(5), 369 (2019)
48. Koumaditis, K., Themistocleous, M., Vassilacopoulos, G.: 'PINCLOUD: Integrated E-Health Services Over the Cloud' SUCRE 2014
49. Kiourtis, A., et al.: Electronic health records at people's hands across Europe: the InteropEHRate Protocols. In: pHealth 2022, pp. 145–150 (2022)
50. Mavrogiorgou, A., Kiourtis, A., Touloupou, M., Kapassa, E., Kyriazis, D., Themistocleous, M.: The road to the future of healthcare: transmitting interoperable healthcare data through a 5G based communication platform. In: Themistocleous, M., Rupino da Cunha, P. (eds.) EMCIS 2018. LNBIP, vol. 341, pp. 383–401. Springer, Cham (2019). https://doi.org/10.1007/978-3-030-11395-7_30

Using Computational Knowledge Extraction Approach to Assess Three Decades of Health Management Information Systems for Informed Actions

Josue Kuika Watat[1](✉) and Ebenezer Agbozo[2]

[1] HISP Centre, University of Oslo, Gaustadalléen 30, 0373 Oslo, Norway
josuekw@ifi.uio.no

[2] Ural Federal University, ul. Mira 32, 620002 Yekaterinburg, Russia
eagbozo@urfu.ru

Abstract. This research explores the role of the District Health Information System (DHIS2) in global health decision-making. Although DHIS2 is widely used in healthcare systems around the world, there is a significant shortage of in-depth scholarly research assessing its effectiveness and scope. This gap is critical, as understanding DHIS2's full potential is central in today's global health context, where data-driven decision-making is fundamental to both managing public health crises and improving healthcare delivery. To address this, our research employs the Latent Dirichlet Allocation (LDA) methodology to analyze a corpus of DHIS2-centric research. By doing so, we aim to uncover the main themes and applications of DHIS2 for informed action, thereby providing a more structured view of its impact in the health data management field. In addition, the study yielded a word cloud which depicted a variety of diseases connected to DHIS2 research, outlining its fundamental value in disease surveillance and management. The results emphasize DHIS2's enabling role within the broader HMIS ecosystem in developing a data-centric approach for effective public health interventions and management of health emergencies. This study develops a blueprint for exploiting HMIS to its maximum potential for informed public health activities by gaining sophisticated knowledge of DHIS2 through the LDA analysis. It guides researchers in understanding DHIS2's applications, facilitating further exploration in areas like disease trend analysis. Practitioners can use these insights to improve health surveillance, for instance, by tracking and managing outbreaks, thereby making more informed decisions in public health scenarios.

Keywords: Topic Modelling · Machine Learning · Evidence-based decision-making · DHIS2 · BERTopic · Informed Actions

1 Introduction

The digital age has marked a profound shift in how information is used across all sectors, including healthcare. Amidst a backdrop of rapidly evolving technology and growing health challenges, the need for efficient and comprehensive information systems has

M. Papadaki et al. (Eds.): EMCIS 2023, LNBIP 501, pp. 248–260, 2024.
https://doi.org/10.1007/978-3-031-56478-9_18

never been greater [1]. The Health Management Information Systems (HMIS), a tool designed to gather, process, and disseminate health data, has gained traction as a powerful force in modern healthcare. HMIS goes beyond the confines of traditional data collection systems. It offers an integrated approach, marrying health metrics with actionable insights [2].

The significance of HMIS in the present-day context is manifold. Firstly, as global health threats become more intricate, from chronic diseases to pandemics, it becomes paramount to have a pulse on health trends at all scales - local, national, and global. A systematized method to capture, analyze, and report health data offers invaluable insights into the state of public health and the efficacy of interventions. Secondly, as healthcare becomes more decentralized and diverse in-service delivery, HMIS enables coordination and coherence. A standardized data structure facilitates communication between different healthcare entities, from primary care providers to specialized health agencies, ensuring that all players in the ecosystem are harmonized in their efforts.

The District Health Information System 2 (DHIS2) serves as a prominent example that highlights the efficacy and adaptability of HMIS. DHIS2 being an open-source platform, has gained significant global adoption as a HMIS being utilized by health ministries across more than 70 nations [3]. The platform's flexibility in responding to different health domains, its user-friendly interface, and its strong analytical capabilities position it as an innovation in the field of health data [4]. As an example, DHIS2 was utilized by Sierra Leone, Sri Lanka, and Uganda to further strengthen their surveillance systems in reaction to COVID-19. This highlights the system's capacity to quickly adapt and its significant influence during the height of the epidemic [5].

As the healthcare industry increasingly embraces digitalization, the growing number of electronic health records, wearable devices, and enhanced medical imaging has given rise to an environment defined by the abundance of data. The sheer volume and complexity of this large amount of data provide hurdles for data storage, interpretation, and extraction of useful insights. For the purpose of maximizing the potential of this data, sophisticated analysis techniques are absolutely necessary. Machine learning techniques, notably Latent Dirichlet Allocation (LDA), are particularly salient in this context. LDA which is a topic modeling method, adeptly navigates through enormous amounts of unstructured data [6]. By identifying intricate patterns and clustering similar data points, it can uncover commonalities in patient histories, decipher trends in disease trajectories, and predict treatment outcomes [6, 7].

The aim of this study is to explore the research agenda of one of the most widely used health management information systems worldwide, by analyzing the abstracts of 253 publications from 1994 to 2023. The emerging literature in information systems shows that various research projects have analyzed previous publications based on metadata from the articles. Schoormann et al. [8] applied machine learning approaches on 95 articles to investigate how information systems research utilizes artificial intelligence to boost sustainable development. Also, abstract data tends to be word-dense, making it more suitable for the LDA algorithm [9]. As such, we follow similar research patterns to conduct our study. The following sections of the paper are structured as follows: in the next section, the methodology employed for this research is described. Next, the results are explained, and a general overview of the article's objectives is presented.

2 HMIS for Informed Actions

HMIS plays a fundamental role in supporting evidence-based policy development, planning, implementation, and evaluation of healthcare programmes by ensuring that appropriate use of resources is made across the entire healthcare system [10]. As an example, HMIS collects, stores, analyses and evaluates health-related data, which is then aggregated into decision-enabling analytical reports and visualizations, from healthcare facilities to district, regional and national administrative levels [11].

This wide-ranging coordination extends across a number of management aspects that are critical to the effective operation of healthcare organizations [12]. A well-functioning HMIS increases both the accessibility and quality of service delivery through evidence-based practices [13]. Moreover, HMIS quality impacts health care delivery. For instance, data quality should be routinely monitored, as high-quality statistics are contingent on data quality assessment and improvement measures [14]. The DHIS2 serves as a prime example of *HMIS for informed actions*. A study exploring the implementation of DHIS2 in Tanzania highlighted its crucial role in evidence-based policy-making, planning, and implementation of health programs [10]. The study highlighted the efficacy of DHIS2 in enhancing the quality and accessibility of data across all tiers of the healthcare system, hence enabling well-informed decision-making and allocation of resources. Additionally, the research also identified the deficiencies and determinants impacting the efficacy of DHIS2 within the framework of the nation's dynamic healthcare industry, emphasizing the necessity for ongoing advancement and tailoring of HMIS systems to accommodate evolving healthcare demands and settings.

HMIS acts as a trigger for informed action in healthcare, by providing an essential infrastructure for data-driven decision-making. Continuous research and improvement of these systems becomes imperative, as they not merely enhance efficiency and effectiveness of healthcare delivery, but also contribute to greater societal development [15, 16]. The literature on HMIS, especially in the context of developing countries, demonstrates the significance of such systems in transforming healthcare practices, policy formulation and resource allocation [17]. The body of knowledge points to an ongoing commitment to innovation, quality assessment and the active engagement of healthcare professionals in the use of HMIS to achieve optimal healthcare outcomes.

3 Methodology

Conducting a comprehensive literature analysis of existing research is a necessary initial phase in knowing and deliberating over an area of area of study. It supports the development of a specific study domain, while offering valuable perspectives on prevalent research patterns and potential avenues for future investigations [18, 19]. Literature review of prior research is crucial in scholarly and scientific research, as it builds a foundation on which new knowledge is derived [20]. The process consists of a thorough study of one or more scientific sources pertaining to a particular question, area of research or theory. The primary objective is to offer a comprehensive overview, pinpoint deficiencies in current research, and outline the context for future study [21]. Furthermore, it also serves as a means of confirming that new research draws on existing knowledge, thus reducing duplication and fostering improved understanding of the field.

Our study adopts a computational literature review (CLR) approach to extract knowledge, trends, and salient themes that will provide an overview of the state of research with regards to DHIS2. Research has identified that while CLR is still in its infancy, with regards to the tools, there has been commendation of its use as it plays a crucial role in the literature review process [22]. In this study, CLR is accomplished through content analysis and topic modelling. Through topic modelling, a statistical technique (using methods such as latent Dirichlet allocation (LDA) and latent semantic analysis (LSA)) for identifying the development of a topic of interest and has been applied extensively in numerous fields of study [23]. Our study used the BERTopic a topic model (neural network-based) which is based on language models (pre-trained transformer-based) that clusters embedding and generates topics [24].

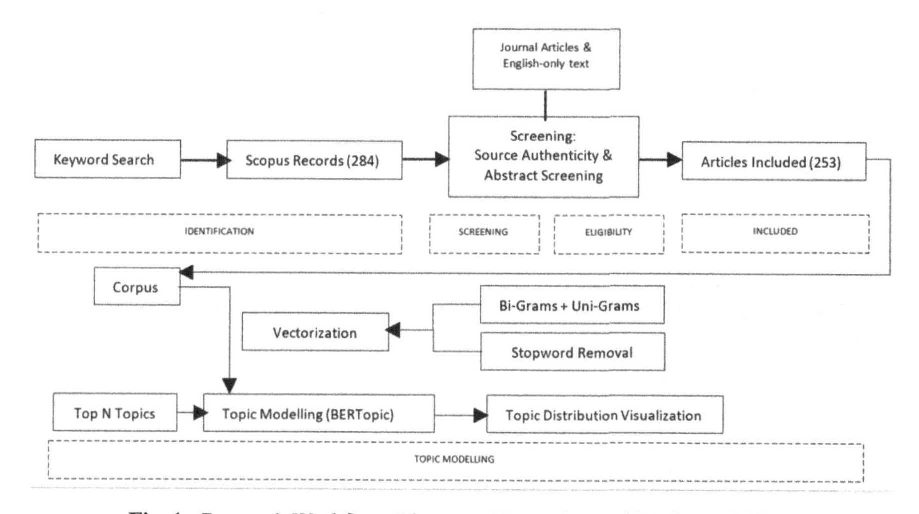

Fig. 1. Research Workflow (Literature Extraction and Topic Modelling)

Figure 1 illustrates the research workflow. To accomplish the set goal, we collated data from Scopus and set the search string as follows: "Decision-Making" OR "Decision" OR "Plan*" OR "Monitor*" OR "Improv*" OR "Development" OR "Surveillance" OR "governance" OR "Manag*" OR "Control*" OR "Direct*" AND "DHIS2" OR "District Health Information Systems". The resultant list of articles was narrowed down to a total of 253 publications. The Inclusion Criteria included: (a) All English publications; (b) Journal Articles. The Exclusion Criteria included: (a) non-English publications; (b) Pre-Prints. We followed the recommendations of Müller & Sæbø regarding the hijacking of journals by predatory organizations so as to ensure the authenticity of the sources used in our study [25]. This verification enabled us to assess the reliability and credibility of sources used in our research.

As highlighted in Table 1, the articles publication period span from 1994 to 2023; the annual growth rate in the theme in focus of this study is 11.74%; authored by 1608 researchers (with an international collaboration rate of 58.1%; and 7.31% co-authorship per document); and 119 scientific journals - Malaria Journal, PLOS One, BMC Health Services Research, South African Medical Journal, Electronic Journal of

Table 1. Data Description.

Description	Results
MAIN INFORMATION ABOUT DATA	
Timespan	1994:2023
Sources (Journals)	119
Documents	253
Annual Growth Rate %	11.74
Document Average Age	3.96
Average citations per doc	9.344
References	7847
DOCUMENT CONTENTS	
Keywords Plus (ID)	1481
Author's Keywords (DE)	716
AUTHORS	
Authors	1608
Authors of single-authored docs	11
AUTHORS COLLABORATION	
Single-authored docs	11
Co-Authors per Doc	7.31
International co-authorships (%)	58.1
DOCUMENT TYPES	
article	253

Information Systems in Developing Countries, BMC Public Health, BMJ Open, BMC Medical Informatics and Decision Making, Pan African Medical Journal, and BMC Pregnancy and Childbirth making up the top 10 journals. The next section discusses the results of applying CLR techniques to our article abstracts.

4 Result

This section presents and discusses results of the LDA topic modelling. It presents a list of words and a generic set of thematic terms. Table 1 gives the main words produced by the modelling along with the topics (themes) that cover them. A consensus on the meanings of 8 topics was established through discussion with the various authors.

As presented in Table 2 (see appendix for acronyms), the 8 topics derived from the analyzed abstracts are presented, as well as the corresponding number of articles that make up those topics. In addition, we deduced the topics based on the collection of keywords.

Table 2. Topic Modelling Results - Salient Topics, and Corresponding Themes.

#	Keywords	Topic	Number of Papers
1	health - data - maternal - intervention - facility - care - delivery - neonatal - service - quality	Data-Driven Interventions for Quality Neonatal and Maternal Healthcare	38
2	information - health - data - dhis - user - process - management - level - technology - hospital	Enhancing User-Centric Health Information Management and Data Processing	36
3	malaria - case - report - irs - data - incidence - control - health - surveillance - test	Health Strategies for Malaria Control	35
4	hiv - test - data - art - facility - estimate - health - pmtct - report - survey	Monitoring and Improving HIV Testing, Treatment, and Prevention of Mother-to-Child Transmission (PMTCT)	34
5	dhis - data - report - health - ri - intervention - facility - immunization - increase - timeliness	Routine Immunization (RI) and Intervention Timeliness Enhancement with Data-Driven Initiatives	23
6	ebola - surveillance - disease - health - outbreak - data - report - case - period - idsr	Epidemiological Disease Surveillance and Response Strategies from Integrated Health Data Reporting and Case Monitoring	20
7	child - service - health - covid - immunition - indicator - coverage - impact - vitamin - vaccine	The Impact of Covid-19 on Child Health Service Provision and Nutrition	20
8	mental - health - service - stroke - visit - cc - disorder - care - ncd - medicine	Safety-Critical Healthcare for Mental Health and Disorders	11

Topic 1 is summarized *as "Data-Driven Interventions for Quality Neonatal and Maternal Healthcare"*. A study conducted within the context of Malawi revealed that maternal healthcare improvement in low-resource settings could be challenged with human-based data entry errors. The Global Action in Nursing (GAIN) project, which is being spearheaded by the University of California San Francisco (UCSF), is providing interventions for managing and monitoring data in an efficient manner [26]. A separate research emphasized the need of providing training to healthcare staff in South Africa's KwaZulu Natal region about the collection of data to avoid the transfer of HIV from mother to child [27]. A subsequent study has reaffirmed the elevated maternal mortality ratio attributed to the limited availability of prenatal care in the Sub-Saharan Africa region. It emphasizes the necessity of implementing social support initiatives to mitigate the social stigma associated with antenatal care in rural settings [28]. Furthermore, Researchers emphasized that disaggregated data is an essential instrument for learning details on patients' and healthcare systems' conditions, which could help decision-making at the local level [28].

Topic 2 is summarized as *"Enhancing User-Centric Health Information Management and Data Processing"* which is a core value of DHIS/DHIS2. Studies have shown that health information systems must be implemented so as assist every tier of healthcare providers make evidence-based decisions and effectively manage patient care [29]. From a governance perspective, when implementing human-centric health information systems, it is essential that there be a team that consists of healthcare experts and managers to provide the right balance which will foster understanding rather than seeking to deploy solutions that meet their respective needs.

Topic 3 which is deduced as *"Health Strategies for Malaria Control"* is comprised of feedback control actions such as monitoring, surveillance, and reporting. A research on malaria approaches in Mali, such as indoor residual spraying, indicated that feedback was necessary to raise the standard of healthcare service and allocate resources more effectively [30]. Another study in Uganda utilised advanced technology and methods, including remote sensing, to collect data on rainfall patterns, environmental factors, and socio-economic indicators. These data were then used to develop models that can assist in making informed decisions, ultimately leading to a decrease in malaria cases [31].

Topic 4 which we coined as *"Monitoring and Improving HIV Testing, Treatment, and Prevention of Mother-to-Child Transmission (PMTCT)"* consists of antiretroviral treatment (ART) concepts associated with surveys and reports as feedback mechanisms to improve healthcare delivery for mothers suffering with HIV. Research conducted in the South African context proposed policy interventions with the objective of enhancing data quality, a critical factor in monitoring and decision-making processes. These interventions encompassed data standardization, data verification, and the standardization of information management systems [32]. Another study highlighted the increasing rate of maternal mortality in South Africa's KwaZulu-Natal and identified the impact of antenatal syphilis point-of-care testing in improving maternal and child healthcare [33].

Topic 5 - *"Routine Immunization (RI) and Intervention Timeliness Enhancement with Data-Driven Initiatives"* highlights the very relevant activities such as routine immunization, data reporting and timeliness to improve quality healthcare delivery. For example, in the case of Nigeria where research highlighted interventions such as on-the-job training, stakeholder engagement, in-facility mentoring and the provisioning of tools to ensure quality job delivery are essential in improving the timeliness of healthcare provision [33, 34]. One study also highlighted factors such as computer literacy, infrastructure, governance and workflow orchestration to improve vaccine information systems and their role in quality decision making [35].

Topic 6 - *"Epidemiological Disease Surveillance and Response Strategies from Integrated Health Data Reporting and Case Monitoring"* highlights a very crucial initiative that has become relevant in recent times – with the epidemics of Ebola and covid-19 pandemics that plagued our world. Investment in surveillance and monitoring systems coupled with open-source technology, sharing and interoperability are very essential to decision support in controlling such epidemics [5]. Research attests to and recommends the leveraging of epidemic surveillance and monitoring technologies as well as regular follow-ups by ministry of health stakeholders for effective healthcare delivery [36].

Topic 7 was deduced as *"The Impact of Covid-19 on Child Health Service Provision and Nutrition"*. One study highlighted the need for resources and promotion of services essential to children to curb adverse deterioration of the health and survival of children [37]. Another study pointed out that dietary supplement improvement strategies are being implemented and monitored in Ethiopia due to the lack of vitamin-rich (primarily vitamin A) meals which are essential for child growth [38].

Finally, Topic 8 is deduced as *"Safety-Critical Healthcare for Mental Health and Disorders"* - a topic of great discussion in recent times. India launched a National Mental Health Program and deployed a web-based information system named i-MANN which is aimed at gathering, managing and analyzing mental health records in the country [39]. Another study revealed the need for mental health facilities to deal with such disorders as there is a lack of focus by healthcare organizations in low and middle-income nations as such cases go untreated [40].

Fig. 2. Diseases associated with DHIS2 research

Figure 2 depicts a word cloud that visually represents the disorders linked to research on DHIS2, as derived from the literature sources utilised in the study. In order to achieve this outcome, the abstracts underwent pre-processing to exclude extraneous keywords such as stop words and non-alphanumeric letters. Subsequently, a corpus of the abstracts was created, and this corpus was tokenized to identify the terms included within it. In addition, a carefully selected compilation of diseases, ailments, and disorders is employed as a vocabulary for conducting comparison searches inside our corpus. If a sickness is identified within our corpus, it is then incorporated into a final list. Ultimately, a wordcloud is generated to emphasize our findings, where the size of the text corresponds to its level of prominence within the corpus. Regarding DHIS2, it can be deduced that malaria, HIV, and COVID-19 have been the main topics of debate.

DHIS2 has served a crucial role in facilitating well-informed healthcare decisions in HIV management. The assessment of health facilities' performance in reporting HIV indicators using DHIS2 was undertaken in Kenya, as evidenced by a research [41].

The project seeks to evaluate the effectiveness of HIV indicator reporting at the national level through the application of machine learning algorithms and data visualization techniques. The utilization of this technique performed an instrumental part in recognizing of trends and execution of timely interventions, illustrating the importance of thorough reporting and data analysis in DHIS2 in directly impacting and improving healthcare policies and actions in the field of HIV. Moreover, the augmentation of HIV preventive programmes in low- and middle-income countries (LMICs) has bolstered health management information systems (HMIS), consequently enhancing the accessibility of routinely reported HIV data. The aforementioned data, commonly derived from monthly reports prepared by several institutions, are consolidated into a condensed format and subsequently sent to systems such as DHIS2. The improved accessibility and excellence of HMIS data, provided by DHIS2, allows for informed policy development, assessment of programme efficacy, advocacy, and allocation of resources in HIV management.

The utilization of DHIS2 in the treatment of malaria has played a crucial role in facilitating well-informed decision-making. A recent research assessed the reporting of malaria using DHIS2, specifically examining nine indicators of malaria in three administrative regions [42]. It emphasized the potential of the system in the context of malaria monitoring and management. However, it also identified challenges related to the completeness, timeliness, and trustworthiness of the data. These difficulties were exacerbated by factors such as insufficient supervision and training, large workloads, and geographical distance. An evaluation of data quality for the initial four years (2014–2017) of malaria reporting in Senegal using DHIS2 indicated a gradual enhancement throughout the period [43]. It showed that the percentage of anticipated malaria cases reported and the thoroughness of facility reports had significantly improved. Nevertheless, a significant disparity was seen between public and private institutions, since a far smaller proportion of private facilities opted to utilise the system. Integrating malaria data from a parallel system created by the National Malaria Control Programme into DHIS2 in Burkina Faso has significantly enhanced the quality of the data. The integration of malaria data with the Health Management Information System (HMIS) through the use of DHIS2 has resulted in enhanced precision, promptness, comprehensiveness, and dependability of the data. The 2017 malaria data quality evaluation reported a considerable increase in data correctness from 43% to 83%, timeliness from 62% to 80%, completeness from 64% to 78%, and dependability from 67% to 87%.

Amidst the COVID-19 epidemic, the use of DHIS2 supplied critical support to nations in effectively handling and addressing the issue. In the context of COVID-19 case notification and response operations, DHIS2 was employed by a total of fifty-five nations, namely Sierra Leone, Sri Lanka, and Uganda, to customize their respective national public health surveillance systems [5]. This adaptation has involved leveraging the platform's community of practice, long-term capacity building and local autonomy to implement effective responses to COVID-19.

The use of DHIS2 in cancer management, specifically in Rwanda, illustrates its significant impact on health data systems. Previously, the administration of cancer data in Rwanda was encumbered by laborious, paper-based procedures that resulted in data loss and delays in reporting. The integration of DHIS2 with CanReg5, a specialist cancer registry system, has transformed the cancer data management landscape[1]. DHIS2 made real-time data capture and preliminary analysis easier, while CanReg5 tackled more complex analyses. The integration has not only streamlined data management but also improved data quality, leading to more effective cancer control strategies.

5 Conclusion

Conducting a review of the extant literature on a research area offers a useful tool for assessing the extent and scope of the research. Traditional literature review approaches, however, entail a lengthy manual review process. In this research we employed a semi-automated approach to perform a concise analysis of literature review on health information systems, and its significance in informed decision making. LDA ensures that themes are extracted quickly and objectively from a scientific standpoint. It enables researchers to perform an in-depth analysis.

Our research identified 8 topics based on an assessment of 253 articles regarding the assessment of health management information systems for informed action. The findings provide suggestions for future research in the global field of health technology decision support systems. DHIS2 are a subset of HMIS and play a vital role in contributing to the healthcare ecosystem primarily in deprived communities. Our research discovered from literature that there is a growing focus on human-centric solutions to healthcare and following the COVID-19 pandemic, there have been analytical perspectives integrated into the DHIS2 architecture. The study highlights DHIS2's impact in streamlining operations, improving patient outcomes, and facilitating strategic health planning, showcasing the practical benefits of HMIS in diverse healthcare settings. Moreover, this highlights the significance of Health Management Information Systems (HMIS) in effectively addressing health inequalities through the implementation of focused interventions and fair allocation of resources. This is especially significant in settings with limited resources, where the effective utilization of data may result in substantial enhancements in health outcomes. Lastly, the findings strongly advocate for continued advancements in HMIS. The integration of cutting-edge technologies such as artificial intelligence and machine learning with systems like DHIS2 can further enhance their efficiency and predictive capabilities. As these technologies evolve, they offer new opportunities for HMIS to provide deeper insights and more efficient healthcare solutions.

Our research has some limitations. First, the focus of the research was narrowed down to journal articles, which reduces the sampling. Conference papers and other articles published in other databases (apart from Scopus) have not been integrated into this analysis, which raises the prospect of further research. Secondly, this study concentrates on specific parts of the articles' metadata. In some cases, however, abstracts are not

[1] https://community.dhis2.org/t/integrated-canreg5-and-dhis2-improving-cancer-data-management-in-rwanda/52854.

always a faithful depiction of article content. Thus, future research could expand upon the complete analysis (full text).

Appendix - Acronyms

- SIA - Supplemental immunization activity
- ART- antiretroviral treatment
- IDSR - Integrated Disease Surveillance and Response
- RI – Routine Immunization
- RMNCH - reproductive, maternal, newborn and child health (RMNCH)
- CHCs - community health centers
- CHPs - community health posts
- MCHPs - maternal and child health posts
- DHIMS - District Health Information Management System
- CC - community clinics
- NCD - non-communicable diseases

References

1. Sahay, S.: Big data and public health: challenges and opportunities for low and middle income countries. Commun. Assoc. Inf. Syst. **39**, 20 (2016)
2. Akpan, G.U., et al.: Leveraging polio geographic information system platforms in the African region for mitigating COVID-19 contact tracing and surveillance challenges. JMIR Mhealth Uhealth **10**, e22544 (2022)
3. Nicholson, B., Nielsen, P., Sahay, S., Sæbø, J.I.: Digital public goods platforms for development: the challenge of scaling. Inf. Soc. **38**, 364–376 (2022)
4. Msiska, B., Nielsen, P.: Innovation in the fringes of software ecosystems: the role of socio-technical generativity. Inf. Technol. Dev. **24**, 398–421 (2018)
5. Kinkade, C., et al.: Extending and strengthening routine DHIS2 surveillance systems for COVID-19 responses in Sierra Leone, Sri Lanka, and Uganda. Emerg. Infect. Dis. **28**, S42 (2022)
6. Gong, J., Abhishek, V., Li, B.: Examining the impact of keyword ambiguity on search advertising performance: a topic model approach. MIS Q. **42**, 805–830 (2018)
7. Yang, Y., Subramanyam, R.: Extracting actionable insights from text data: a stable topic model approach. MIS Q. **47** (2023)
8. Schoormann, T., Strobel, G., Möller, F., Petrik, D., Zschech, P.: Artificial intelligence for sustainability—a systematic review of information systems literature. Commun. Assoc. Inf. Syst. **52**, 8 (2023)
9. Gatti, C.J., Brooks, J.D., Nurre, S.G.: A historical analysis of the field of OR/MS using topic models. arXiv preprint arXiv:1510.05154 (2015)
10. Nyamtema, A.S.: Bridging the gaps in the health management information system in the context of a changing health sector. BMC Med. Inform. Decis. Mak. **10**, 36 (2010)
11. Lippeveld, T., Azim, T., Boone, D., Dwivedi, V., Edwards, M., AbouZahr, C.: Health management information systems: backbone of the health system. In: Macfarlane, S.B., AbouZahr, C. (eds.) The Palgrave Handbook of Global Health Data Methods for Policy and Practice, pp. 165–181. Palgrave Macmillan, London (2019)

12. Kasambara, A., et al.: Assessment of implementation of the health management information system at the district level in Southern Malawi. Malawi Med. J. **29**, 240–246 (2017)
13. Kebede, M., Adeba, E., Chego, M.: Evaluation of quality and use of health management information system in primary health care units of east Wollega zone, Oromia regional state, Ethiopia. BMC Med. Inform. Decis. Mak. **20**, 107 (2020)
14. Endriyas, M., et al.: Understanding performance data: health management information system data accuracy in Southern Nations Nationalities and People's Region Ethiopia. BMC Health Serv. Res. **19**, 175 (2019)
15. Khan, A.Z., Mahmood, F., Bokhari, R.H., Mushtaq, R., Abbas, R.: Challenges of e-government implementation in health sector: a step toward validating a conceptual framework. Digit. Policy Regul. Gov. **23**, 574–597 (2021)
16. Bernardi, R., Sarker, S., Sahay, S.: The role of affordances in the deinstitutionalization of a dysfunctional health management information system in Kenya: an identity work perspective. MIS Q. **43** (2019)
17. Uneke, C.J., Sombie, I., Keita, N., Lokossou, V., Johnson, E., Ongolo-Zogo, P.: Improving maternal and child health policymaking processes in Nigeria: an assessment of policymakers' needs, barriers and facilitators of evidence-informed policymaking. Health Res. Policy Syst. **15**, 125–138 (2017)
18. Keyworth, C., Hart, J., Armitage, C.J., Tully, M.P.: What maximizes the effectiveness and implementation of technology-based interventions to support healthcare professional practice? A systematic literature review. BMC Med. Inform. Decis. Mak. **18**, 1–21 (2018)
19. Snyder, H.: Literature review as a research methodology: an overview and guidelines. J. Bus. Res. **104**, 333–339 (2019)
20. Smith, V., Devane, D., Begley, C.M., Clarke, M.: Methodology in conducting a systematic review of systematic reviews of healthcare interventions. BMC Med. Res. Methodol. **11**, 1–6 (2011)
21. Tate, M., Furtmueller, E., Evermann, J., Bandara, W.: Introduction to the special issue: the literature review in information systems. Commun. Assoc. Inf. Syst. **37**, 5 (2015)
22. Principe, V.A., et al.: A computational literature review of football performance analysis through probabilistic topic modeling. Artif. Intell. Rev. **55**, 1351–1371 (2022)
23. Li, X., Lei, L.: A bibliometric analysis of topic modelling studies (2000–2017). J. Inf. Sci. **47**, 161–175 (2021)
24. Grootendorst, M.: BERTopic: neural topic modeling with a class-based TF-IDF procedure. arXiv preprint arXiv:2203.05794 (2022)
25. Müller, S.D., Sæbø, J.I.: The 'hijacking' of the Scandinavian Journal of Information Systems: implications for the information systems community. Inf. Syst. J. n/a (2023)
26. Blair, A.H., et al.: Assessing combined longitudinal mentorship and skills training on select maternal and neonatal outcomes in rural and urban health facilities in Malawi. J. Transcult. Nurs. **33**, 704–714 (2022)
27. Mphatswe, W., et al.: Improving public health information: a data quality intervention in KwaZulu-Natal, South Africa. Bull. World Health Organ. **90**, 176–182 (2012)
28. Kim, K.H., Choi, J.W., Oh, J., Moon, J., You, S., Woo, Y.: What are the barriers to antenatal care utilization in Rufisque district, Senegal?: a bottleneck analysis. J. Korean Med. Sci. **34** (2019)
29. Chikumba, P.A.: Management of health information in Malawi: role of technology. Adv. Sci. Technol. Eng. Syst. J. **2**, 157–166 (2017)
30. Wagman, J., et al.: Rapid reduction of malaria transmission following the introduction of indoor residual spraying in previously unsprayed districts: an observational analysis of Mopti Region, Mali, in 2017. Malar. J. **19**, 1–10 (2020)
31. Ssempiira, J., et al.: Interactions between climatic changes and intervention effects on malaria spatio-temporal dynamics in Uganda. Parasite Epidemiol. Control **3**, e00070 (2018)

32. Kaposhi, B.M., Mqoqi, N., Schopflocher, D.: Evaluation of antiretroviral treatment programme monitoring in Eastern Cape South Africa. Health Policy Plan. **30**, 547–554 (2015)
33. Mashamba-Thompson, T.P., Drain, P.K., Kuupiel, D., Sartorius, B.: Impact of implementing antenatal syphilis point-of-care testing on maternal mortality in KwaZulu-Natal, South Africa: an interrupted time series analysis. Diagnostics **9**, 218 (2019)
34. Ohiri, K., Makinde, O., Ogundeji, Y., Mobisson, N., Oludipe, M.: Strengthening routine data reporting in private hospitals in Lagos Nigeria. Health Policy Plan. **38**, 822–829 (2023)
35. Rahmadhan, M.A.W.P., Handayani, P.W.: Challenges of vaccination information system implementation: a systematic literature review. Hum. Vaccin. Immunother. **19**, 2257054 (2023)
36. Reynolds, E., et al.: Implementation of DHIS2 for disease surveillance in Guinea: 2015–2020. Front. Public Health **9**, 761196 (2022)
37. Jensen, C., McKerrow, N.H.: Child health services during a COVID-19 outbreak in KwaZulu-Natal Province South Africa. South Afr. Med. J. **111**, 114–119 (2021)
38. Laillou, A., Baye, K., Zelalem, M., Chitekwe, S.: Vitamin A supplementation and estimated number of averted child deaths in Ethiopia: 15 years in practice (2005–2019). Matern. Child Nutr. **17**, e13132 (2021)
39. Kaur, J., et al.: I-MANN: a web-based system for data management of mental health research in India. Indian J. Psychol. Med. **42**, S15–S22 (2020)
40. Burns, J.K.: The burden of untreated mental disorders in KwaZulu-Natal Province-mapping the treatment gap. S. Afr. J. Psychiatry **20**, 6–10 (2014)
41. Gesicho, M.B., Were, M.C., Babic, A.: Evaluating performance of health care facilities at meeting HIV-indicator reporting requirements in Kenya: an application of K-means clustering algorithm. BMC Med. Inform. Decis. Mak. **21**, 6 (2021)
42. Wangdi, K., Sarma, H., Leaburi, J., McBryde, E., Clements, A.C.A.: Evaluation of the malaria reporting system supported by the district health information system 2 in Solomon Islands. Malar. J. **19**, 372 (2020)
43. Muhoza, P., et al.: A data quality assessment of the first four years of malaria reporting in the Senegal DHIS2, 2014–2017. BMC Health Serv. Res. **22**, 18 (2022)

Artificial Intelligence

The Role of Artificial Ethics Principles in Managing Knowledge and Enabling Data-Driven Decision Making in Supply Chain Management

Saeeda Alhaili(✉) and Farzana Mir

British University in Dubai, Dubai, United Arab Emirates
22001638@student.buid.ac.ae, farzana.mir@buid.ac.ae

Abstract. In today's data-driven business environment, the ethical management of knowledge and data utilization for decision-making in supply chain management has become increasingly vital. This study explores how artificial ethics principles can guide businesses in managing knowledge ethically and enable data-driven decision-making in supply chain management. The study specifically looks into two key areas: establishing moral standards for handling data and knowledge throughout the supply chain and incorporating artificial ethics principles into data analytics systems to support fairness and impartiality. The study follows a semi-systematic review approach. The findings show the importance of ethical considerations and their contributions to knowledge management and data-driven decision-making in supply chain management. By integrating artificial ethics principles, organizations can uphold ethical values such as accountability, fairness, and transparency in their decision-making procedures. Moreover, integrating these principles into data analytics systems ensures unbiased and equitable decision-making. This study emphasizes the value of integrating ethics into supply chain operations and provides advice for businesses looking to use data ethically and efficiently.

Keywords: Knowledge Management · Supply Chain Management · Artificial Intelligence · Machine Learning · Artificial Ethics Principles

1 Introduction

Recently, data-driven decision-making in supply chain management has gained significant importance as organizations seek to gain a competitive edge in an ever-changing business landscape. However, along with the increased reliance on data and technology, ethical considerations surrounding these practices have become a growing concern [1]. As organizations continue to collect and utilize vast amounts of data, there is an emerging need to develop and implement artificial ethics principles to ensure the fairness and impartiality of data-driven decision-making processes. By integrating ethical considerations into their knowledge management and data-driven decision-making processes, organizations can improve transparency, build stakeholder trust, and mitigate the risks of unethical or biased autonomous decision-making [2].

M. Papadaki et al. (Eds.): EMCIS 2023, LNBIP 501, pp. 263–277, 2024.
https://doi.org/10.1007/978-3-031-56478-9_19

The ethical concerns surrounding data-driven autonomous decision-making are not limited to a single industry or sector. Supply chain operations, in particular, are facing increased scrutiny as organizations seek to streamline their operations and improve efficiency [3]. From procurement and transportation to production and distribution, supply chain operations heavily rely on data and analytics to make informed decisions that impact their bottom line and the well-being of their stakeholders [50]. This research explores how organizations can leverage artificial ethics principles to effectively manage knowledge and data-driven decision-making across their supply chain. This paper addresses the primary research question: "How might organizations use artificial ethics principles to manage knowledge and achieve data-driven decision-making across their supply chain?" In addition to the main research question, the following secondary questions are examined:

1. How can organizations effectively address the ethical concerns related to artificial intelligence and data-driven decision-making in supply chain management?
2. What are organizations' key considerations and challenges when implementing artificial ethics principles in their supply chain decision-making processes?

Through a comprehensive semi-systematic review of existing literature, this study critically analyzes and synthesizes the findings from previous research to provide insights and recommendations for organizations seeking to integrate artificial ethics principles into their supply chain operations. By answering these research questions, this study aims to understand how organizations can navigate the ethical challenges associated with artificial intelligence and data-driven decision-making in supply chain management (The study contributes to the growing body of knowledge on the ethical implications of AI and data-driven decision-making in supply chain management and provides practical insights for organizations seeking to adopt these technologies ethically. By doing so, organizations can ensure that they are sustainable socially and economically, contributing to a more ethical and equitable future for all [32].

2 Literature Review

Supply Chain Management (SCM) is a complex field that requires organizations to make critical decisions to ensure efficient operations and gain a competitive edge [1, 13, 46]. Artificial intelligence (AI) and data-driven decision-making integration has become a transformative strategy in SCM in recent years [11]. This integration can improve performance, achieve strategic goals across various functions, and improve supply chain operations. Nevertheless, ethical issues and implications are raised by adopting AI and data-driven approaches in SCM. In this literature review, we examine how organizations can use artificial ethics principles to manage knowledge and achieve data-driven decision-making across their supply chain.

2.1 AI and Data-Driven Decision Making in Supply Chain Management

In Supply Chain Management (SCM), the integration of artificial intelligence (AI) and data-driven decision-making has emerged as a crucial strategy for businesses to gain a competitive advantage, reduce risks, and make informed choices [11]. This integration plays a crucial role, as it holds the potential to improve SCM operations and achieve strategic goals across various functions, including procurement, shipping, warehousing, and inventory management. AI and machine learning (ML) technologies enable organizations to process vast amounts of data, generate insights, and automate decision-making procedures, improving supply chain visibility, demand forecasting accuracy, and inventory optimization [40]. For example, companies can utilize AI algorithms to make real-time inventory decisions based on demand patterns, optimize route planning for transportation, and automate warehouse operations for improved efficiency. Effective knowledge management (KM) is integral to the success of SCM, enabling organizations to maximize their operations and make informed judgments [16]. Knowledge Management (KM) is crucial for implementing AI and data-driven methodologies in SCM, as it involves acquiring, analyzing, and using explicit and tacit knowledge [41]. Knowledge management models, such as the Socialization, Externalization, Combination, and Internalization (SECI) Model by Nonaka and Takeuchi [32] or the Knowledge Management Value Chain Model, offer insights into how organizations can effectively manage their information assets to support ethical decision-making. These models emphasize the importance of Knowledge Management in providing ethical guidance by effectively leveraging information resources.

Organizations must harness explicit and tacit knowledge to enhance the integration of AI and data-driven decision-making in SCM. Explicit knowledge refers to codified, recorded, and easily shared information, while tacit knowledge encompasses implicit expertise and know-how that is challenging to formalize and communicate [41]. By leveraging both types of knowledge, organizations can develop more robust AI models, improve decision-making capabilities, and optimize supply chain performance.

By integrating knowledge management models into AI and data-driven methodologies, organizations can effectively manage their knowledge assets, make informed judgments, and achieve operational excellence in SCM [34]. Numerous studies have shown the practical applications and benefits of AI and data-driven approaches in SCM showing how organizations have successfully implemented AI-based demand forecasting models, improving accuracy and significantly reducing inventory holding costs (e.g. [21, 42]. A study by [12] in the healthcare sector focuses on implementing Business intelligence (BI) in healthcare industry processes and emphasizes the importance of leveraging knowledge management practices and BI tools in supply chain decision-making. These examples provide evidence of the advantages of integrating AI and data driven decision making in SCM operations. However, it is essential to consider the challenges and limitations associated with implementation, including the need for skilled personnel, data quality issues, and resistance to change.

By effectively harnessing KM and incorporating AI and data-driven techniques, organizations can navigate the complexities of supply chain management, optimize their operations, and make well-informed decisions. This integration empowers organizations to enhance their competitive edge, improve operational efficiency, and achieve strategic objectives in an increasingly AI-driven landscape.

2.2 Artificial Ethics Principles in Supply Chain Management

Artificial Ethics Principles (AEPs) aim to ensure that decision-making processes in SCM align with ethical values and standards, promoting fairness, transparency, and accountability in using data and algorithms. Several studies have highlighted the potential benefits of AEPs in the context of supply chain management, including improved transparency, trust, and stakeholder engagement [19].

However, implementing AEPs poses significant challenges, such as managing the complexity of providing ethical guidance, lack of standardized frameworks for AEPs, and the potential for unforeseen consequences [19]. Knowledge Management (KM) is crucial in SCM. Integrating KM with AEPs in SCM is a relatively new area of exploration aimed at examining the benefits and challenges associated with this approach. While AI and ML ethics have been extensively studied across various domains, the specific focus on AI and ML ethics in supply chain management is a more recent investigation [19]. The rapid advancement of AI and ML in supply chain management (SCM) has brought attention to the ethical implications of these technologies. It raises concerns such as potential biases, discrimination risks, and privacy infringement. It is crucial to address these ethical challenges to ensure responsible and transparent decision-making processes.

Ethical Concerns in SCM: One of the primary ethical concerns in SCM is the potential for biases in AI algorithms. When trained on partial data, these algorithms may perpetuate unfair hiring practices, procurement decisions, and resource allocation [36].

ML algorithms can unintentionally discriminate against specific individuals or groups, resulting in unfair treatment or exclusionary practices. Furthermore, using vast amounts of data in AI and ML technologies raises concerns about privacy rights and protecting personal and sensitive information [8].

Consequences of Unethical Decision-Making: Unethical decision-making in SCM can affect organizations negatively. It can lead to reputational damage, erode customer trust, and result in financial losses. Moreover, unethical practices of decision-making can negatively impact stakeholders, including customers, employees, and the wider community. It is crucial to prioritize ethical considerations to mitigate these risks and maintain sustainable and responsible SCM practices [30].

The (AEPs) concept has emerged to address these ethical concerns. Ethical decision-making models, such as the Ethical Decision-Making Model by [39] or the Four-Component Model by Badaracco [7], provide a structured approach for organizations to consider moral values, principles, and guidelines throughout the supply chain. These models guide decision-makers in assessing the potential ethical implications of AI and ML technologies, including bias, fairness, transparency, and privacy issues.

Organizations should actively engage stakeholders and consider their perspectives and interests to address these challenges. Stakeholder Theory, proposed by [18], serves as a theoretical foundation for understanding the significance of stakeholder engagement and accountability in integrating AI and ethical principles in SCM. By involving suppliers, customers, employees, and regulatory bodies in developing and implementing AI-driven ethical practices, organizations can ensure transparency, address ethical concerns, and be accountable to relevant stakeholders.

2.3 Ethical Implications of AI and Data-Driven Decision Making

The ethical implications of incorporating (AI) and data-driven decision-making in SCM have gained significant attention in recent years. Integrating AI and machine learning (ML) into SCM brings ethical challenges such as prejudice, discrimination, and privacy invasion [27]. To ensure responsible and ethical decision-making, standardized frameworks and a shared understanding of ethical norms among stakeholders are crucial [19]. To guide the integration of AI and ML into SCM, researchers have developed frameworks and recommendations that emphasize the importance of addressing ethical issues [30]. Comprehensive risk assessments are essential to identify potential ethical obstacles and opportunities throughout the supply chain [34]. By integrating the Data-Driven Decision-Making Model by [24] and the Six-Step Decision-Making Model by Davenport [17] with the existing literature, organizations can effectively manage the ethical implications of AI and data-driven decision-making in SCM. These models offer structured approaches to leveraging data and insights from AI and ML technologies in decision-making processes while considering ethical concerns [17, 24].

Ultimately, integrating AI and data-driven decision-making in SCM offers immense potential for organizations to optimize their operations and achieve competitive advantages [11]. However, ethical issues must be carefully considered to ensure responsible and accountable practices. This literature review has highlighted the ethical implications of AI and data-driven decision-making in SCM, including concerns related to bias, fairness, transparency, privacy, and stakeholder engagement [9, 25, 38].

3 Methodology

The systematic review approach identifies, evaluates, and analyzes all available research relevant to a research question, topic area, or phenomenon of interest [48]. This research investigates the integration of artificial ethics principles, knowledge management, and data-driven decision-making in supply chain management. This paper follows a systematic approach to search for and analyze relevant articles, providing a comprehensive overview of the existing literature on the topic. The search for relevant articles was conducted using multiple databases, including ProQuest, JSTOR, ScienceDirect, and online resources such as Google Scholar and BUID library. The search terms used for this semi-systematic review paper were "Artificial Ethics principles," "Knowledge Management," "Data-Driven Decision making," and "Supply Chain Management." The inclusion criteria for article selection were limited to peer-reviewed journal publications in English published between 2011 and 2022 [14].

This research aims to provide historical context and demonstrate the field's evolution over time by including older models. While the primary focus is on recent publications, incorporating earlier works allows for a more comprehensive examination and a broader understanding of the subject matter [23]. During the article selection process, a systematic screening approach was employed. The identified articles were evaluated for relevancy and quality, considering the research questions and objectives of the study. Measures were taken to minimize bias and ensure consistency in the selection process [23].

The data extraction and analysis process involved a content analysis approach. A coding template was developed based on the research questions and objectives. The selected articles were carefully reviewed, and relevant information on the study's questions and objectives was extracted [35]. The analysis primarily employed a qualitative approach, identifying key themes and findings from the literature [11].

It is important to acknowledge certain limitations associated with the research approach. The reliance on secondary data may introduce biases, and the exclusion of non-English publications and studies published before 2011 may limit the scope of the review. Additionally, the semi-systematic review approach may reach a different level of rigor than a full systematic review [11]. However, the strength of the semi-systematic review lies in its ability to provide a systematic and comprehensive overview of the existing literature on integrating artificial ethics principles, knowledge management, and data-driven decision-making in supply chain management [28]. It allows for a structured approach to gathering and analyzing secondary data from diverse sources, contributing valuable insights to the field.

In summary, this research paper involved the analysis of 51 relevant articles and aims to investigate the integration of artificial ethics principles, knowledge management, and data-driven decision-making in supply chain management. The methodology employed in this semi-systematic review paper follows a systematic approach involving a thorough search and analysis of relevant articles. While acknowledging its limitations, this research provides a comprehensive overview of the existing literature and contributes to understanding the topic [43] (Fig. 1).

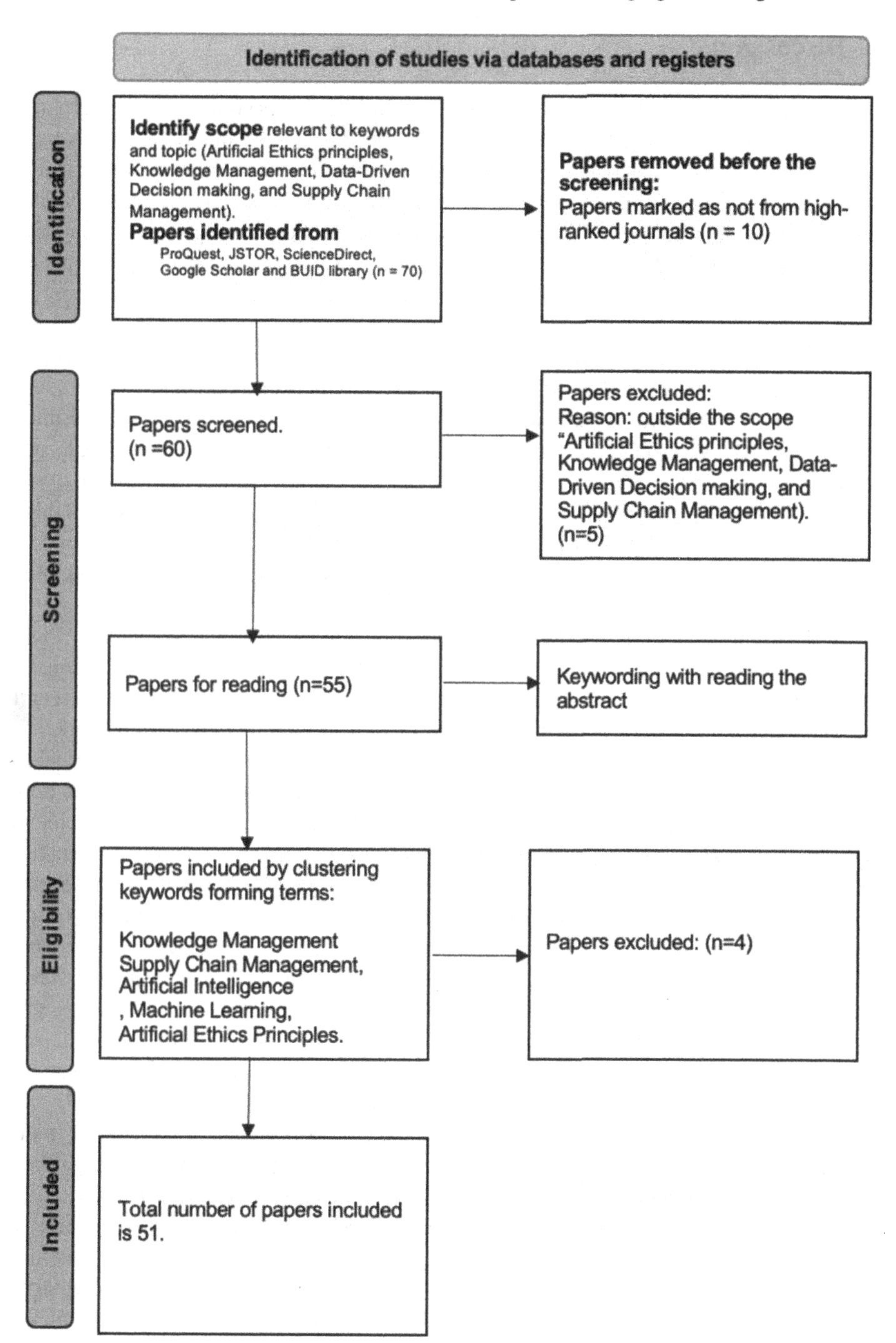

Fig. 1. PRISMA Flow Diagram

4 Discussion

The literature review that supply chain management (SCM) has significantly progressed as AI and data-driven decision-making have developed [11]. Organizations increasingly utilize artificial intelligence and machine learning innovations to increment efficiency, save expenses, and gain a competitive edge. However, these advantages raise ethical concerns regarding adopting AI and data-driven approaches in supply chain management. Bias, discrimination, and privacy infringement are significant issues that must be addressed [11]. AI algorithms may perpetuate biases in the training data, resulting in unfair hiring practices, procurement decisions, and resource allocation [36]. Discrimination against specific individuals or groups can occur, leading to unfair treatment or exclusionary practices [9]. Additionally, using vast amounts of data, including personal and sensitive information, raises concerns about privacy rights [25]. The "Artificial Ethics Principles" (AEPs) concept has emerged to mitigate these ethical challenges. AEPs are designed to promote fairness, transparency, and accountability in decision-making processes, aligning them with ethical values and principles [22]. Artificial ethics principles (AEPs) must be implemented to manage information across the supply chain properly and guarantee that decision-making processes align with ethical norms and principles [11].

AEPs implementation in SCM can enhance transparency, increase stakeholder trust, and improve partner engagement [45]. However, implementing AEPs poses challenges, such as the complexity of ethical guidance, the absence of standardized frameworks, and the potential for unforeseen consequences. Establishing a comprehensive framework that promotes ethical behavior and responsible use of AI and ML technologies is essential to answer the main research question of this study. The research question focuses on how organizations can utilize Artificial Ethics Principles to manage knowledge and achieve data-driven decision-making in their supply chain. This framework should encompass guidelines for effectively gathering, evaluating, and utilizing explicit and tacit information. Effective knowledge management is vital in Supply Chain Management, as explicit knowledge refers to formal, codified information, while tacit knowledge is subjective and context-dependent [52]. So, to improve decision-making skills and optimize overall supply chain operations, the framework should include instructions for successfully gathering, assessing, and using explicit and tacit knowledge. The ability of firms to use their knowledge assets to make informed choices makes effective knowledge management essential in the context of supply chain management (SCM) [52].

The findings from the literature review indicate the importance of knowledge management (KM) in supporting ethical decision-making. Effective KM enables organizations to maximize operations and make informed judgments [16]. Knowledge acquisition is the foundation of knowledge management and data-driven decision-making in the supply chain. Organizations should establish processes for gathering, organizing, and storing explicit and tacit knowledge. This can be achieved through knowledge-sharing sessions, cross-functional collaboration, and technological platforms that facilitate knowledge capture and transfer. By collecting explicit and tacit knowledge, organizations can comprehensively understand supply chain dynamics and make well-informed decisions. Analysis and interpretation of knowledge follow the acquisition phase. AI and ML technologies are crucial in identifying large datasets' patterns, trends, and correlations, extracting valuable insights for data-driven decision-making [20]. However, using these technologies ethically and responsibly is essential, ensuring the detection and elimination of biases to prevent unfair or discriminatory outcomes.

Efficient utilization of information requires organizations to disseminate and share knowledge across the supply chain. This involves developing platforms and knowledge-sharing mechanisms like communication networks, collaboration tools, and online repositories. By facilitating the flow of knowledge, organizations can ensure that decision-makers at all levels can access relevant data and insights, enabling them to make choices aligned with strategic goals and ethical standards. Additionally, fostering a culture of continuous learning and information sharing within the supply chain through training programs, mentoring, and communities of practice enhances group decision-making and promotes supply chain management [4].

Stakeholder engagement and effective communication are critical to successfully integrating AEPs into supply chain decision-making processes. It is essential to involve key stakeholders, including suppliers, customers, and employees, in the development and execution of AEPs in order to guarantee transparency, reliability, and accountability within the supply chain. The engagement of Stakeholders in the development process of Artificial Ethics Principles should be from the early stages. Thus, organizations can gain valuable insights, various perspectives, and a more in-depth understanding of the specific ethical concerns associated with their supply chain [11, 20, 52]. It is crucial to emphasize that stakeholder engagement should be continuous and iterative. AEPs should be continuously assessed and modified in response to stakeholder feedback, adjusting ethical landscapes, and emerging challenges. This dynamic approach addresses ethical supply chain issues in a way that keeps AEPs valuable and efficient [20]. Additionally, decision-making procedures must be transparent and inclusive, allowing stakeholders to understand the rationale behind choices and providing avenues for accountability and constructive criticism.

Effective knowledge management and adherence to ethical standards enable organizations to enhance the efficiency of their supply chain operations. Data-driven decision-making facilitates the identification of inefficiencies, streamlines processes, and enables prompt responses to changing market demands. This comprehensive approach supports improved inventory control, supplier relationship management, and demand forecasting. Furthermore, addressing ethical concerns fosters cooperation and trust among supply chain participants, leading to better coordination, reduced risks, and enhanced sustainability [8].

To ensure transparency, trustworthiness, and accountability within the supply chain, it is essential to involve all key stakeholders, including suppliers, clients, and employees, in the planning and execution of Artificial Ethics Principles (AEPs). Effective communication and stakeholder involvement is essential for addressing ethical issues and promoting a collaborative environment. By incorporating stakeholders from the outset of AEP development, organizations can gain valuable insights, diverse viewpoints, and a deeper understanding of the ethical challenges specific to their supply chain environment [31]. This inclusive approach facilitates the identification of potential ethical risks and opportunities, allowing for the customization of AEPs to address the specific concerns of different stakeholders.

Particular emphasis should be placed on including suppliers in the creation and execution of Artificial Ethics Principles (AEPs). Suppliers play a critical role in determining a firm's ethical compliance. Collaborative efforts to promote ethical practices can help suppliers align with the organization's ethical principles and adhere to the defined AEPs. This can be achieved through initiatives such as supplier training, moral audits, and regular communication to address potential issues or challenges. Furthermore, involving consumers in AEP development can enhance transparency and instil confidence. With increasing consumer awareness and expectations regarding ethical practices, businesses must demonstrate their commitment to ethical behavior to attract and retain customers [20]. By including consumers in discussions about AEPs, businesses can gain valuable insights into their expectations, concerns, and values. This knowledge can then be used to effectively convey the organization's ethical beliefs and design AEPs that align with consumer expectations. By ensuring the active involvement of both suppliers and consumers, organizations can make more ethical decisions throughout the supply chain. This comprehensive approach, driven by stakeholder engagement, can help foster trust, enhance transparency, and promote responsible business practices [30].

It is essential to recognize workers as stakeholders and include them in the formulation and execution of AEPs. Workers often encounter moral dilemmas in their daily tasks as essential participants in supply chain operations. By involving workers in creating AEPs, employers can empower their workforce to act ethically and contribute valuable insights. Workers can provide input on potential ethical issues and help ensure that AEPs align with the organization's values and guiding principles. This can be achieved through training programs, open discussion forums, and regular communication channels, fostering increased employee involvement and ownership of ethical decision-making [1].

Transparency is a vital outcome of effective stakeholder participation and communication during AEP implementation. By involving workers and other stakeholders in the decision-making process and addressing ethical concerns transparently, organizations demonstrate their commitment to ethical standards. Open communication about AEPs enhances stakeholders' confidence by enabling them to understand the rationale behind actions and the safeguards to mitigate ethical risks [1]. This comprehensive approach ensures that workers' perspectives are considered and contributes to developing an ethical workplace culture. By involving stakeholders in the creation and execution of AEPs, organizations demonstrate their willingness to listen and value diverse viewpoints, considering the impact of choices on different stakeholders. This inclusive approach promotes trust and accountability as natural outcomes [30]. When stakeholders actively engage in decision-making and have a voice, they are more likely to hold the organization responsible for its conduct. This responsibility extends beyond the organization and benefits the entire supply chain ecosystem. Stakeholders ensure adherence to ethical standards throughout the supply chain by providing input, voicing concerns, and monitoring the organization's compliance with the established AEPs [1].

Organizations must consider the potential unintended consequences of AI and data-driven decision-making. These technological advancements can reinforce prejudices, discriminate against specific individuals or groups, and violate privacy rights [8]. Organizations should prioritize justice, diversity, and inclusion in their decision-making processes to mitigate these risks. Regular monitoring and auditing of AI algorithms and data sources are crucial to identify and correct any biases or discriminatory patterns that may arise [11].

The literature review and research topic emphasize the importance of integrating AEPs into supply chain management (SCM) and effectively managing knowledge in order to address the ethical issues related to AI and data-driven decision-making. Organizations must create elaborate structures encouraging moral conduct, openness, and accountability. By collaborating with stakeholders, carrying out thorough risk assessments, and establishing unique ethical norms, organizations can overcome the ethical issues that arise from integrating AI and machine learning (ML) in SCM. In the data-driven and digital business environment, this proactive approach promotes stakeholder trust while improving operational effectiveness, giving it a competitive edge [15]. Businesses can pursue sustainability and succeed in the fiercely competitive global market by integrating AEPs and ethical decision-making into supply chain management.

To sum up, integrating AI and data-driven decision-making in SCM offers significant potential, but addressing the ethical implications of bias, discrimination, and privacy infringement is crucial. By implementing Artificial Ethics Principles, involving stakeholders, and integrating personalized ethical rules, organizations can navigate these challenges and optimize their SCM practices while ensuring transparency, fairness, and accountability. Effective knowledge management further supports ethical decision-making by maximizing operations and leveraging information assets. Overall, a holistic approach is required to realize the benefits of AI and data-driven decision-making in SCM while upholding ethical standards.

5 Conclusion and Recommendations

In conclusion, the integration of AI and data-driven decision-making in supply chain management (SCM) offers significant potential for improved performance, strategic goal achievement, and enhanced operational excellence. However, addressing the ethical implications arising from this integration is crucial, including bias, discrimination, and privacy infringement. These concerns can lead to unfair practices, exclusionary behavior, and privacy rights violations. To mitigate these ethical challenges, the concept of Artificial Ethics Principles (AEPs) has emerged, aiming to promote fairness, transparency, and accountability in decision-making processes. Ethical decision-making models, such as the Ethical Decision-Making Model by Rest and the Four-Component Model by Badaracco, provide a structured approach for organizations to consider moral values, principles, and guidelines throughout the supply chain. Implementing AEPs can enhance transparency, increase stakeholder trust, and improve partner engagement. However, implementing AEPs poses challenges, including the complexity of ethical guidance, the absence of standardized frameworks, and the potential for unforeseen consequences. Therefore, organizations must embrace stakeholder engagement and accountability, as emphasized by Stakeholder Theory, to ensure the successful integration of AI and ethical principles in SCM. By involving suppliers, customers, employees, and regulatory bodies in developing and implementing AI-driven ethical practices, organizations can ensure transparency, address ethical concerns, and be accountable to relevant stakeholders.

In addition to these considerations, the following recommendations are proposed to effectively manage the ethical implications of AI and data-driven decision-making in Supply chain management. Firstly, organizations should integrate personalized ethical rules that align with ethical norms and principles to guide decision-making processes and ensure fairness, transparency, and accountability. Secondly, conducting comprehensive risk assessments is crucial to identify and mitigate potential ethical risks associated with AI and data-driven decision-making. This includes assessing biases in training data, evaluating impacts on diverse stakeholder groups, and addressing privacy concerns. Furthermore, fostering knowledge management is vital in supporting ethical decision-making in SCM. Organizations should leverage explicit and tacit knowledge, utilizing knowledge management models such as the SECI and the Knowledge Management Value Chain Model to maximize operations, make informed judgments, and support ethical practices. Finally, establishing collaborative partnerships with suppliers, customers, employees, and regulatory bodies is essential. These partnerships facilitate the co-creation and implementation of AI-driven ethical practices, ensuring diverse perspectives, enhancing transparency, and fostering accountability throughout the supply chain. By implementing these recommendations, organizations can navigate the complexities of ethical decision-making and optimize their SCM practices while upholding ethical standards. Recognizing that a holistic approach is required, considering both technical and ethical aspects, is essential to fully realize the benefits of AI and data-driven decision-making in SCM.

6 Limitations and Areas of Future Research

To advance this field, future research can focus on several key areas. Firstly, exploring the practical applications of Artificial Ethics Principles in Supply Chain Management will provide valuable insights into how organizations have successfully integrated ethical principles into their decision-making processes. This can involve conducting case studies or empirical investigations to understand the challenges faced, best practices employed, and outcomes associated with incorporating ethical considerations into SCM. Secondly, investigating the potential trade-offs between moral considerations and performance outcomes in data-driven decision-making is essential. Future research can explore the complex relationship between ethical decision-making and organizational performance. By studying scenarios where organizations face dilemmas between maximizing performance through data-driven approaches and adhering to ethical principles, researchers can provide guidance on balancing ethics and performance in SCM decision-making.

Thirdly, the development of frameworks or standards is necessary to guide the successful incorporation of ethical principles into knowledge management (KM) and decision-making processes in SCM. Future research can focus on designing guidelines or protocols that facilitate the integration of AEPs into existing KM practices, ensuring ethical considerations are embedded in information sharing, decision-making, and knowledge utilization throughout the supply chain. Such frameworks or standards will provide organizations with practical tools to navigate the ethical complexities of SCM decision-making. Lastly, exploring the role of AEPs in sustainability and ethical SCM practices is a promising avenue for future research. This can involve investigating how integrating AEPs in SCM practices contributes to sustainable supply chain operations, responsible sourcing, and environmentally friendly practices. Understanding the linkages between AEPs and sustainability will further enhance supply chain activities' ethical and social impact. Future research can also explore how AEPs can support the implementation of ethical SCM practices, such as fair trade, supplier diversity, and social responsibility initiatives.

By addressing these research gaps, future studies can provide worthwhile insights into the practical applications of the Artificial Ethics principle in Supply Chain management, the trade-offs involved in data-driven decision-making, the development of frameworks or standards for ethical principles, and the role of AEPs in promoting sustainability and ethical SCM practices. These research directions will contribute to a more in-depth understanding of integrating AEPs into SCM and foster adopting ethical and responsible decision-making procedures throughout the supply chain.

References

1. Akkermans, H.A.: Responsible supply chain management: the role of transparency and trustworthiness. J. Bus. Ethics (2016)
2. Alreshidi, E.: Smart sustainable agriculture (SSA) solution underpinned by internet of things (IoT) and artificial intelligence (AI). arXiv preprint (2019)
3. Badii, A.C.: Ethical considerations for intelligent agents in supply chain management. J. Bus. Ethics (2016)
4. Baker-Brunnbauer, J.: Management perspective of ethics in artificial intelligence. AI Ethics **1**(2), 173–181 (2021)
5. Berkowitz, E.N., Williams, C.K.: The strategic use of knowledge management. J. Knowl. Manag. **3**(4), 303–313 (1999)
6. Bosch, R., et al.: A case study of Volkswagen unethical practice in diesel emission scandal. Int. J. Sci. Res. Publ. **6**(9), 225–229 (2016)
7. Badaracco, J.L.: The discipline of building character. Harv. Bus. Rev. **76**(2), 115–124 (1998)
8. Bechtsis, D., Tsolakis, N., Iakovou, E.V.: Data-driven secure, resilient and sustainable supply chains: gaps, opportunities, and a new generalised data sharing and data monetisation framework. Int. J. Prod. Res. 4397–4417 (2022)
9. Buolamwini, J., Gebru, T.: Gender shades: intersectional accuracy disparities in commercial gender classification. In: Proceedings of the 1st Conference on Fairness, Accountability, and Transparency, pp. 77–91 (2018)
10. Belhadi, A., Kamble, S., Fosso Wamba, S., Queiroz, M.M.: Building supply-chain resilience: an artificial intelligence-based technique and decision-making framework. Int. J. Prod. Res. **60**(14), 4487–4507 (2022)
11. Brendel, A.B., Mirbabaie, M., Lembcke, T.B., Hofeditz, L.: Ethical management of artificial intelligence. Sustainability 1974 (2021)
12. Basile, L.J., Carbonara, N., Pellegrino, R., Panniello, U.: Business intelligence in the healthcare industry: the utilization of a data-driven approach to support clinical decision making. Technovation **120**, 102482 (2023). https://doi.org/10.1016/j.technovation.2022.102482
13. Chen, H., Themistocleous, M., Chiu, K.: Approaches to supply chain integration followed by SMEs: an exploratory case study. In: Proceedings of Tenth American Conference on Information Systems (AMCIS), New York, USA, pp. 2610–2620 (2004)

14. Connelly, L.M.: Inclusion and exclusion criteria. Medsurg Nurs. **29**(2) (2020)
15. Cunningham, E.: Artificial intelligence-based decision-making algorithms, sustainable organizational performance, and automated production systems in big data-driven smart urban economy. J. Self-Governance Manag. Econ. 31–41 (2021)
16. Du, Y.L.: Sustainable Supply Chain Management in the Sharing Economy: A Resource-Based View Perspective. Resources, Conservation and Recycling (2016)
17. Davenport, T.H.: Process Innovation: Reengineering Work through Information Technology. Harvard Business Press (2013)
18. Freeman, R.E.: Strategic management: a stakeholder approach. Pitman (1984)
19. Gava, O., Bartolini, F., Venturi, F., Brunori, G., Pardossi, A.: Improving policy evidence base for agricultural sustainability and food security: a content analysis of life cycle assessment research. Sustainability 1033 (2020)
20. Hussain, M., et al.: Blockchain-based IoT devices in supply chain management: a systematic literature review. Sustainability (2021)
21. Johnson, M., Brown, K., Davis, L.: Implementing AI-driven demand forecasting in Company B: a case study. Int. J. Logist. Supply Chain Manag. **15**(2), 87–100 (2020)
22. Jobin, A., Ienca, M., Vayena, E.: The global landscape of AI ethics guidelines. Nat. Mach. Intell. **1**(9), 389–399 (2019)
23. Johnson, D.S., Sihi, D., Muzellec, L.: Implementing big data analytics in marketing departments: Mixing organic and administered approaches to increase data-driven decision making. Informatics **66** (2021)
24. Johnson, P.F., Leauby, B.A., Klassen, R.D.: The data-driven decision-making model: a data-driven approach to student success. J. Student Affairs Res. Pract. **53**(1), 55–68 (2016)
25. Kroll, J.A., et al.: Accountable algorithms. Univ. Pa. Law Rev. **165**(3), 633–705 (2017)
26. Keele, S.: Guidelines for performing systematic literature reviews in software engineering (2007)
27. Lee, M.H.: An ethical decision-making framework for supply chain management. Sustainability (2017)
28. Mishra, S.B., Alok, S.: Handbook of research methodology (2022)
29. Nolan, C., et al.: The Volkswagen emissions scandal: a case study in corporate misbehaviour. J. Manag. Policy Pract. **18**(1), 44–56 (2017)
30. Nasim, S.F., Ali, M.R., Kulsoom, U.: Artificial intelligence incidents & ethics a narrative review. Int. J. Technol. Innov. Manag. (2022)
31. Nitsche, A.-M.N., Matthias, C.L.: Technological and Organisational Readiness in the Age of Data-Driven Decision Making: A Manufacturing Perspective. Leeds Beckett Repository (2020)
32. Nonaka, I., Takeuchi, H.: The Knowledge-Creating Company: How Japanese Companies Create the Dynamics of Innovation. Oxford University Press (1995)
33. Peppoloni, S., Di Capua, G.: Geoethics as global ethics to face grand challenges for humanity. Geological Society, London, Special Publications, 13–29 (2021)
34. Pournader, M., Ghaderi, H., Hassanzadegan, A., Fahimnia, B.: Artificial intelligence applications in supply chain management. Int. J. Prod. Econ. **241**, 108250 (2021)
35. Patino, C.M., Ferreira, J.C.: Inclusion and exclusion criteria in research studies: definitions and why they matter. J. Bras. Pneumol. **44**, 84 (2018)
36. Rajagopal, P.R.: Fuzzy logic-based approach for managing ethical issues in supply chain management. J. Bus. Ethics (2017)
37. Rong, P., Liu, S.: The impact of the ethical CEO on top management team's creativity from the perspective of knowledge management: the moderating role of psychological distance. Curr. Psychol. 1–15 (2022)

38. Ribeiro, M.T., Singh, S., Guestrin, C.: "Why should I trust you?": explaining the predictions of any classifier. In: Proceedings of the 22nd ACM SIGKDD International Conference on Knowledge Discovery and Data Mining, pp. 1135–1144 (2016)
39. Rest, J.R.: Moral development: advances in research and theory. Praeger (1986)
40. Saghiri, S.: A hybrid model for green supply chain management under uncertainty. J. Clean. Prod. (2016)
41. Scheibe, K.P., Mukandwal, P.S., Grawe, S.J.: The effect of transactive memory systems on supply chain network collaboration. Int. J. Phys. Distrib. Logist. Manag. (2020)
42. Smith, J., Adams, R., Johnson, M.: AI-based demand forecasting models in supply chain management: a case study of Company A. J. Supply Chain Manag. **25**(3), 123–135 (2019)
43. Seif El-Nasr, M., Kleinman, E.: Data-driven game development: ethical considerations. In: Proceedings of the 15th International Conference on the Foundations of Digital Games, pp. 1–10 (2020)
44. Singh, R.: An overview of artificial intelligence and its application in supply chain management. Int. J. Emerg. Technol. Adv. Eng. (2016)
45. Sleep, S., Gala, P.: Removing silos to enable data-driven decisions: the importance of marketing and IT knowledge, cooperation, and information quality. J. Bus. Res. (2022)
46. Soratana, K., Landis, A.E., Jing, F., Suto, H.: Supply Chain Management of Tourism Towards Sustainability. Springer, Cham (2020)
47. Themistocleous, M., Irani, Z., Love, P.E.D.: Evaluating the integration of supply chain information systems: a case study. Eur. J. Oper. Res. **159**(2), 393–405 (2004)
48. Themistocleous, M., Cunha, P., Tabakis, E., Papadaki, M.: Towards cross-border CBDC interoperability: insights from a multivocal literature review. J. Enterprise Inf. Manag. **36**(5), 1296–1318 (2023). https://doi.org/10.1108/JEIM-11-2022-0411
49. Tseng, M.L., Ha, H.M., Tran, T.P., Bui, T.D., Chen, C.C., Lin, C.W.: Building a data-driven circular supply chain hierarchical structure: resource recovery implementation drives circular business strategy. Bus. Strateg. Environ. **31**(5), 2082–2106 (2022)
50. Tseng, M.-L., Tran, T.P.: Sustainable industrial and operation engineering trends and challenges Toward Industry 4.0: a data driven analysis. J. Ind. Prod. Eng. 581–598 (2021)
51. Tseng, M.L., Ha, H.M., Tran, T.P.T., Bui, T.D., Chen, C.C., Lin, C.W.: Building a data-driven circular supply chain hierarchical structure: Resource recovery implementation drives circular business strategy. Bus. Strategy Environ. **31**, 2082–2106 (2022)
52. Tsolakis, N., Iakovou, E.: Data-driven secure, resilient and sustainable supply chains: gaps, opportunities, and a new generalised data sharing and data monetisation framework. Int. J. Prod. Res. **60**, 4397–4417 (2021)
53. Wang, X., Chen, X., Tian, F.: An economic analysis of the 2008 milk scandal in China. J. Public Econ. **95**(11–12), 1253–1262 (2011)
54. Zekhnini, K., Cherrafi, A., Bouhaddou, I., Benghabrit, Y., Garza-Reyes, J.A.: Supply chain management 4.0: a literature review and research framework. Benchmarking Int. J. 465–501 (2020)

Fine-Tuning Large-Scale Project Scheduling

George Sklias[1,3(✉)], Socratis Gkelios[2,3], and Dimitrios Dimitriou[1]

[1] Department of Economics, Democritus University of Thrace, Komotini, Greece
{gsklias,ddimitri}@econ.duth.gr

[2] Department of Electrical and Computer Engineering, Democritus University of Thrace, Komotini, Greece
sgkelios@ee.duth.gr

[3] Intelligent Systems Laboratory, Neapolis University Pafos, Paphos, Cyprus

Abstract. This paper explores the integration of artificial intelligence (AI) into project management, proposing a decision support system that optimizes project timelines and resources. The pilot study focuses on the Port of Agios Konstantinos in Greece. The methodology section introduces dual annealing as a stochastic optimization method and explains the use of a customizable cost function with overlap calculation to prioritize project aspects. An objective function is defined to maximize task alignment with optimal scheduling periods. The experimental results section presents three optimization cases, adjusting schedules for critical tasks in the pilot project based on different weightings of budget and weather considerations.

Keywords: Artificial Intelligence in Project Management · Decision Support Systems · Resource Allocation Optimization

1 Introduction

The use of artificial intelligence has increased exponentially in a variety of scientific fields, from computer vision and robotics [2,3,9,10], to real estate evaluations [6,7]. In recent years, the pervasive integration of artificial intelligence into information systems has led to a significant paradigm shift in the realm of business process optimization [1,18]. The recognition of the manifold benefits associated with AI implementation has propelled its adoption across industries. In particular, the infusion of AI techniques into traditional project management practices has emerged as a powerful catalyst for enhancing operational efficiency and amplifying Key Performance Indicators (KPIs), particularly within the domain of infrastructure projects. Given the paramount significance of critical infrastructure in attaining objectives like the EU Green Transition deal, the utilization of AI-based information systems to bolster the project management processes of such infrastructure ventures is not only beneficial but strongly recommended [8].

M. Papadaki et al. (Eds.): EMCIS 2023, LNBIP 501, pp. 278–290, 2024.
https://doi.org/10.1007/978-3-031-56478-9_20

As expounded by Miller and Gloria J. [15], projects invariably grapple with the constraints of time and budget, which inherently limit the availability of resources. However, the algorithms derived from AI projects possess the potential to exert profound and enduring impacts. Therefore, it becomes imperative to adopt a comprehensive perspective on success when planning and executing such projects, with a dedicated focus on addressing their broader societal implications as explicit critical success factors.

Furthermore, Vărzaru et al. [21] underline the pivotal role of customization as the most influential characteristic contributing to the enhanced efficiency and effectiveness of project management activities, closely followed by innovation.

Taking a step further, AI seamlessly integrates with Building Information Modeling (BIM), offering rapid information retrieval and in-depth analysis capabilities across the entire lifecycle of construction projects. The transformative potential of AI lies in its capacity to provide intelligent, data-driven insights into the digitally constructed environments forged by BIM. This empowers project stakeholders to make more informed decisions, optimize various project stages, and ultimately attain higher levels of efficiency, superior quality, enhanced collaboration, and increased sustainability [17].

Moreover, Chang et al. [5] illustrate how AI techniques in project management can significantly aid organizations in mitigating risks associated with ecological agriculture projects. Their study introduces an indicator system and a novel heuristic algorithm designed to promote risk reduction. The risk assessment index system encompasses six critical dimensions, including nature, society, market economy, management, technology, and finance, incorporating a total of 23 indicators. This systematic and comprehensive approach effectively reflects the multifaceted factors influencing the risk profile of ecological agriculture projects.

AI also proves instrumental in the real-time evaluation of infrastructure project health, optimal project execution, and the seamless integration of personnel with project deliverables, as demonstrated by Lokhande [12]. These factors are deemed paramount for ensuring effective and smooth project management processes.

Furthermore, as asserted by Ke et al. [11], manual data screening and collection remain mandatory processes to ensure the accuracy of information systems, enabling AI in project management to wield a profound impact on the thinking, organizational structure, and methodologies of enterprise management. This sentiment is echoed by Mohammad I. et al. [13], whose research suggests that the technical framework of intelligent systems should be scalable and flexible, accommodating additional data sources, attributes, and dimensions to maximize the impact of AI techniques in project management. In addition to validating its findings, the author further extends their research by introducing additional factors and subjecting them to a thorough mathematical assessment, as expounded upon in [14].

Additionally, Wei Yin et al. [22] and Nusen et al. [16] developed a Genetic Algorithm for Multi-Objective Optimization of Construction Schedules, culminating in the conclusion that project schedules can exert a significant influence on project costs. In a similar vein, Berezka [4] posited that Decision Support System (DSS) models can be broadly applied as tools for decision support in the planning of investment construction projects. These models facilitate the estimation and optimization of schedules, ultimately enhancing the selected parameters of economic efficiency of investments, all through the lens of multi-objective schedule optimization.

The burgeoning need for infrastructure projects, driven by climate crises and burgeoning population demands, underscores the critical importance of effective project management in the construction domain. Timely and precise delivery of infrastructure projects holds paramount significance for public agencies. Over recent years, continuous project delays have been observed to adversely impact the quality of services offered at regional levels. These delays, coupled with friction in project delivery, have consequential effects on resource allocation and total project budgets.

In light of these considerations, the fusion of optimized project timelines and efficient resource allocation holds substantial potential for elevating Key Performance Indicators associated with project delivery. An integrated support system, incorporating these two critical factors, can prove invaluable for public agencies. Such a system, designed as a decision support tool, can aid in optimizing project timelines based on the unique needs of individual project activities and facilitate more effective monitoring of activity progress throughout the project lifecycle. This multifaceted decision support tool can accomplish the following: a) optimize resource allocation and budget; b) enhance regional office visibility regarding project progress; and c) assist top-level executives in identifying the root causes of any delays and actively engage with responsible parties to mitigate any setbacks in the timely delivery of projects.

To realize these objectives, the decision support system must perform both global optimization of Gantt chart activities and local optimization of these activities, all while factoring in external variables such as budget and duration. With this overarching goal in mind, this paper lays the foundation for a comprehensive decision support system aimed at enhancing the Key Performance Indicators of construction projects, minimizing resource utilization, and optimizing project durations through the application of AI techniques. Additionally, the system, facilitated by AI techniques, can globally optimize Gantt Chart activities based on preferred time parameters for each project activity, thus enabling public agencies to tailor project Gantt Charts based on specific assumptions for each activity.

For the purposes of this research paper, the pilot use case will focus on the Port of Agios Konstantinos in the central region of Greece. This ambitious port project encompasses the construction of four terminals, a surrounding parking facility for vehicles and motorcycles, and the establishment of ferry routes to the Sporades region. The port can accommodate up to four ferry boats and two large fishing boats simultaneously, involving 35 distinct activities with a total projected duration of 24 months.

The remainder of this paper is structured as follows: In Sect. 2, we lay the foundation by introducing the proposed methodology—a flexible framework designed to address the complexities of project scheduling. Moving forward to Sect. 3, we delve into practicality through a pilot study, providing tangible insights into the real-world applicability and effectiveness of our approach. The heart of our research is unveiled in Sect. 4, where we offer a comprehensive presentation and analysis of the experimental results. Lastly, Sect. 5 encapsulates our journey, summarizing the conclusions drawn from these findings and outlining promising avenues for future research endeavors.

2 Methodology

The primary objective of this optimization approach is to craft an intricately efficient schedule for the Port of Ag. Konstantinos project, harmonizing the intricate interplay between budgetary constraints and the influence of weather conditions. This endeavor unfolds through a multifaceted and innovative procedure, harnessing the power of the dual-annealing algorithm. Dual annealing, a well-regarded stochastic optimization method, distinguishes itself by its adeptness at comprehensively exploring solution spaces. Unlike rigid, deterministic algorithms, dual annealing introduces a calculated element of randomness into its search process, endowing it with the unique ability to not only navigate toward improved solutions but also to venture into uncharted territories, uncovering hidden gems of optimization that might otherwise elude conventional approaches.

In essence, this optimization approach acts as a dynamic orchestrator, balancing the delicate interplay between budget constraints and unpredictable weather patterns, ultimately sculpting a schedule that optimizes resource utilization and promotes efficient project execution. Its stochastic nature imbues it with the agility to adapt to changing conditions and unearth innovative solutions, making it a potent tool for complex real-world project scheduling challenges such as the Port of Ag. Konstantinos.

2.1 Dual-Annealing

Dual annealing represents an innovative fusion of two powerful optimization methodologies: simulated annealing and genetic algorithms. These two approaches, simulated annealing and genetic algorithms, each bring unique strengths to the table.

Simulated annealing, inspired by the real-world annealing process, introduces an element of adaptability into the optimization process. Unlike rigid algorithms, simulated annealing occasionally accepts suboptimal solutions. This feature is crucial as it prevents the algorithm from getting stuck in local minima, enabling it to explore a wider solution space and potentially discover more favorable configurations.

On the other hand, genetic algorithms draw inspiration from nature's principles of evolution and natural selection. They employ mechanisms such as mutation and crossover to traverse the solution space efficiently. This adaptability allows genetic algorithms to consider a diverse range of potential solutions, mimicking the process of genetic variation and selection in nature.

Within this optimization framework, dual annealing plays a central role. It serves as the engine that systematically explores the solution space in search of the most favorable configuration for project scheduling. What sets dual annealing apart from deterministic algorithms is its controlled injection of randomness into the optimization process. This controlled randomness not only helps refine existing solutions but also allows the algorithm to venture into uncharted territories within the solution space, potentially revealing superior task arrangements.

In particular, dual annealing focuses on maximizing a score, which in this context, quantifies the degree of alignment between the optimal time frame for each task and their respective start and end boundaries. By prioritizing this alignment, the algorithm aims to find an arrangement that optimally synchronizes tasks with their ideal time frames. This synchronization is pivotal for achieving a delicate balance between budget constraints and weather considerations in project management.

This approach demonstrates how the fusion of simulated annealing and genetic algorithms, with the aid of dual annealing, can significantly enhance project scheduling by dynamically optimizing task arrangements and adapting to project-specific priorities. This innovative optimization methodology not only refines existing solutions but also explores new horizons, promising more efficient and effective project management.

2.2 Utilization of a Customizable Cost Function

The use of a versatile and user-defined cost function is a fundamental aspect of our optimization process. This unique feature empowers project managers by giving them the flexibility to assign precise weights to various project aspects, including financial considerations and weather-related factors. These customizable weights serve as critical decision-making parameters, allowing project managers to emphasize specific aspects in alignment with the project's specific requirements.

For example, assigning a higher weight to the budget within the cost function indicates a strong emphasis on rigorous cost control. Consequently, this strategic choice can result in the creation of a project schedule that is carefully designed to minimize expenses while adhering to predefined financial constraints.

This adaptable cost function demonstrates the flexibility and customization capabilities inherent in our optimization process. By enabling project managers to customize the optimization criteria based on the project's priorities, it facilitates the creation of schedules that are not only efficient but also perfectly aligned with the overarching goals and constraints of the project. This adaptability is a valuable asset in the realm of project management, allowing for the optimization of diverse projects, each with its unique set of demands and objectives.

2.3 Overlap Calculation

The "Overlap" metric, denoted as Overlap(i), is a crucial component of our optimization process. It quantifies the degree of overlap between a specific task and its optimal scheduling period. The formula for calculating the overlap is as follows:

$$\text{Overlap}(i) = \frac{\text{Optimal Period End} - \text{Optimal Period Start}}{\min(\text{Late Finish}_i, \text{Optimal Period End}) - \max(\text{Early Start}_i, \text{Optimal Period Start})}$$

Here's the explanation of the variables involved: - Optimal Period End and Optimal Period Start represent the end and start dates of the optimal scheduling period for the task, respectively. - Late Finish_i is the latest possible finish date for the ith task. - Early Start_i is the earliest possible start date for the ith task.

2.4 Objective Function

Our optimization objective is to maximize the following function:

$$\text{Maximize} \sum_i (\text{Overlap}(\text{Task}_i, \text{OptimalPeriod}_i) \times \text{Weight}_i)$$

where: - Task_i represents the ith task in the project schedule. - OptimalPeriod_i denotes the optimal time period for task i. - Weight_i signifies the weight assigned to task i, which can be related to factors such as budget or weather considerations.

This objective function aims to find an optimal schedule that maximizes the overlap of tasks with their respective optimal periods, taking into account the assigned weights.

3 Pilot

For the purpose of this research paper, we have chosen the port of Agios Konstantinos, located in the central region of Greece, as our pilot use case. This port is a significant maritime transportation hub, boasting four terminals and a parking facility for vehicles and motorcycles.

The primary function of this port is to operate ferry routes serving the Sporades region. Remarkably, the port has the capability to simultaneously accommodate up to four ferry boats and two large fishing boats, making it a vital component of the local transportation infrastructure.

3.1 Project Overview

The construction project at the port of Agios Konstantinos encompasses 35 distinct activities and is projected to span 24 months.

A noteworthy aspect of this endeavor is that approximately 80% of the construction work is carried out underwater, introducing unique challenges and considerations into the project management process.

Fig. 1. Back view of the port of Agios Konstantinos

Fig. 2. Front view of the port of Agios Konstantinos

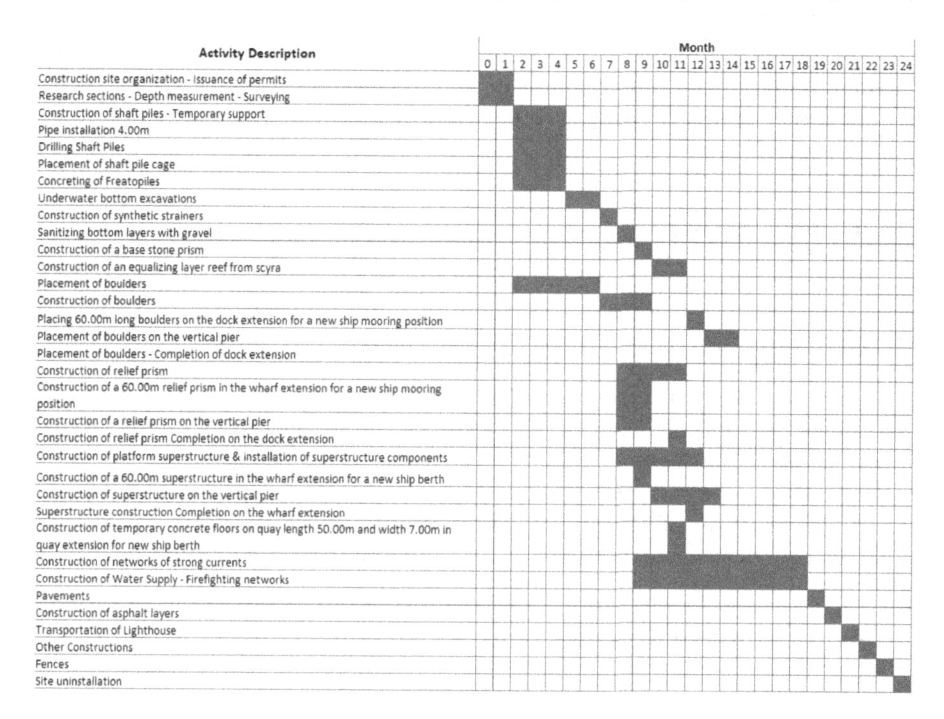

Fig. 3. A Gantt chart of the Project

3.2 Key Adjustable Parameters

Within the context of this project, two primary parameters can be adjusted to influence the course of activities: weather conditions and budget allocations. These factors are crucial in shaping the project's progress and outcomes.

3.3 Project Contract

The Agios Konstantinos port construction project contract was formally executed on February 26, 2020, with a predetermined total duration of 24 months. The financial resources allocated for this project amount to a total budget of €4,058,732. Importantly, the project remains in progress today, with various phases and activities actively underway. The port is illustrated in Fig. 1 and Fig. 2, while the Gantt chart is depicted in Fig. 3.

3.4 Project Milestones

This project is structured around three significant milestone dates, each of which marks a critical phase in its execution timeline:

1. Setting of Boulders - Foundation (Scheduled for 6 months after the initiation of the project)

2. Construction & Installation of Cobblestones (Scheduled for completion at the 9-month mark from project initiation)
3. Completion of the Superstructure (Anticipated at the 12-month milestone from project initiation)

3.5 Crucial Activities

Several key project activities have been identified as particularly sensitive to external factors, necessitating close monitoring and management. These activities fall into two distinct categories of constraints:

1. Underwater Bottom Excavations (Activity 8) and Construction of synthetic strainers (Activity 9), both of which are heavily influenced by unpredictable weather conditions.
2. Construction of Synthetic Strainers (Activity 14), Construction of Boulders, and Construction of Pavements (Activity 29), all of which are subject to budgetary constraints that must be carefully managed and optimized.

4 Experimental Results and Analysis

The experiments conducted to optimize the schedule at the Port of Agios Konstantinos are pivotal in demonstrating the remarkable capabilities of our algorithm. In the context of a complex project involving various tasks, our meticulous analysis has enabled us to identify the critical tasks most profoundly impacted by the schedule. Specifically, we have identified four tasks for optimization: Subsea bottom excavations, Road pavements, Construction of synthetic drainage systems, and an additional significant task. The proposed schedule for these tasks, along with the project's initial start and end dates as outlined in the Gantt chart before any optimization efforts, is presented in Table 1.

Table 1. Task Schedule (Before Optimization)

Task	Early Start	Early Finish	Late Start	Late Finish
Subsea Bottom Excavations	6/3/2020	6/11/2020	4/14/2020	8/1/2020
Road Pavements	9/22/2021	10/25/2021	9/22/2021	10/25/2021
Construction of Synthetic Drainage Systems	7/3/2020	7/12/2020	7/3/2020	11/26/2020
Construction of boulders	8/3/2020	2/20/2021	2/20/2021	2/20/2021

In the sequel, a detailed analysis of the results for each optimization case is given.

Table 2. Task Schedule (Case 1: Weight - {Budget = 0.8, Weather = 0.2})

Task	Early Start	Early Finish	Late Start	Late Finish
Subsea Bottom Excavations	6/19/2020	9/7/2020	8/16/2020	10/6/2020
Road Pavements	11/27/2021	11/27/2021	12/30/2021	12/30/2021
Construction of Synthetic Drainage Systems	9/7/2020	10/27/2020	9/16/2020	1/31/2020
Construction of boulders	10/8/2020	4/27/2021	4/27/2021	4/27/2021

4.1 Case 1: Weight - {Budget = 0.8, Weather = 0.2}

In this case, we assigned a higher weight to the budget (0.8) compared to weather considerations (0.2), indicating a stronger emphasis on cost control. As a result, the schedule was adjusted by 66 days:

This case demonstrates that by prioritizing budget constraints, we were able to optimize the schedule while keeping the impact on weather-related delays relatively low. The project's completion date was reduced by 66 days, potentially resulting in cost savings due to earlier completion.

4.2 Case 2: Weight - {Budget = 0.5, Weather = 0.5}

In this scenario, we assigned equal weight to both budget (0.5) and weather considerations (0.5), indicating a balanced approach. The schedule was adjusted by 83 days:

Table 3. Task Schedule (Case 2: Weight - {Budget = 0.5, Weather = 0.5})

Task	Early Start	Early Finish	Late Start	Late Finish
Subsea Bottom Excavations	7/6/2020	9/24/2020	9/2/2020	10/23/2020
Road Pavements	12/14/2021	12/14/2021	1/16/2021	1/16/2021
Construction of Synthetic Drainage Systems	9/24/2020	11/13/2020	10/3/2020	2/17/2021
Construction of boulders	10/25/2020	5/14/2021	5/14/2021	5/14/2021

In this case, we achieved a balance between budget control and weather-related delays. The project's completion date was reduced by 83 days compared to the initial schedule. This approach ensures cost control while also considering weather-related uncertainties (Table 2).

4.3 Case 3: Weight - {Budget = 0.2, Weather = 0.8}

In Case 3, we assigned a higher weight to weather considerations (0.8) compared to the budget (0.2), prioritizing schedule optimization to minimize weather-related delays. This resulted in a schedule adjustment of 121 days:

In this case, by giving higher importance to weather-related delays, we achieved a significant schedule adjustment of 121 days. This approach aims to reduce weather-related disruptions (Tables 3 and 4).

Table 4. Task Schedule (Case 3: Weight - {Budget = 0.2, Weather = 0.8})

Task	Early Start	Early Finish	Late Start	Late Finish
Subsea Bottom Excavations	8/13/2020	11/1/2020	10/10/2020	11/30/2020
Road Pavements	1/21/2022	1/21/2022	2/23/2022	2/23/2022
Construction of Synthetic Drainage Systems	11/1/2020	12/21/2020	11/10/2020	3/27/2021
Construction of boulders	12/2/2020	6/21/2021	6/21/2021	6/21/2021

5 Conclusion and Future Work

The optimization approach employed in this study represents a sophisticated blend of advanced optimization techniques. Dual annealing's capacity for global exploration is harnessed to navigate the complex solution space effectively. The inclusion of weights in the evaluation function allows for customization of the optimization process to align with project-specific priorities. This approach, by maximizing the overlap of task durations with their optimal periods, offers a powerful tool for efficiently managing the construction schedule of large-scale projects like the Port of Ag. Konstantinos. It stands as a testament to the potential of combining advanced optimization algorithms with thoughtful problem formulation to address complex real-world challenges.

Furthermore, the high degree of flexibility and adaptability intrinsic to this optimization approach leaves ample room for significant enhancements. This adaptability opens the door to tackling even more intricate and multifaceted problems, paving the way for further advancements in the field of large-scale project management.

In the pursuit of refining the optimization approach for project scheduling, several promising directions emerge. One avenue is the integration of deep learning methodologies to bolster forecasting accuracy, particularly for variables like weather conditions and material prices. Advanced neural network architectures and predictive models can leverage historical data to offer more precise projections as showcased by [19] and [20]. Moreover, tailoring optimization strategies to specific task categories within the project presents an opportunity for fine-tuned scheduling. Tasks susceptible to weather fluctuations or constrained by tight budgets could benefit from specialized optimization techniques. Embracing adaptive optimization techniques that dynamically adjust in response to real-time project data and evolving conditions represents another progressive stride. Incorporating uncertainty modeling, such as Monte Carlo simulations, can further fortify the optimization framework against unforeseeable events. Additionally, integrating machine learning for automated task dependency analysis and IoT-driven real-time data feeds for dynamic scheduling could revolutionize construction project management. Lastly, creating a user-friendly interface and visualization platform enables project managers to interact seamlessly with the optimization results, fostering wider adoption in real-world construction endeavors. These potential future advancements hold the promise of significantly enhancing the efficacy and applicability of the optimization approach in large-scale project scheduling. They address diverse challenges and offer substantial benefits in terms of efficiency, cost-effectiveness, and sustainability.

References

1. Ahmad, T., Van Looy, A.: Business process management and digital innovations: a systematic literature review. Sustainability **12**(17), 6827 (2020)
2. Apostolidis, S.D., Vougiatzis, G., Kapoutsis, A.C., Chatzichristofis, S.A., Kosmatopoulos, E.B.: Systematically improving the efficiency of grid-based coverage path planning methodologies in real-world UAVs' operations. Drones **7**(6), 399 (2023)
3. Bakas, N., Langousis, A., Nicolaou, M., Chatzichristofis, S.: Gradient free stochastic training of ANNs, with local approximation in partitions. Stochast. Environ. Res. Risk Assess. 1–15 (2023)
4. Berezka, V.: Application of the integrated decision support system for scheduling of development projects. In: MATEC Web of Conferences, vol. 251, p. 05033. EDP Sciences (2018)
5. Chang, Y., Liang, Y.: Intelligent risk assessment of ecological agriculture projects from a vision of low carbon. Sustainability **15**(7), 5765 (2023)
6. Dimopoulos, T., Bakas, N.: Sensitivity analysis of machine learning models for the mass appraisal of real estate. Case study of residential units in Nicosia, Cyprus. Remote Sens. **11**(24), 3047 (2019)
7. Dimopoulos, T., Tyralis, H., Bakas, N.P., Hadjimitsis, D.: Accuracy measurement of Random Forests and Linear Regression for mass appraisal models that estimate the prices of residential apartments in Nicosia, Cyprus. Adv. Geosci. **45**, 377–382 (2018)
8. Gailhofer, P., et al.: The role of artificial intelligence in the European green deal. European Parliament Luxembourg, Belgium (2021)
9. Gkelios, S., Kastellos, A., Boutalis, Y., Chatzichristofis, S.A.: Universal image embedding: retaining and expanding knowledge with multi-domain fine-tuning. IEEE Access (2023)
10. Gkelios, S., Sophokleous, A., Plakias, S., Boutalis, Y., Chatzichristofis, S.A.: Deep convolutional features for image retrieval. Expert Syst. Appl. **177**, 114940 (2021). https://doi.org/10.1016/j.eswa.2021.114940. https://www.sciencedirect.com/science/article/pii/S095741742100381X
11. Ke, J.: Design and research of economic management problem fusion method based on decision information system. Secur. Commun. Netw. **2022** (2022)
12. Lokhande, A.: Use of artificial intelligence smart tools in projects. In: 2022 8th International Conference on Smart Structures and Systems (ICSSS), pp. 1–6. IEEE (2022)
13. Merhi, M.I.: Evaluating the critical success factors of data intelligence implementation in the public sector using analytical hierarchy process. Technol. Forecast. Soc. Chang. **173**, 121180 (2021)
14. Merhi, M.I.: An evaluation of the critical success factors impacting artificial intelligence implementation. Int. J. Inf. Manag. **69**, 102545 (2023)
15. Miller, G.J.: Artificial intelligence project success factors—beyond the ethical principles. In: Ziemba, E., Chmielarz, W. (eds.) FedCSIS-AIST ISM 2021. LNBIP, vol. 442, pp. 65–96. Springer, Cham (2022). https://doi.org/10.1007/978-3-030-98997-2_4

16. Nusen, P., Boonyung, W., Nusen, S., Panuwatwanich, K., Champrasert, P., Kaewmoracharoen, M.: Construction planning and scheduling of a renovation project using BIM-based multi-objective genetic algorithm. Appl. Sci. **11**(11), 4716 (2021)
17. Pan, Y., Zhang, L.: Integrating BIM and AI for smart construction management: current status and future directions. Arch. Comput. Methods Eng. **30**(2), 1081–1110 (2023)
18. Paschek, D., Luminosu, C.T., Draghici, A.: Automated business process management-in times of digital transformation using machine learning or artificial intelligence. In: MATEC Web of Conferences, vol. 121, p. 04007. EDP Sciences (2017)
19. Salinas, D., Flunkert, V., Gasthaus, J., Januschowski, T.: DeepAR: probabilistic forecasting with autoregressive recurrent networks. Int. J. Forecast. **36**(3), 1181–1191 (2020)
20. Shi, X., et al.: Deep learning for precipitation nowcasting: a benchmark and a new model. In: Advances in Neural Information Processing Systems, vol. 30 (2017)
21. Vărzaru, A.A.: An empirical framework for assessing the digital technologies users' acceptance in project management. Electronics **11**(23), 3872 (2022)
22. Yin, W.: Multi-objective optimization of construction schedule based on artificial intelligence optimization algorithm. In: 2021 3rd International Conference on Artificial Intelligence and Advanced Manufacture, pp. 1621–1624 (2021)

Integrating LLMs in Higher Education, Through Interactive Problem Solving and Tutoring: Algorithmic Approach and Use Cases

Nikolaos P. Bakas[1,2(✉)], Maria Papadaki[3,5], Evgenia Vagianou[1], Ioannis Christou[1], and Savvas A. Chatzichristofis[4]

[1] School of Liberal Arts and Sciences, Department of Information Technology & AI Lab, American College of Greece, Deree, Athens, Greece
{nbakas,jes,ichristou}@acg.edu
[2] National Infrastructures for Research and Technology - GRNET, 7 Kifisias Avenue, 11523 Athens, Greece
[3] British University in Dubai, Dubai, United Arab Emirates
maria.papadaki@buid.ac.ae
[4] Intelligent Systems Laboratory, Department of Computer Science, Neapolis University Pafos, Paphos, Cyprus
s.chatzichristofis@nup.ac.cy
[5] Department of Mechanical, Aerospace and Civil Engineering, The University of Manchester, Manchester M139PL, UK

Abstract. Despite the concerns that recent developments in Large Language Models (LLMs) have raised, they undoubtedly revealed a novel potential of Artificial Intelligence (AI) algorithms in educational environments. Whether they are used for tutoring, in a manner similar to that of Intelligent Tutoring Systems (ITS), or to support assessment design and delivery, their impact in a learning setting is remarkable. In this paper, we propose an interactive tutoring approach, utilizing ChatGPT's API. By exploiting ChatGPT's programming interface, we can develop customized interactive problem-solving and tutoring sessions on specific topics of interest. The API's versatility allows for dynamic interactions, fostering a deeper understanding of subjects taught and effective problem-solving skills. We demonstrate the application of the developed code in an applied educational setting with specific use cases.

Keywords: Large Language Models · Intelligent Tutoring Systems · Adaptive Learning

1 Introduction

The idea that machines can imitate humans is not new. In ancient Greek mythology, Talos [9], was a giant bronze robot-like built by Hephaestus to protect Europa in Crete from Pirates and invaders, crossing the island's shores three times a day. Nevertheless, the definition of AI was most clearly stated in 1950

M. Papadaki et al. (Eds.): EMCIS 2023, LNBIP 501, pp. 291–307, 2024.
https://doi.org/10.1007/978-3-031-56478-9_21

by Alan Turing, one of the founding fathers of artificial intelligence. In [27], Turing assumes a human "interrogator" asking questions to two hidden entities, a *human* and a *machine*. If the "interrogator", based on the responses provided, is not able to distinguish between the human and the machine, then we have an "intelligent machine". Following the above interpretation and using the (1931) proven "Incompleteness Theorem" by Kurt Gödel [11], he seems to have the following chain of thoughts:

- in any sufficiently strong and consistent logical system, there can be formulated statements which can be neither proved or disproved within the system,
- a computing machine is assumed as a logical system,
- a digital computer with infinite capacity is considered,
- there are certain things that such a machine cannot do.

Eventually, he argues that machines have certain limitations compared to human intelligence.

With the substantial increase of computational power [1,16] and the recent developments in LLMs, the unfeasible, for a machine, operations have been narrowed down significantly, and ChatGPT can certainly be viewed as the proof. Cognitive comprehension, language generation, multilingual competency, scale and complexity adaptability are only a few of the impressive capabilities of the tool [10]. The diverse domains of ChatGPT's applicability produce equally diverse areas of research. Naturally, there is extensive criticism regarding the use of such a tool, ranging from effectiveness to safety and trust, topped by moral and ethical concerns. In a recent interview, Chomsky, who formulated the theory of *Universal grammar* [4], postulating the existence of a *biological structure* of language in humans, states that ChatGPT is basically a high-tech plagiarism machine, which can mobilize neither the will to learn nor understanding [8]. In the industry, in July 2022 Google fired an engineer, who claimed AI technology had sentiments [17]. Specifically, Lemoine, in an interview with Bloomberg, stated that when he asked the LLM a vague question, it replied with a sense of humor [24].

The scientific community was quick to react to ChatGPT entering the Internet. In [22], it is reported that ChatGPT has been utilized even as a co-author in scientific papers, an act which was considered unacceptable by many researchers. A few days later (24-Jan-2023), Nature, issued an updated [7] policy for papers to be published, with two main elements:

1. LLMs are not accepted as certified authors on research papers because authorship involves responsibility for the work, which an AI model cannot undertake.
2. LLMs' utilization should be documented in an appropriate section of the manuscript.

Science Magazine is stricter by clearly prohibiting AI-generated text in scientific papers [25]. Nevertheless, the research community, though quite concerned about the use of such tools, is also intrigued by the potential positive impact [15,21,23]. Along these lines, Nature published a new article in which AI is considered as potentially capable of providing the "gift of time" to researchers [26].

Hence, soon after, a new aspect of LLM tools was considered, and in specific, save time for researchers, who can now engage in more challenging tasks of their research.

Recent research also reveals concepts such as self-education and self-assessment, where the utilization of LLMs may play a central role. Boubker [2] conducted a survey on a sample of Moroccan higher education institutions, regarding the utilization of ChatGPT from students. The analysis of the questionnaires revealed statistical significance on perceived usefulness and student satisfaction, which is compatible with the fundamental principle "Will to Learn" [3]. On another occasion, ChatGPT was found capable of answering 51% of Otolaryngology questions [20], while being constantly re-trained with new data, making it a useful tool in education; however, caution is advised due to the possibility of generating incorrect or "hallucinated" results.

The purpose of this paper is to present an algorithmic approach for integrating LLMs in a higher education framework. In Sect. 2, we introduce the "embeddings-based" approach and an overview of potential capabilities. In Sect. 3, we present the algorithm of such integration and the computer code for specific use cases, such us interactive problem-solving, and tutoring. Finally, in Sect. 6 we conclude with the main findings and suggestions for future research directions.

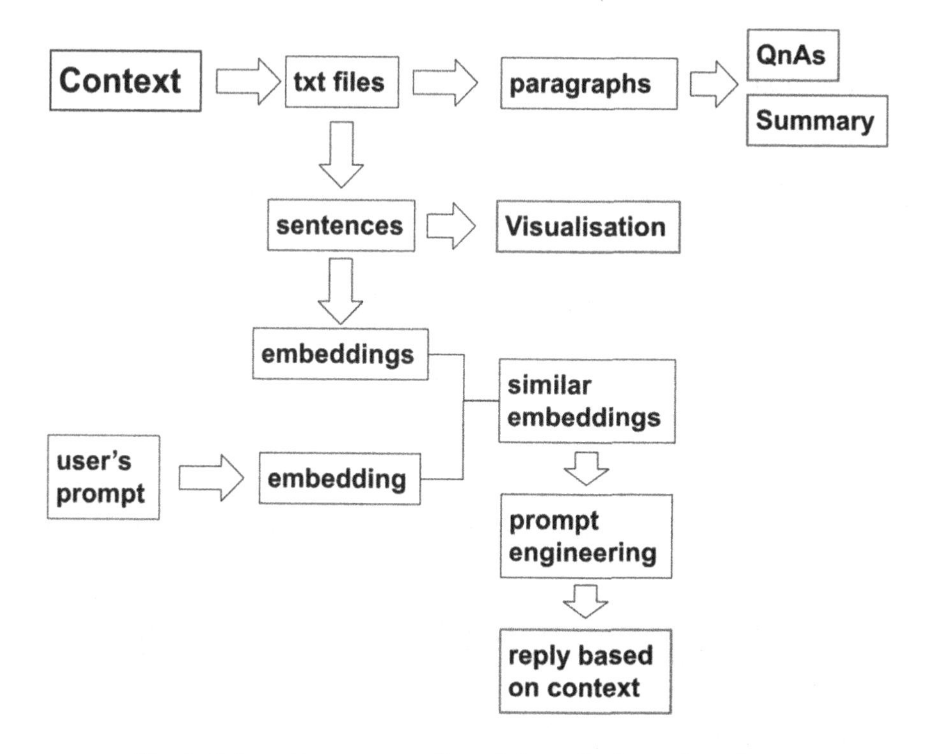

Fig. 1. Schematic representation of the preliminary formulation

2 Preliminary Formulation and ChatGPT's Capabilities

The initial phase of this work regarded the development of a custom chatbot to act like an interactive tutor. We used word vectors to extract text similar to a student's question, formulate a context, and then ask the model to reply based on that context. Figure 1 depicts the workflow for in-context gathering, processing, and employing data to produce replies tailored to the user queries. The data is subjected to preprocessing and conversion into `txt files`, which are subsequently divided into paragraphs, then used to create a pool of QnAs and text summarization, and eventually segmented into individual sentences. The extracted individual sentences, along with the user's prompt, are converted into embeddings-vector representations that encapsulate the semantics of the input sentences. Following, we find sentences of the initial text that are similar to the user's prompt and formulate a particular context, which with appropriate prompt engineering is fed to the model to produce the final result.

The approach worked well for small-scale models, as the replies were based on the initial context provided. However, occasionally the selected cases were not as relevant to the user's prompt, while sometimes significant other sentences were omitted. On top of that, recent versions of ChatGPT [18] are capable of handling thousands of tokens. Hence, as we describe in the next Sect. 3, we can now feed the model with the complete initial context without the need to pre-process it.

3 Algorithmic Approach

AI has the potential to revolutionize education through its multifaceted roles as an **enabler** [12], **cognitive partner** [13], and **mediator** [28] for critical thinking. These multifaceted roles lead to the development of adaptive learning approaches [14]. As an enabler, AI can personalize learning experiences, adapting to the unique needs and preferences of individual students. It can identify gaps in their knowledge and suggest appropriate resources to fill them. AI can act as a cognitive partner by providing instant feedback, answering questions, and facilitating interactive learning, fostering a collaborative and engaging educational environment. Furthermore, AI can serve as a mediator to encourage critical thinking by presenting students with diverse perspectives and challenging questions, prompting them to analyze, evaluate, and synthesize information from various sources. In this way, AI not only enhances the efficiency of education but also empowers students to become independent, analytical thinkers, preparing them for the demands of the 21st century.

In the following sub-sections, we demonstrate the prompts used to instruct the LLM to act as an interactive tutor, along with sample responses. We use code written in Python [19] and OpenAI's API [18].

3.1 Framework Text

In the following code snippet, we start by defining the Framework Text, i.e. what the teacher is willing to teach in class. We present a case from the *Project*

Management field and, particularly the initiation phase of project management. The same process applies to any other field, and the text stored in variable "my_framework_text", will be used later in the prompts sent to ChatGPT.

```
1  my_framework_text = r"""
2  Initiation phase of Project Management
3
4  Theory
5
6  The Initiation Phase is the foundational step in the lifecycle of a project.
   ↪ It involves the following critical components:
7
8  1. Define Project Goals: Before embarking on any project, it's essential to
   ↪ have clear and well-defined goals. What are you trying to achieve?
9  These goals will provide direction throughout the project's lifecycle and
   ↪ serve as a benchmark for its success.
10
11 2. Stakeholders: Identifying the stakeholders is crucial. These are
   ↪ individuals or groups who have an interest in the project's outcome.
12 Their inputs, needs, and concerns can influence the direction and execution
   ↪ of the project.
13
14 3. Create a Business Case: The business case is the justification for the
   ↪ project. It answers key questions such as:
15     Why is this project necessary?
16     What are the expected outcomes?
17     How will it be achieved?
18     Who will be involved?
19
20 4. Impact: Every project has consequences, both intended and unintended.
   ↪ Evaluating these impacts helps in understanding the potential benefits
21 and pitfalls. It will also aid in gaining stakeholder buy-in and ensuring
   ↪ the project aligns with the organization's strategic objectives.
22
23 5. Is the Project Worth Doing?: This is one of the most critical questions
   ↪ to answer before proceeding.
24 By doing a preliminary cost estimation and evaluating the associated risks,
   ↪ you can determine if the project's potential outcomes justify
25 the required investments.
26
27 6. Project Charter: The project charter is a formal document that describes
   ↪ the project in detail. It contains:
28     Objectives: What the project aims to achieve.
29     Constraints: Any limitations or boundaries that the project may have,
   ↪ such as budget or time.
30     How It Will Be Carried Out: The approach or methodology that will be
   ↪ used.
31     Stakeholders: A list of all parties involved, detailing their roles and
   ↪ responsibilities.
32
33 In essence, the Initiation Phase sets the stage for the project. It provides
   ↪ clarity, direction, and the necessary justification for proceeding.
34 Without a strong initiation, projects can lack focus, direction, and the
   ↪ support they need to succeed.
35 """
```

3.2 Question Answering and Feedback

In the following code snippet, we provide the model with the *framework text* and ask it to create a pool of 10 Questions based on the provided text:

```
my_prompt = f"""
    You will be provided with some text in triple quotes.
    Create 10 Questions based on the text in triple quotes.
    The text is: \n\n\"\"\"{my_framework_text}\"\"\"
"""
my_messages = [{"role": "system", "content": "You are an interactive
    tutor"}]
my_messages.append({"role": "user", "content": my_prompt})
t0 = time()
response_plain = openai.ChatCompletion.create(model="gpt-3.5-turbo",
    messages=my_messages, top_p = 0.91)
t1 = time()
response_plain =
    response_plain.choices[0]["message"]["content"].replace("\n\n","")
print(response_plain)
```

For example, one automatically generated set of Questions is:

```
1. What is the purpose of the Initiation Phase in project
   management?
2. Why is it important to define project goals before starting a
   project?
3. Who are stakeholders in a project, and why is it crucial to
   identify them?
4. What is a business case, and what questions does it answer?
5. Why is it important to evaluate the impact of a project?
6. What is the significance of determining if a project is worth
   doing before proceeding?
7. What is a project charter, and what information does it
   contain?
8. How does the Initiation Phase set the stage for a project?
9. What are the potential consequences of not having a strong
   initiation phase in a project?
10. Why is it necessary for a project to align with an
   organization's strategic objectives?
```

Accordingly, we select one random Question:

```
i = np.random.randint(10)
gpt_question = response_plain.split("?")[i].replace("\n","") +
    "?"
gpt_question
```

and provide it to the student:

```
'4. What is a business case, and what questions does it answer?'
```

And let's assume that the student answered "a business case, answers key questions such as why is this project necessary, and what are the expected outcomes". We store the student's answer in the variable `my_answer`:

```
1 my_answer = "a business case, answers key questions such as
  ↪ why is this project necessary, and what are the expected
  ↪ outcomes"
```

Then, we give the model student's answer, the automatically generated question stored in the variable `gpt_question`, along with the *framework text*, stored in the variable `my_framework_text`. Correspondingly, we ask the model to evaluate the student's response and provide feedback.

```
1  my_prompt = f"""
2  You gave a student the Question: {gpt_question}, and the student
   ↪ Answered: {my_answer}.
3  Give the student feedback, what they should change in their Answer,
4  based on the text below in triple quotes.
5  Rate the student from 1 to 10. Be strict in your evaluation!
6
7  The text is: \n\n\"\"\"{my_framework_text}\"\"\"
8  """
9  my_messages = [{"role": "system", "content": "You are a helpful
   ↪ assistant."}]
10 my_messages.append({"role": "user", "content": my_prompt})
11 t0 = time()
12 gpt_response = openai.ChatCompletion.create(model="gpt-3.5-turbo",
   ↪ messages=my_messages)
13 t1 = time()
14 gpt_response_choices =
   ↪ gpt_response.choices[0]["message"]["content"].replace("\n\n","")
15 print("Question>>",gpt_question,"\n")
16 print("Your Answer>>",my_answer,"\n")
17 print("ChatGPT's Evaluation>>",gpt_response_choices,"\n")
18 print("Time elapsed: %.3f seconds." % (t1-t0))
```

Interestingly, ChatGPT correctly identified the parts of the "business case" that the student answered (*why*, and *what*), and suggested that the student should also include in their answer the *how*, and *who* parts. Also, the evaluation 5.5 out of 10 is close to an instructor's evaluation.

```
Question>> 4. What is a business case, and what questions does it
↪ answer?

Your Answer>> a business case, answers key questions such as why
↪ is this project necessary, and what are the expected outcomes

ChatGPT's Evaluation>> The student's answer is partially correct
↪ but missing some key points. They mentioned that a business
↪ case answers the questions of why the project is necessary and
↪ what are the expected outcomes. However, they didn't mention
↪ other important questions that a business case should answer,
↪ such as how will it be achieved and who will be involved.
↪ Based on their answer, I would rate the student a 5.5 out of
↪ 10. They understood the concept of a business case but missed
↪ some important details.

Time elapsed: 2.180 seconds.
```

3.3 Assess the Level of Understanding

In this Section, we ask the student to write below briefly their understanding of the topic. For example, let's assume the following answer:

```
my_understanding_of_the_topic = f"""
The Initiation Phase is the critical first step in project
↪ management, where goals are defined to guide the project
↪ and measure its success. Stakeholders, whose interests and
↪ input can shape the project, are identified. A business
↪ case justifies the project by outlining necessity,
↪ expected outcomes, and involved parties. Impact assessment
↪ ensures alignment with strategic objectives and
↪ stakeholder support. A cost-risk evaluation ascertains if
↪ the project's benefits outweigh its investments. A project
↪ charter formalizes all this information, detailing
↪ objectives, constraints, methodology, and stakeholder
↪ roles. This phase is essential for providing the project
↪ with a clear focus and direction.
"""
```

Accordingly, we ask the model to compare the initial "framework text", with the corresponding understanding by the student and evaluate it. We use this prompt:

```
1  my_prompt = f"""
2  You will be provided with a topic provided below in triple quotes.
3  The Topic is:
4  \n\n\"\"\"{my_framework_text}\"\"\"
5
6  You will also be provided with a student's Understanding Level of the
   ↪ topic in triple parentheses.
7  The Understanding Level is:
8  \n\n\(\(\({my_understanding_of_the_topic}\)\)\)
9
10 Evaluate the student's Understanding Level. Highlight any gaps and
   ↪ clarify what they might have missed.
11 """
12 my_messages = [{"role": "system", "content": "You are a helpful
   ↪ assistant."}]
13 my_messages.append({"role": "user", "content": my_prompt})
14 t0 = time()
15 gpt_response = openai.ChatCompletion.create(model="gpt-3.5-turbo",
   ↪ messages=my_messages)
16 t1 = time()
17 gpt_response_choices =
   ↪ gpt_response.choices[0]["message"]["content"].replace("\n\n","")
18 print("ChatGPT's Evaluation>>",gpt_response_choices)
19 print("Time elapsed:
```

In this particular case, through the model's response, which is presented in the following frame, we may notice that the model identified missing parts from student's understanding, such as "evaluating the impacts of the project", "assessing if the project is worth doing", the "impacts", "reliminary cost estimation and risk assessment", as well as the "the purpose and components of a project charter".

```
ChatGPT's Evaluation>> The student's Understanding Level is quite
↪ good as they have captured the key concepts of the Initiation
↪ Phase of Project Management. They mention that this phase
↪ involves defining project goals, identifying stakeholders,
↪ creating a business case, and providing clear focus and
↪ direction to the project. However, there are a few minor gaps
↪ in their understanding. They did not mention evaluating the
↪ impacts of the project, assessing if the project is worth
↪ doing by considering cost estimation and associated risks, and
↪ the importance of a project charter that describes the project
↪ in detail.To further enhance their understanding, the student
↪ could explore the importance of evaluating the impacts of the
↪ project to understand potential benefits and pitfalls. They
↪ could also learn about conducting preliminary cost estimation
↪ and risk assessment to determine if the project's potential
↪ outcomes justify the required investments. Additionally,
↪ understanding the purpose and components of a project charter,
↪ which includes objectives, constraints, approach, and
↪ stakeholders, would provide them with a more comprehensive
↪ understanding of the Initiation Phase.
Time elapsed: 4.622 seconds.
```

3.4 Interactive Problem Solving

In this Section, we ask the LLM to create a use case for the student to solve. Based on the initial "framework text", we use the following prompt to develop an applied problem on the provided topic, and ask students to provide their solution and rationale.

```
1  my_prompt = f"""
2  You are an interactive tutor.
3  Provide a student with an applied problem on the topic provided below
   ↪ in triple quotes.
4  The problem should apply in the field of ICT or Cybersecurity.
5  Ask students to explain their thinking.
6  The problem that you will provide should be solvable based on
   ↪ knowledge directly stemming from the topic provided below,
7  and no other background information. Provide the problem in 100 words
   ↪ at maximum.
8
9  Topic:
10 \n\n\"\"\"{my_framework_text}\"\"\"
11 """
12 my_messages = [{"role": "system", "content": "You are a helpful
   ↪ assistant."}]
13 my_messages.append({"role": "user", "content": my_prompt})
14 gpt_response = openai.ChatCompletion.create(model="gpt-3.5-turbo",
   ↪ messages=my_messages)
15 gpt_response_problem =
   ↪ gpt_response.choices[0]["message"]["content"].replace("\n\n","")
16 print("Problem Statement>>",gpt_response_problem)
```

Accordingly, we get the following auto-generated problem in the initiation phase of project management:

```
Problem Statement>> Problem:You are the project manager for a
↪ software development project for a client in the banking
↪ industry. Your client wants to develop a new mobile banking
↪ application that will allow their customers to perform various
↪ banking transactions, such as balance inquiries, fund
↪ transfers, and bill payments, conveniently from their
↪ smartphones. Your client has approached your company with this
↪ idea, and they are considering implementing this project. As
↪ the project manager, you are responsible for the initiation
↪ phase of the project. Using the knowledge from the initiation
↪ phase theory, explain the key components and steps you would
↪ undertake to initiate this project effectively.
```

Let's now assume, that the student provided the following answer:

```
my_solution = f"""
To initiate the project effectively, I would start by defining
    the project's purpose and objectives,
ensuring they align with the client's strategic goals. Next,
    I'd identify key stakeholders, including banking
    officials,
IT staff, and end-users, and establish their needs and
    expectations.
Finally, I'd appoint a skilled project team, setting the stage
    for detailed planning.
Effective communication and stakeholder engagement would be
    crucial throughout this phase.
"""
```

We now ask the model to provide feedback on the student's answer, and what they should change in their answer, in order to match the provided topic, as defined in the "framework text". We use this prompt:

```
my_prompt = f"""
You are an interactive tutor who helps students understand concepts.
You provided a student with the following problem:
    {gpt_response_problem}.
They replied: {my_solution}.
Give the student feedback, what they should change in their Answer,
    in order to match the provided topic below in triple quotes.
Rate the student from 1 to 10. Be strict in your evaluation!

Topic:
\n\n\"\"\"{my_framework_text}\"\"\"
"""
my_messages = [{"role": "system", "content": "You are a helpful
    assistant."}]
my_messages.append({"role": "user", "content": my_prompt})
gpt_response = openai.ChatCompletion.create(model="gpt-3.5-turbo",
    messages=my_messages)
gpt_response_choices =
    gpt_response.choices[0]["message"]["content"].replace("\n\n","")
print("ChatGPT's Evaluation>>",gpt_response_choices)
```

Accordingly, we get the following response from the model. We observe, that the model identified that the student did not mention "creating a business case", "evaluating the impact of the project", as well as the "project charter":

```
ChatGPT's Evaluation>> The student's answer is quite good, but
↪   there are some areas that could be improved to better align
↪   with the provided topic. 1. The student mentioned defining the
↪   project's purpose and objectives, which is similar to defining
↪   project goals. However, it would be more accurate to
↪   specifically mention the project goals as one of the key
↪   components in the initiation phase.2. The student correctly
↪   identified the stakeholders and mentioned the need to
↪   establish their needs and expectations. However, they did not
↪   mention the significance of stakeholders influencing the
↪   direction and execution of the project, which is an important
↪   aspect to include.3. The student did not explicitly mention
↪   creating a business case, which is an essential component of
↪   the initiation phase as it justifies the project and answers
↪   key questions about its necessity, expected outcomes, and how
↪   it will be achieved.4. The student did not mention evaluating
↪   the impact of the project, including potential benefits and
↪   pitfalls. This step is crucial for gaining stakeholder buy-in
↪   and ensuring alignment with strategic objectives.5. The
↪   student briefly mentioned appointing a skilled project team,
↪   but it would be beneficial to emphasize the importance of
↪   assembling a capable team for successful project initiation.6.
↪   The student did not mention the project charter, which is a
↪   formal document that describes the project in detail,
↪   including objectives, constraints, approach/methodology, and
↪   stakeholders. This document is a critical aspect of the
↪   initiation phase.Taking these points into consideration, I
↪   would rate the student's answer a 6 out of 10. The student has
↪   a good understanding of the initiation phase, but there are
↪   some key components and steps that should be included or
↪   expanded upon to fully match the provided topic.
```

4 Analyzing the Modules

In this section, we delve into the analysis of the three modules and explore how they align with AI's roles as an enabler, cognitive partner, and mediator for critical thinking. We aim to establish a meaningful correlation between these modules and these multifaceted roles. The results of this analysis are presented in Table 1.

The AI-based "Question answering and feedback" module can be best described as a cognitive partner in the educational context. This module excels in providing instant feedback and answers to students' questions, which, in turn, greatly enhances their understanding and engagement with the learning material.

Table 1. Correlation of AI Roles with Educational Modules

	Question answering and feedback	Assess the level of understanding	Interactive problem solving
Enabler		X	
Cognitive Partner	X		X
Mediator for Critical Thinking	X	X	X

By facilitating interactive learning and offering immediate assistance, it fulfills the key characteristics of a cognitive partner. While it also generates questions and promotes learning, its primary function lies in providing valuable support and feedback, making it a strong match for the cognitive partner role.

Moving on, the AI-based "Assess the level of understanding" module plays a crucial role as an enabler in the educational landscape. This module encourages students to engage in thoughtful self-reflection about their grasp of the topic and express their insights in writing. By doing so, it enables a highly personalized learning experience that caters to the unique needs and preferences of individual students. Furthermore, it assists in pinpointing knowledge gaps and motivates students to delve deeper into the subject matter. Although it involves the assessment of understanding, its primary function revolves around supporting and enabling students to enhance their comprehension and learning. This aligns it closely with the characteristics of an enabler in AI's educational applications.

Finally, the AI-based "Interactive problem solving" module assumes the role of a mediator for critical thinking. This module presents students with real-world problems that require them to think critically and apply their knowledge to practical situations. By offering these challenging use cases, it encourages students to analyze, evaluate, and synthesize information from diverse sources in order to arrive at solutions. In essence, this module actively fosters and nurtures critical thinking skills. Its function as a mediator prompts students to engage in higher-level cognitive processes, making it a fitting candidate for the mediator role in the context of AI's contribution to education.

In summary, these three modules, each in its unique way, contribute to the overarching goal of enhancing the educational experience through AI. They align well with the roles of cognitive partner, enabler, and mediator, respectively, showcasing the versatility and potential of artificial intelligence in education.

5 Further Educational Uses

Besides their potential as intelligent tutors, soon LLMs will undoubtedly find their place in higher education as intelligent advisors. Of the main tasks of an academic advisor is to provide guidance in the course selection process, aiming for the formulation of student personalized study plans, which will facilitate the advises progress. At ACG, we have created an Open-Source software tool called SCORER that can produce personalized study plans for the students of the college, based on the their history of completed courses, their desires, and of course, curriculum constraints (see [6] for more details). SCORER creates a Mixed Integer Programming model (see [5]), the solution of which provides an optimal study plan that integrates both student preferences and curricula constraints. While LLMs are not yet capable of solving complex optimization problems themselves, they can connect to the Wolfram Alpha servers to solve complex mathematical problems involving calculus, linear algebra, and various other branches of mathematics. We anticipate the release of a plugin between ChatGPT and the GUROBI (or GAMS or AMPL) servers, which will enable ChatGPT to tackle complex optimization problems as effectively as it currently tackles complex mathematical problems. Once such a connection is available, ChatGPT will be able to function as a true intelligent tutor by creating optimal study plans for students after engaging in a natural language dialogue with them, rather than resorting to the limitations of a classical software Graphical User Interface for the acquisition of the necessary information.

6 Conclusions

We presented an AI-based tutoring approach utilizing state-of-the-art LLMs and particularly ChatGPT's API. Using prompt engineering and based on "framework text" provided by a human tutor, we focused on three areas:

1. Question answering with automatic evaluation from the model.
2. Assessing the student's level of understanding of the provided topic.
3. Automatic extraction of an applied problem based on the topic, and evaluation of student's response by the model.

We observed that the model was capable of correctly identifying missing parts in the student's responses, assessing their level of understanding, and proposing enhancements in their responses. The entire process is based on interactive tutoring sessions, where the student can interact with the model, to self-evaluate their understanding throughout the learning process, and get feedback on what they should improve on. The flexibility of the usage of such tools offers a new learning context, with numerous capabilities and corresponding tools to develop.

Acknowledgments. ChatGPT [18] has been utilized for the development of the interactive tutoring sessions presented.

Funding. This work received financial support from the EuroCC 2 Project (101101903) of the European Commission.

References

1. Bakas, N., Markou, G., Charmpis, D., Hadjiyiannakou, K.: Performance and scalability of deep learning models trained on a hybrid supercomputer: application in the prediction of the shear strength of reinforced concrete slender beams without stirrups. In: 8th International Conference on Computational Methods in Structural Dynamics and Earthquake Engineering. ECCOMAS Thematic Conference and IACM Special Interest Conference. COMPDYN 2021 (2021). https://2021.compdyn.org/
2. Boubker, O.: From chatting to self-educating: can AI tools boost student learning outcomes? Expert Syst. Appl. **238** (2024). https://doi.org/10.1016/j.eswa.2023.121820
3. Bruner, J.S.: Toward a Theory of Instruction. Harvard University Press (1966)
4. Chomsky, N.: Tool module, chomskys universal grammar (1960). https://thebrain.mcgill.ca/flash/capsules/outil_rouge06.html
5. Christou, I.: Quantitative Methods in Supply Chain Management: Models and Algorithms. Springer, Cham (2011)
6. Christou, I.T., Vagianou, E., Vardoulias G.: Planning courses for student success at the American College of Greece. INFORMS J. Appl. Anal. (2024). https://doi.org/10.1287/inte.2022.0083
7. EDITORIAL: Tools such as ChatGPT threaten transparent science; here are our ground rules for their use. Nature (2023). https://www.nature.com/articles/d41586-023-00191-1
8. EduKitchen: Chomsky on ChatGPT, education, Russia and the unvaccinated (2023). https://www.youtube.com/watch?v=IgxzcOugvEI&list=WL, youTube video
9. Gill, D.: The red and the black: studies in Greek pottery. Antiquity **70**(270), 988–992 (1996)
10. Gill, S.S., Kaur, R.: ChatGPT: vision and challenges. Internet Things Cyber-Phys. Syst. **3**, 262–271 (2023)
11. Gödel, K.: Über formal unentscheidbare sätze der principia mathematica und verwandter systeme i. Monatshefte für mathematik und physik **38**, 173–198 (1931)
12. Jaakkola, H., Henno, J., Lahti, A., Järvinen, J.P., Mäkelä, J.: Artificial intelligence and education. In: 2020 43rd International Convention on Information, Communication and Electronic Technology (MIPRO), pp. 548–555. IEEE (2020)
13. Kadel, R.S., Gazi, Y., Harmon, S., Goel, A., Courville, T.: Toward a road map for scalable advanced learning ecosystems (SALEs). Int. J. Innov. Online Educ. **3**(1) (2019)
14. Koutsantonis, D., Koutsantonis, K., Bakas, N.P., Plevris, V., Langousis, A., Chatzichristofis, S.A.: Bibliometric literature review of adaptive learning systems. Sustainability **14**(19), 12684 (2022)
15. Liebrenz, M., Schleifer, R., Buadze, A., Bhugra, D., Smith, A.: Generating scholarly content with ChatGPT: ethical challenges for medical publishing. Lancet Digit. Health **5**(3), e105–e106 (2023)
16. Markou, G., Bakas, N., Chatzichristofis, S., Papadrakakis, M.: A general framework of high-performance machine learning algorithms: application in structural mechanics. Comput. Mech. (2023). https://doi.org/10.1007/s00466-023-02386-9, accepted August 2023
17. Maruf, R.: Google fires engineer who contended its AI technology was sentient (2022)

18. OpenAI: GPT-4 architecture (2023). https://platform.openai.com/docs/guides/gpt
19. Python: Python programming language. https://www.python.org/
20. Revercomb, L., Patel, A., Choudhry, H., Filimonov, A.: Performance of ChatGPT in otolaryngology knowledge assessment. Am. J. Otolaryngol. - Head Neck Med. Surg. **45**(1) (2024). https://doi.org/10.1016/j.amjoto.2023.104082
21. Stahl, B., Eke, D.: The ethics of ChatGPT - exploring the ethical issues of an emerging technology. Int. J. Inf. Manag. **74** (2024). https://doi.org/10.1016/j.ijinfomgt.2023.102700
22. Stokel-Walker, C.: ChatGPT listed as author on research papers: many scientists disapprove. Nature (2023). https://www.nature.com/articles/d41586-023-00107-z
23. Sullivan, M., Kelly, A., McLaughlan, P.: ChatGPT in higher education: considerations for academic integrity and student learning. J. Appl. Learn. Teach. **6**(1), 31–40 (2023). https://doi.org/10.37074/jalt.2023.6.1.17
24. Bachelor of Technology: Google engineer on his sentient AI claim (2022). https://www.youtube.com/watch?v=kgCUn4fQTsc, youTube video
25. Thorp, H.H.: ChatGPT is fun, but not an author (2023)
26. Tregoning, J.: AI writing tools could hand scientists the 'gift of time'. Nature (2023). https://www.nature.com/articles/d41586-023-00528-w
27. Turing, A.M.: Computing Machinery and Intelligence. Springer, Heidelberg (1950). https://academic.oup.com/mind/article/LIX/236/433/986238
28. Xia, Q., Chiu, T.K., Chai, C.S., Xie, K.: The mediating effects of needs satisfaction on the relationships between prior knowledge and self-regulated learning through artificial intelligence chatbot. Br. J. Educ. Technol. (2023)

Author Index

M. Papadaki et al. (Eds.): EMCIS 2023, LNBIP 501, pp. 309–310, 2024.
https://doi.org/10.1007/978-3-031-56478-9

Printed in the United States
by Baker & Taylor Publisher Services